Education Innovation Series

Series Editors

Education holds the key to unlock human resources that a society needs to survive and flourish. This is particularly salient in a borderless knowledge economy. For the past decades, the sterling performance of economies such as Hong Kong, Finland, Japan, Singapore and Taiwan in international studies (e.g., TIMSS, PIRLS and PISA) has attracted much attention internationally.

Researchers, policy makers and practitioners all over the world wish to understand how education innovations propel the emerging systems from good to great to excellent, and how their trajectories will provide insights for reforms in the education system, schooling innovation, and classroom practices. The Education Innovation Book Series delves into education innovations enacted by these emerging systems and situates them in both the local and the broader international contexts. The series primarily focuses on pedagogy and classroom practices; education policy formulation and implementation; school and instructional leadership; and the context and interface between education research, policy and practice. We believe that the latter is critical in making education innovations come to bear. Each volume documents insights and lessons learned based on empirical research (both quantitative and qualitative) and theoretical analyses. Implications to research, policy and professional practice are surfaced through comparing and synthesizing their experience in the process of comparative studies on successful reforms around the world.

The audience of the edited volumes and monographs published in this series includes researchers, policy makers, practitioners and students in the fields of education and teacher education, and public policies related to learning and human resources.

Please contact Grace Ma (e-mail: grace.ma@springer.com) for submitting book proposals for this series.

Hiroki Fujii • Sun-Kyung Lee
Editors

Science Education for Sustainable Development in Asia

Springer

Editors
Hiroki Fujii
Okayama University
Okayama, Japan

Sun-Kyung Lee
Cheongju National University of Education
Cheongju, Korea (Republic of)

ISSN 2211-4874 ISSN 2211-4882 (electronic)
Education Innovation Series
ISBN 978-981-99-8710-8 ISBN 978-981-99-8711-5 (eBook)
https://doi.org/10.1007/978-981-99-8711-5

This Springer imprint is published by the registered company Springer Nature Singapore Pte Ltd.
The registered company address is: 152 Beach Road, #21-01/04 Gateway East, Singapore 189721, Singapore

Series Editors' Foreword

Science Education for Sustainable Development in Asia, co-edited by Hiroki Fujii and Sun-Kyung Lee, is a timely book to highlight the role science education plays in supporting and promoting sustainable development. As the editors mention in their introduction chapter:

> Education for Sustainable Development (ESD) occupies a prominent place in the 2030 Agenda for Sustainable Development and the Sustainable Development Goals (SDGs) adopted by the United Nations. It is a "vital means of implementation for sustainable development" and "a key enabler" of all the SDGs. (United Nations General Assembly, 2091, cited by editors)

As such, a book of this nature which explores the potential and limitations of how science education can play a significant role that can contribute well towards the ESD represents an invaluable signpost in this journey. In fact, the United Nations published *The Global Sustainable Development Report 2023* (United Nation, 2023) last year urging for strong science education participation. This report was preamble with the following title:

Times of Crisis, Times of Change, Science for Accelerating Transformations to Sustainable Development (Front Page).

The report refers to a document entitled "What scientists need to do to accelerate progress on the SDGs: Nature Commentary" (Malekpour et al., 2023), contributed by 19 authors. This report cautions that "At the current rate of progress, the world will not eradicate poverty, end hunger or provide quality education for all by 2030 – to name just some, central, aspirations of the SDGs. Instead, by the end of this decade, our world will have 575 million people living in extreme poverty, 600 million people facing hunger, and 84 million children and young people out of school.... And, at the current rate, it will take 300 years to attain gender equality" (Malekpour et al., Sept 2023). This group of authors concede that there are urgent needs for the scientists to (1) identify tailored approaches to remove key roadblocks to progress, such as forcing to adopt net-zero emissions in service-based economics (citing Switzerland and Singapore as an example); (2) find feasible and cost-effective pathways, such as investments in health and education to improve economic productivity and growth, people's incomes and government revenue that will increase consumption but reduce resource demand, environmental degradation and pollution, as well as exacerbating inequalities; (3) scientists need to develop criteria to assess the impact of different SDG governance processes. Researchers should also provide guidance to enhance accountability to ensure that funded projects will contribute to enhancing sustainability.

In response to the call for proposals and actions to make the SDGs achievable, as a matter of urgency, this book is a demonstration of how the Asian academics and practitioners can make a difference in raising awareness, designing applicable pedagogies to teach science for sustainability and making good use of Asia's cultural contexts to promote sustainability in the formal curriculum through classroom practices. To add a note, our years of Asian studies and direct experiences have led us to understand that in general, across countries in Asia and despite cultural differences, the belief and pursuit of harmony between human and nature seems to be ingrained within Asian societies. Assuming that this is a reasonably accurate observation, it can be argued that Asia has both the cultural (or even spiritual) tradition and motivation to make a successful go at achieving the SDGs directly or indirectly. At the same time, with a generally positive view in the value of education, Asian environments are conducive in promoting sustainable development not only through the science curriculum but through schooling at large.

If you are looking for literature on science education and sustainable development in Asia, this book comes highly recommended. As it covers a large number of Asian countries, the diversity in perspectives and the generalizable lessons learnt can prove to be an invaluable resource. Besides, it can be most interesting to see how different Asian countries work towards this SDGs agenda through science education within their respective contexts. This will provide a rich ground for understanding how different countries in Asia adopt different pedagogical approaches to achieve the same goal, i.e. to promote SDG goals through science education.

We are sure this will be a significant resource book on the topic, providing information from Asia. This book also shows how Asian countries have made attempts to attain the SDG goals in their contextualized ways.

Institute for Adult Learning
Singapore University of Social Sciences
Singapore, Singapore
Singapore University of Social Sciences
Singapore, Singapore

Wing On Lee
Cheah Horn Mun
Rebekah Weiying Lim

References

Malekpour et al. (2023). What scientists need to do to accelerate progress on the SDGs: Nature commentary. *Nature*, *621*, 250–254.

United Nations. (2023). *The global sustainable development report*. United Nations.

Acknowledgements

This book is based on a research project titled "Teacher Education for ESD in the Asia-Pacific Region" from 2017 to 2019, supported by the Japan Society for the Promotion of Science and the Japanese National Commission for United Nations Educational, Scientific and Cultural Organization (UNESCO).

The project involved the participation and collaboration of 34 teacher education institutions in 16 Asia-Pacific countries, UNESCO Bangkok and Asia-Pacific Cultural Centre for UNESCO. It was a challenging initiative to mainstream the idea of sustainability in science education of each individual, institution and country.

Contents

Chapter 1
Introduction: Challenges of Science Education in the Era of Sustainable Development Goals

Hiroki Fujii and Sun-Kyung Lee

1.1 Introduction

Education for sustainable development (ESD) occupies a prominent place in the 2030 Agenda for Sustainable Development and the Sustainable Development Goals (SDGs) adopted by the United Nations. It is "a vital means of implementation for sustainable development" and "a key enabler" of all the SDGs (United Nations General Assembly, 2019). To date, ESD has broadly affected school curricula, pedagogy, educational evaluations, and educational environments worldwide.

Science education is required to strengthen the concept of sustainability and contribute to the achievement of sustainable development. This request is based on the common understanding that science education plays a crucial role in education for sustainability. First, scientific knowledge of climate change, renewable energy, biodiversity, disaster risk reduction, and sustainable consumption and supply is the basis for considering sustainability-related issues. Second, the scientific processes emphasized in science education are essential for the development of key competencies for sustainability, such as systems thinking, anticipatory, normative, strategic, collaboration, critical thinking, self-awareness, integrated problem-solving competencies (UNESCO, 2017), and guiding rational and reasonable behavior. Considering this crucial aspect, this book presents a perspective on science education in the era of the SDGs.

H. Fujii (✉)
Okayama University, Okayama, Japan
e-mail: fujii-hi@okayama-u.ac.jp

S.-K. Lee
Cheongju National University of Education, Cheongju, Republic of Korea

H. Fujii, S.-K. Lee (eds.), *Science Education for Sustainable Development in Asia*, Education Innovation Series, https://doi.org/10.1007/978-981-99-8711-5_1

The contributors to this book researched and practiced science education for sustainability in a research project titled "Teacher Education for ESD in the Asia-Pacific Region" from 2017 to 2019, supported by the Japan Society for the Promotion of Science and the Japanese National Commission for United Nations Educational, Scientific and Cultural Organization (UNESCO). The project was a challenging initiative and breathed new life into science education for each individual, institution, and country. In fact, the contributors' research and practices have triggered innovation in science education for sustainability. Excellent examples of school teaching and learning, science curricula and assessments, science education in collaboration with local communities, and science teacher education have been successfully accumulated. While embodying Asia's sustainability education paradigms, policies, and practices, these examples further depict the diversity and uniqueness of Asia's natural, social, and cultural contexts. The book vividly paints this diversity and highlights some commonalities.

Science teachers, educators, curriculum designers, and teachers are leaders in shaping a sustainable society. They are required to connect, collaborate, engage in, and create science education for sustainability with students, colleagues, and people from different sectors. Simultaneously, they are expected to deepen their learning in this process, cultivate their own interest in science education for sustainability, and change their own way of life and action. The chapters included in this book will help fulfill these expectations by presenting useful examples and prospects for new science classes, curricula, and teacher education.

The realization of a new era based on a sustainable society begins with education. Expectations are high in science education, which plays an important role. This book aims to be a milestone in Asia in meeting these expectations.

1.2 Connections Between School Science and ESD

1.2.1 Learners' Competencies for Sustainability To Be Acquired Through ESD

ESD initiated from the requirement for education to address growing sustainability challenges. ESD employs action-oriented, innovative pedagogy to enable learners to develop knowledge and awareness and transform society into a more sustainable one. UNESCO has been leading and coordinating ESD globally since 2005 and, more recently, promoting it worldwide through its new roadmap, ESD for 2030, as well as through the Sustainable Development Goal 4 on education under Target 4.7 (UNESCO, 2020). ESD is also a vital means of implementation for the 2030 Agenda for Sustainable Development and a key enabler of all Sustainable Development Goals, as reaffirmed by the United Nations General Assembly in its resolution 72/222.

To achieve the SDGs, UNESCO (2017) advocated a new idea, Education for Sustainable Development Goals, which is an expansion of the conventional ESD. The idea is based on the assumption that the SDGs constitute a sizeable agenda, which cannot be achieved without education, thereby implying that an ideal system of education for achieving the SDGs is with ESD at the core. The aim of Education for Sustainable Development Goals is to develop "sustainability citizens" as an image of humanity who can connect, communicate, create, and engage in the transformation required in today's complex and uncertain world. Furthermore, citizens must acquire key competencies for sustainability mentioned earlier. The list of those competencies has been discussed over the past decades (e.g., Wiek et al., 2011).

Many of these competencies have been incorporated into school ESD policies in many countries. In Japan, for example, they overlap with the "Competencies and attitudes to be emphasized in ESD," as indicated by the Japanese National Commission for UNESCO (2021): specifically, (1) ability to think critically; (2) ability to plan with anticipation of a future scenario; (3) ability to think in multidimensional and integrative ways; (4) ability to communicate; (5) ability to cooperate with others; (6) attitude to respect relations and connections; and (7) attitude to participate proactively. Furthermore, these competencies overlap with "learners' competencies" drawn by the Korean National Curriculum (Ministry of Education, 2015) in schools: (1) self-management ability; (2) knowledge information processing competency; (3) creative thinking ability; (4) aesthetic emotional competency; (5) communication ability; and (6) community competency.

More recently, a framework of sustainability competencies titled "GreenComp" has been presented by the European Commission (Bianch et al., 2022). It includes three competencies in each of the four areas: (1) embodying sustainability values (valuing sustainability, supporting fairness, and promoting nature); (2) embracing complexity in sustainability (systems thinking, critical thinking, and problem framing); (3) envisioning sustainable futures (futures literacy, adaptability, and exploratory thinking); and (4) acting for sustainability (political agency, collective action, and individual initiative). Thus, the discussion regarding sustainability competencies to be acquired by learners is active, and frameworks and models are evolving further.

1.2.2 Framing of School Science Incorporating ESD

The connection between science education and ESD has rapidly expanded over the past two decades. High-quality science education is required not only to sustain a lively scientific community that is able to address global problems, such as global warming and pandemics, but also to bring about and maintain a high level of scientific literacy in the general population (Eijck & Roth, 2008). Therefore, science education must be configured for ESD to address these issues. Moreover, if we consider the argument that ESD should focus more on ecological perspectives

(Kopnina, 2020), science education that can make use of such perspectives will play a more important role in ESD.

Internationally, science education, which is linked to ESD, has predominantly been about empowering students to analyze and discuss issues rather than providing them the experiences they require to actively participate in social change and take a proactive role (Bezeljak et al., 2020). Science education in Asia is no exception to this trend. If the goal of science education is to engage students in science in a manner that is relevant and empowers them as future citizens, the curricula and lessons for science education should be properly framed (Tytler, 2012). This remains a challenge in science education.

Framing science education should emphasize at least the following three competencies for students to acquire, based on the aforementioned competency-driven discussion that has been popular in ESD: first, the competency to envision and create the future, which is called future thinking (Buntting & Jones, 2015; Laherto & Rasa, 2022), corresponding to the behavioral domain; second, the competency to solve problems scientifically, corresponding to the cognitive domain; and third, the competency to develop empathy and collaboration, corresponding to the social-emotional domain. This frames the objectives of science education.

With regard to framing the science content connected to ESD, the focal themes are climate change, renewable energy, biodiversity, natural disaster prevention and mitigation, sustainable consumption and production, and health promotion. These issues are distributed in terms of coordinating the certainty of required and available knowledge and agreement on norms and values (Block et al., 2019). For example, climate change is located in the far direction on both coordinates and is a wicked problem. Therefore, school science should consider the characteristics of each of these issues.

In the Asian context, emphasis on any of these issues varies. In Japan, Taiwan, the Philippines, Indonesia, and other countries and regions severely affected by volcanic activity, earthquakes, and typhoons, natural disaster prevention and mitigation are treated considerably in school science. In Mongolia, climate change is an important educational topic in science and geography in schools as the country is vulnerable to climate change. In addition, certain studies have explored the potential of ESD in school science in relation to indigenous knowledge in the East, especially in Indonesia (Zidny et al., 2020), which is suggestive for teaching topics such as biodiversity conservation.

Framing appropriate pedagogical approaches to science education connected to ESD is in accordance with an action-oriented, innovative pedagogy that calls for a shift from teaching to learning, including self-directed learning, participation and collaboration, problem-orientation, inter- and transdisciplinarity, and the linking of formal and informal learning (UNESCO, 2017). Especially as it relates to the theme of climate change, Perkins et al. (2018) mention recent scholars' visions regarding education for sustainability revolving around five main aspects: (1) teaching all forms of scientific knowledge as rooted in some level of uncertainty, complexity, and nuance; (2) comprehensive empirical knowledge of climate change that includes its main principles, myths, and debates; (3) critical inquiry, integration, and engagement with global and cross-disciplinary perspectives; (4) transformative and learner-centered pedagogies conducive to climate change and sustainability curricula; and (5) authentic student-learner participation in the study and mitigation of climate change. These visions are also reflected in the discussion of ESD pedagogy and are expected to be applied in teaching and learning in school practices.

However, teachers express a difference between the teaching of sustainable development and that of science and geography (Summer et al., 2003), and science teachers are not always familiar with ESD approaches. For example, Mongolian reports (Tugjamba et al., 2018; Yembuu & Getsel, 2020) indicate that the knowledge and skills of Mongolian teachers have been limited in teaching ESD and climate change education; that is, they lack teaching methodologies, and therefore, they remain unsure how to transfer the knowledge to students. A report from Malaysia (Nayan et al., 2020) indicates that the problem is not only in teachers' lack of pedagogies in climate change education but also in their lack of climate literacy, especially in terms of skills and practices in mitigating and adapting to climate change.

Science teachers may have knowledge of problem-solving and inquiry-based pedagogical approaches; however, they are not adept at implementing them in class in reality. Moreover, addressing ESD's emphasis on participatory, collaborative, action-oriented, and trans- and interdisciplinary approaches is a major challenge for them. Several chapters in this book present the setting of competency-oriented learning objectives and the selection of themes that address local and global issues in ESD-infused school science curricula and lessons. Simultaneously, the chapters will approach the solutions through case studies to make the pedagogical approaches effective.

1.3 Science Teacher Education and ESD

1.3.1 Teachers' Competencies of ESD

Globally, one of the essential competencies for teachers is the competencies to practice ESD. The development of these competencies is consistent with the aims of UNESCO's ESD for 2030 – Priority Action Area 3 on building capacities of educators (UNESCO, 2020). To date, various frameworks and models have been proposed for teachers' competencies for practicing ESD (Sleurs, 2008; Rauch et al., 2008; UNECE, 2012; Bertschy et al., 2013; UNESCO, 2018). Derived from them, teachers are required not only to acquire competencies for sustainability citizens (e.g., knowledge, skills, attitudes, values, motivation, and commitment) but also to obtain profession-general competencies in order: (a) to help students acquire those competencies as a result of innovative teaching practices and (b) to expand these practices to people around the students and teachers.

Based on these competency models, the authors of this book have been promoting a project in the Asia-Pacific region that creates a competency framework for teachers practicing ESD in collaboration with 34 teacher education institutions

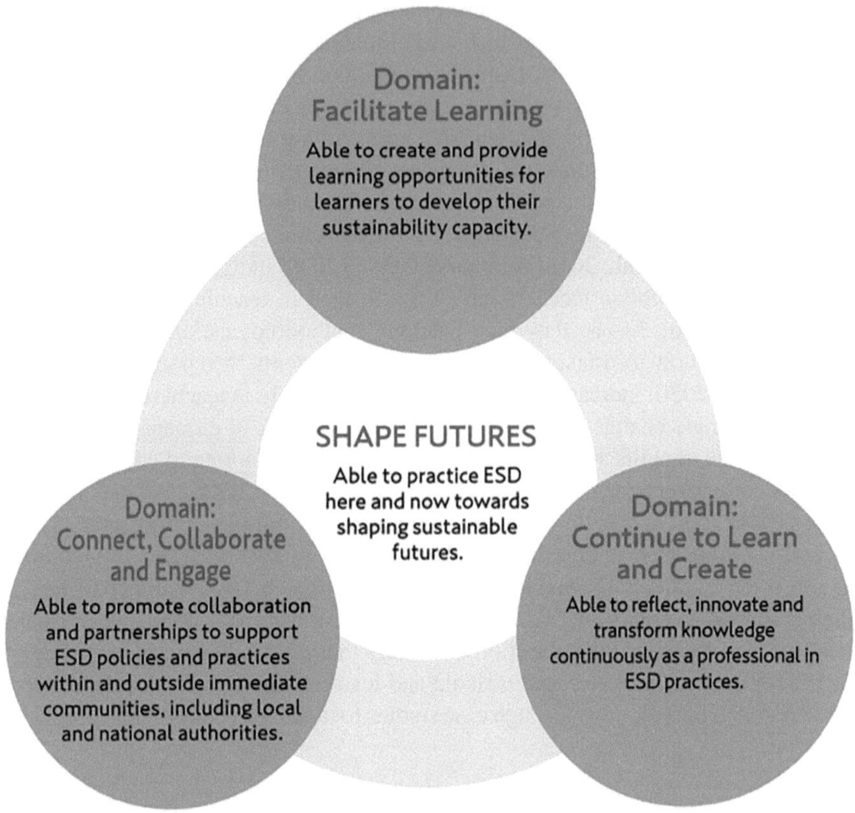

Fig. 1.1 Overview of the Asia-Pacific ESD Teacher Competency Framework

in 16 countries in the Asia-Pacific region, UNESCO Bangkok, and Asia-Pacific Cultural Centre for UNESCO. The framework centered on shaping the future comprises three domains: (a) facilitate learning, (b) continue to learn and create, and (c) connect, collaborate, and engage (Okayama University ESD Promotion Center, 2020) (Fig. 1.1). This was the first Asia-Pacific framework for ESD teacher competency, and it is significant that the development of the framework was accomplished in a region having remarkable diversity of nature, society, and culture. Examples of science teacher education programs and courses based on this framework will be discussed in several chapters of this book.

More recently, the European Union (EU) funded project "A Rounder Sense of Purpose" linked to the SDGs has proposed 12 competencies across three domains (think holistically, envisioning change, and achieving transformation) for teachers and other educators who contribute to sustainable development (Millican, 2022). Although these competencies are assigned to four phases (integration, involvement, practice, and reflexivity) per domain, they can be likened to the 12 colors of an artist's palette and act and interact in various combinations according to the context.

Interestingly, teacher education for sustainable development, including the development of ESD competencies, has the potential for innovation in teacher education in general. Fischer et al. (2022) identified the emphasis on the socio-environmental challenges of our time, methodologies to engage with knowledge diversity (e.g., action research orientation and inter- and transdisciplinarity research modes), and sustainability science learning approaches (e.g., scenario learning and forecasting/backcasting).

1.3.2 Developing Science Teachers to Practice ESD

Regarding science teacher education for sustainable development, Jegstad et al. (2018) with respect to a Norwegian teacher education program with student teachers indicated that an emphasis on experiential learning and sociocultural learning theory builds a thorough foundation for ESD. They argue that science student teachers with a fact-based teaching tradition that emphasizes knowledge transfer to students require more explicit guidance to reveal ESD expressions and contribute with a teaching tradition more in accordance with ESD.

Program development that explicitly addresses ESD is accumulating in Asia. A case of teacher education programs introduced by Fujii (2021) in Japan is a 2-month pre-service science teachers' course that targets ESD based on Lesson Study methodology – a rigorous sequential process for developing a lesson plan collaboratively with colleagues. A trial of the course presented diverse evidence suggesting that it promoted pre-service teachers' understanding of what science teachers should do to teach science lessons incorporating ESD and the capacity to accomplish this. Focusing on the theme of climate change, Chang (2014) organized two professional development workshops in Singapore, for in-service geography teachers, focusing on both their subject matter knowledge and pedagogical knowledge of climate change education. Through the workshop, the teachers were able to critique existing lesson plans and improve them to teach conceptual change and help students clarify misconceptions. The two aforementioned programs developed by Fujii and Chang focus on teachers acquiring general professional competencies to (1) help students acquire sustainability competencies as a result of innovative teaching practices and (2) extend these practices to their future colleagues.

Furthermore, a course for training pre- and in-service geography teachers in Mongolia (Yembuu & Getsel, 2020) offered three climate change education modules: (1) climate change and ecosystem, (2) climate change and society, and (3) ecosystem-based adaptation and mitigation. The aim of this course was to provide teachers not only pure scientific or factual knowledge but also to make an organic awareness of the importance of everyday sustainable practices. In addition, the course implemented a learner-centered pedagogical approach, such as inquiry- and project-based learning, as well as problem-solving skills. Early observations indicated that the course and other connected activities drove behavioral changes in pre-service teachers, such as saving water and paper. This case corresponds to the teachers themselves being required to acquire competencies as sustainability citizens.

Highlighting the characteristics of science teachers, many of them emphasize fact-based teaching, believing that once students receive the correct scientific knowledge, they will automatically take the right actions regarding environmental issues (Borg et al., 2012). However, sustainability issues, including environmental issues, involve a variety of perspectives, views, values, and norms and are more than a matter of knowledge. Therefore, these characteristics can act as obstacles for science teachers in implementing ESD. Additionally, science teachers view their role as that of communicating facts (McGinnis, 2003) and are reluctant to deal with problems that are far from the required and available knowledge and/or from agreement on norms and values – the very topics that ESD is supposed to address. Certainly, as Sjöström et al. (2017) mentioned, the predominance of a scientific literacy mindset that emphasizes scientific content and knowledge of scientific processes, as well as the usefulness and meaningfulness of science, may have influenced these characteristics of science teachers.

In light of the above, the challenges for science teachers practicing ESD are as follows. First, even if science teachers' primary interest is an ecological perspective on sustainable development themes, they must also have an economic and social perspective that is inseparable from the ecological perspective. Second, while science teachers, following the fact-based teaching tradition, tend to use lectures as the most common teaching method, it is important for teachers to engage students in sustainable development themes based on student-centered problem-solving and inquiry. Moreover, it is important for teachers to work in a more pluralistic manner. Third, teachers' positive attitudes toward ESD promote students' learning about sustainable development issues. The challenge, therefore, is to positively enhance science teachers' beliefs and attitudes toward ESD practices.

1.4 Keywords in Science Education Connected with ESD

As discussed, the connection between science education and ESD has become stronger. In charting that connection, Jegstad et al. (2018) presented a model that realizes elements of ESD by focusing on "teaching through science," which emphasizes how general education skills are promoted through the subject of science. The model is characterized by a core of students' acquiring scientific content knowledge, expanding to understanding science in context, understanding sustainable practices of science through nature of science, and thereafter acquiring ESD competencies and ultimately students and teachers developing the skills to participate and engage in "lived ESD" as a whole-school climate and environment.

Based on the concept of *Bildung*, Sjöström et al. (2017) also classified the scientific literacy that science education aims for into three categories: vision I, learning for individual skill development and personal growth; vision II, learning individual social participation by understanding science and its applications; and vision III, science education for values-driven transformation of both individual learners and society. In addition, the curriculum types for each vision are traditional structure-of-the-discipline- and history-of-science-driven curricula for vision I, context-based science education and classical STS curricula for vision II, and socio-scientific issue (SSI)-based science education and other curricula oriented toward critical sustainability for vision III. While these visions of scientific literacy coexist today, vision III, which is oriented toward critical sustainability, has been explored for individual and social actions and transformation through science education.

With reference to these models and classifications, we organize the keywords in science education connected with ESD based on the descriptions in the previous sections (Fig. 1.2). The objective of science education is to cultivate scientific literacy as an umbrella, ultimately transforming individuals and society and promoting social participation and civic engagement. The key to achieving this is the competencies of future, systems, and critical thinking, as well as action competencies. Furthermore, facilitating decision-making as a sustainability citizen necessarily considers norms and values. With respect to its themes, the focus will be on issues involving science, technology, society, and the environment (STSE), particularly complex SSI, based on addressing local and global issues. Specifically, issues related to climate change, renewable energy, biodiversity, disaster risk deduction, sustainable consumption and supply, and health promotion are diverse in terms of the certainty of required and available knowledge and agreement on norms and values. The pedagogical approach to science education is learner-centered,

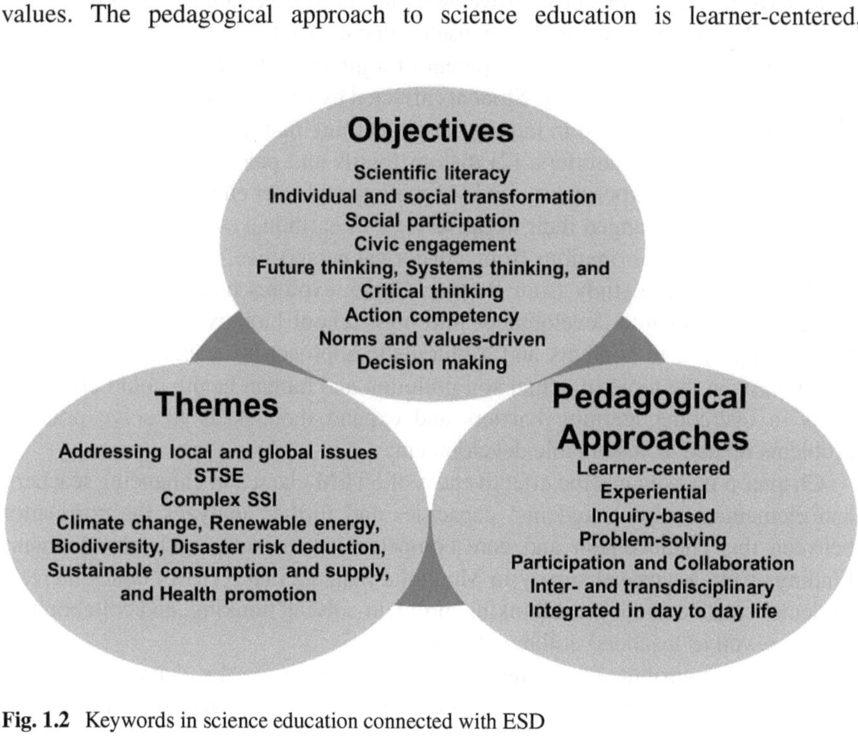

Fig. 1.2 Keywords in science education connected with ESD

inquiry-based problem-solving, and experiential learning. In response to the aforementioned objectives, the emphasis is on participation and collaboration with an inter- and transdisciplinary orientation. Consequently, local and global issues are integrated into daily life.

1.5 Structure of This Book

This book is arranged in 20 chapters in four parts: science teaching and learning for sustainable development in formal schooling, science curriculum for sustainable development in formal schooling, informal science education for sustainable development, and science teacher education for sustainable development. The relationship between the chapters and the keywords mentioned above is as follows.

Part I: Science Teaching and Learning for Sustainable Development in Formal Schooling

Part I covers a variety of research and practices in school science teaching and learning connected to ESD. Chapter 2, jointly by Japan and Korea, focuses on an 8-year science exchange program between Japanese and Korean high school students on the theme of sustainability. The chapter discusses the requirements for program effectiveness, indicating that through bilateral partnership and collaboration, students have made significant improvements in multifaceted understanding, critical thinking, decision-making, and response to global society.

Chapter 3 introduces the development of a junior high school science class for post-disaster recovery in Japan. Students affected by an earthquake discussed in this chapter have been engaged in inquiry-based learning that includes four themes: (1) volcanic deposits and aquifers, (2) diatom fossils and paleoenvironment, (3) lava plateaus and model experiments, and (4) geology and river channels. Consequently, the students have changed their negative view of earthquakes and are able to view the local geological formations from a historical perspective.

Chapter 4, a case study from Beijing, China, explores how to incorporate the concept of sustainable development into high school biology classes based on a holistic approach to contents and a pluralistic approach to teaching. An inquiry-based class on the topic of actual soil pollution and human health could allow students to overcome learning barriers and expand their skills to solve practical problems related to sustainable development.

Chapter 5 investigates the effectiveness of STEM classes in enhancing teachers' and elementary school students' capacities and further analyzes the correlation between this enhancement and core competencies for sustainable development. Findings from a national survey in Mongolia indicate that STEM classes increase students' creative and logical thinking, problem-solving thinking, and collaboration skills, as well as teachers' collaboration skills.

Chapter 6, contributed from the Philippines, proposes the "head, hand, and heart" approach to ESD for science teaching. The chapter emphasizes that science education has great potential for transformative learning by connecting with ESD as the head, referring to competencies from global and local perspectives; ESD as the hand, incorporating a skills-oriented teaching model; and ESD as the heart, emphasizing a value-driven approach to achieving holistic attitudes toward the environment.

Chapter 7 from Thailand highlights the activities of the Southeast Asian Ministers of Education Organization (SEAMEO) with its secretariat in Thailand to promote ESD in Southeast Asia. The SEAMEO-Japan ESD Awards, as one of the activities, identifies best practices of ESD in schools in Southeast Asian countries and has included many of the topics covered in science education, such as climate change, biodiversity, and sustainable consumption and production.

Part II: Science Curriculum for Sustainable Development in Formal Schooling

The part discussing the science curriculum for sustainable development in formal schooling begins with Chap. 8, which assesses the Korean national science curricula using an environmental education (EE) analysis framework. Findings from the analysis indicate that environmental knowledge features heavily in the curricula compared to environmental attitudes and values, competencies, and participation and action. The author argues that learning units related to EE should focus on improving cognitive leaning aspects, attitudes/values, and action via research activities for achieving well-rounded ESD.

Chapter 9 from Indonesia delves three approaches for infusing ESD into science curricula: ESD on the curriculum, utilizing ESD characteristics as a basis for problem-solving; ESD in the curriculum, incorporating ESD content into the curriculum or establishing ESD as part of the subject content; and curriculum of ESD, infusing ESD into all aspects of the curriculum. Subsequently, this chapter demonstrates that a volcano learning program employing the last approach developed for prospective science teachers enhances their problem-solving competencies.

Chapter 10 performs a multiple-case study, based on interviews with Indonesian science teachers, for examining their strategies for integrating sustainability into science curricula. Drawing on examples of community-based nature and cultural studies, this chapter illustrates how the local context can be used to complement ESD and the SDGs approach to ensure "sustainability as policy" by developing "sustainability as a frame of mind."

Chapter 11 presents a case study of the development of a Japanese elementary school science curriculum on biodiversity conservation and invasive alien species, one of the main themes of ESD. The intervention lessons discussed in this chapter have been successful in transforming students' perceptions from a unidimensional view to a neutral, multidimensional view that alien species could be "bad" and "good," depending on how humans dealt with them.

Part III: Informal Science Education for Sustainable Development

Transformative learning through ESD is not confined to in-school education, but extends to out-of-school education. One of the challenges, therefore, is how to develop science education connected to ESD in collaboration with the local community.

Chapter 12 introducing the collaboration between universities and social education facilities in Japan focuses on ways to attain the SDG 15 "life on land" by developing the teaching material *Kamishibai* based on dissemination activities held in zoo. This chapter suggests that observing animals in zoos and learning about their stories through *Kamishibai* can stimulate empathic responses in individuals, thus ultimately inculcating a sense of responsibility for the animal's well-being in people.

Chapter 13 from Japan reports that learning of junior high school students become a multifaceted and collaborative initiative between the teachers, students, and the local community members, giving rise to a multilayered communities of practices (CoP). It suggests that a transformational approach to learning is the core of ESD, and CoP for ESD comprise an aggregate of diverse and small communities that are overlapping and mutually inspired. The findings can be applied to science education for sustainable development collaborating with local communities.

Chapter 14 describes the United Nations University-registered ESD's Regional Center of Expertise in Penang, Malaysia, that is developing a university-school partnership project to enhance the quality of science education in the region. The project, a competition for young inventors to promote sustainable development, proves to be an effective platform for the exchange of knowledge, wherein students and teachers improve their scientific skills and apply their learnings to issues stretching beyond the boundaries of the school curriculum.

Chapter 15 focuses on the ASEAN Sustainability Leadership Program that involves researchers from various disciplines of higher education institutions and the local community, Siem Reap, Cambodia. Through this program with collaborative learning, participants are able to increase their awareness of learning about and engaging in discourse of sustainability, as well as explore ways to promote the study of sustainable development. It is clear that this program provides a platform for experiential and transformative learning by employing an inter- and transdisciplinary approach.

Part IV: Science Teacher Education for Sustainable Development

The last part of this book focuses on educators, especially teachers, who are key actors in facilitating the transformation of learners into sustainable ways of life. First, Chap. 16 reviews the development of science education in the twenty-first century China, analyzes the current situation and training in terms of science education, and reflects on Chinese science teachers' course of training in the context of ESD. This chapter illustrates that the implementation of new school curricula that emphasizes an active learning approach to solve practical issues can prove beneficial by improving the overall quality of science education.

Chapter 17 from Indonesia presents a faculty-wide science teacher training program that incorporates sustainable development. Students who enroll new compulsory ESD courses and existing elective courses with the addition of ESD demonstrate a strong sustainability awareness and improved competencies to practice ESD. This chapter exemplifies a whole-institution approach to ESD teacher education.

Chapter 18 focuses on the third domain "continue to learn and create: reflect, innovate, and transform knowledge continuously as a professional in ESD practices" of the Asia-Pacific ESD Teacher Competency Framework mentioned earlier.

The findings from this chapter are that application-based workshops and trainings including components of the domain have the potential to promote a more reflection-focused disposition in science teachers.

Chapter 19 delves into various challenges and potential solutions to improve ESD, in Kazakhstan, and pitches steps to achieve ESD goals in the country, some of which include private investment in education, digitalization of education, and the creation of an educational hub. These steps are certainly a prerequisite in educating and inspiring science teachers and other teachers to infuse ESD into their educational practices.

Chapter 20 from Laos outlines specific examples of the integration of ESD into the professional education of science teachers. Teachers who have improved their professional skills by engaging in numerous professional development activities have begun to improve their school curriculum and incorporate interactive teaching strategies into their classes. They are improving students' scientific understanding of the insect world, which is the basis for the formation of a view of nature.

Finally, the study from Myanmar in Chap. 21 aims to develop an in-service science teacher training program on ESD based on Lesson Study. The findings indicate an improvement of the science teaching-learning process in terms of the teachers' content knowledge, teaching methodology, and their perspective toward students' participation in science activities. With such training, science teachers will be able to create lesson plans to improve students' competencies, thereby equipping students to become sustainability inclined global citizen of the twenty-first century.

A tenable link between science education and sustainability is critical for achieving a sustainable society. These links appear in competencies for sustainability, themes and topics in educational content, pedagogical approaches, evaluation and assessment, and whole-institution/community approach discussed in the aforementioned chapters. By reading through the diverse chapters, we will receive a realistic picture of the progress and challenges of science education for sustainable development in Asia.

References

Bertschy, F., Künzli, C., & Lehmann, M. (2013). Teachers' competencies for the implementation of educational offers in the field of education for sustainable development. *Sustainability, 5*(12), 5067–5080.

Bezeljak, P., Scheuch, M., & Torkar, G. (2020). Understanding of sustainability and education for sustainable development among pre-service biology teachers. *Sustainability, 12*(17), 6892.

Bianch, G., Pisiotis, U., & Cabrera Giraldez, M. (2022). *GreenComp. The European sustainability competence framework.* Publications Office of the European Union.

Block, T., Van Poeck, K., & Östmann, L. (2019). Tackling wicked problems in teaching and learning. Sustainability issues as knowledge, ethical and political challenges. In K. Van Poeck, L. Östmann, & J. Öhman (Eds.), *Sustainable development teaching: Ethical and political challenges* (pp. 28–39). Routledge.

Borg, C., Gericke, N., Höglund, H., & Bergman, E. (2012). The barriers encountered by teachers implementing education for sustainable development: Discipline bound differences and teaching traditions. *Research in Science & Technological Education, 30*(2), 185–207.

Buntting, C., & Jones, A. (2015). Futures thinking in the future of science education. In D. Corrigan, C. Buntting, J. Dillon, A. Jones, & R. Gunstone (Eds.), *The future in learning science: What's in it for the learner?* (pp. 229–244). Springer International Publishing.
Chang, C. (2014). *Climate change education: Knowing, doing and being*. Routledge.
Eijck, M., & Roth, W.-M. (2008). Improving science education for sustainable development. *PLoS Biolology, 5*(12), 2763–2769.
Fischer, D., King, J., Rieckmann, M., Barth, M., Büssing, A., Hemmer, I., & Lindau-Bank, D. (2022). Teacher education for sustainable development: A review of an emerging research field. *Journal of Teacher Education, 73*(5), 509–524.
Fujii, H. (2021). Towards the identification of ESD competencies required for pre-service science teachers. In T. Isozaki & M. Sumida (Eds.), *Science education research and practice from Japan* (pp. 155–171). Springer Nature Singapore.
Japanese National Commission for UNESCO. (2021). *Guide to promoting education for sustainable development*. Japanese National Commission for UNESCO. (in Japanese).
Jegstad, K. M., Sinnes, A. T., & Gjøtterud, S. M. (2018). Science teacher education for sustainable development: From intensions to realisation. *NorDiNa (Nordic Studies in Science Education), 14*(4), 350–367.
Kopnina, H. (2020). Education for the future? Critical evaluation of education for sustainable development goals. *The Journal of Environmental Education, 51*(4), 280–291.
Laherto, A., & Rasa, T. (2022). Facilitating transformative science education through futures thinking. *On The Horizon: The International Journal of Learning Futures, 30*(2), 96–103.
McGinnis, J. R. (2003). The morality of inclusive verses exclusive settings. In D. L. Zeidler (Ed.), *The role of moral reasoning on socioscientific issues and discourse in science education* (pp. 195–216). Kluwer Academic Publishers.
Millican, R. (2022). A rounder sense of purpose: Competences for educators in search of transformation. In P. Vare, N. Lausselet, & M. Rieckmann (Eds.), *Competences in education for sustainable development* (pp. 35–43). Springer Nature Switzerland.
Ministry of Education. (2015). *2015 revised national curriculum overview*. Korean Ministry of Education. (in Korean).
Nayan, N., Mahat, H., Hashim, M., Saleh, Y., & Norkhaidi, S. B. (2020). Climate literacy awareness among preservice teachers in Malaysia. *Cakrawala Pendidikan, 39*(1), 89–101.
Okayama University ESD Promotion Center. (2020). *Guide for the effective dissemination of the Asia-Pacific ESD teacher competency framework*. Okayama University.
Perkins, K. M., Munguia, N., Moure-Eraso, R., Delakowitz, B., Giannetti, B. F., Liu, G., Nurunnabi, M., Will, M., & Velazquez, L. (2018). International perspectives on the pedagogy of climate change. *Journal of Cleaner Production, 200*, 1043–1052.
Rauch, F., Streissler, A., & Steiner, R. (2008). *Kompetenzen für Bildung für nachhaltige Entwicklung (KOM-BiNE). Konzept und Anregungen für die Praxis*. BMUKK.
Sjöström, J., Frerichs, N., Zuin, V. G., & Eilks, I. (2017). Use of the concept of Bildung in the international science education literature, its potential, and implications for teaching and learning. *Studies in Science Education, 53*(2), 165–192.
Sleurs, W. (Ed.). (2008). *Competencies for ESD (Education for Sustainable Development) teachers. A framework to integrate ESD in the curriculum of teacher training institutes*. Comenius 2.1 Project.
Summers, M., Corney, G., & Childs, A. (2003). Teaching sustainable development in primary schools: An empirical study of issues for teachers. *Environmental Education Research, 9*(3), 327–346.
Tugjamba, N., Yembuu, B., Gantumur, A., & Gezel, U. (2018). Research study on climate change education for sustainable development in Mongolia. In P. Ordóñez de Pablos (Ed.), *Management strategies and technology fluidity in the Asian business sector* (pp. 192–214). IGI Global.
Tytler, R. (2012). Socio-scientific issues, sustainability and science education. *Research in Science Education, 42*, 155–163.

UNECE. (2012). *Learning for the future. Competences in education for sustainable development*. UNECE.
UNESCO. (2017). *Education for sustainable development goals: Learning objectives*. UNESCO.
UNESCO. (2018). *Integrating Education for Sustainable Development (ESD) in teacher education in South-East Asia: A guide for teacher educators*. UNESCO.
UNESCO. (2020). *Education for sustainable development: A roadmap. ESD for 2030*. UNESCO.
United Nations General Assembly. (2019). *United Nations general assembly resolution 74/223. Education for sustainable development in the framework of the 2030 agenda for sustainable development*. United Nations General Assembly.
Wiek, A., Withycombe, L., & Redman, C. L. (2011). Key competencies in sustainability: A reference framework for academic program development. *Sustainability Science., 6*(2), 203–218.
Yembuu, B., & Getsel, U. (2020). Curriculum development on climate change adaptation: Preservice teacher training in Mongolia. In C. Chang, G. Kidman, & A. Wi (Eds.), *Issues in teaching and learning of education for sustainability: Theory into practice* (pp. 85–93). Routledge.
Zidny, R., Sjöström, J., & Eilks, I. (2020). A multi-perspective reflection on how indigenous knowledge and related ideas can improve science education for sustainability. *Science & Education, 29*, 145–185.

Part I
Science Teaching and Learning for Sustainable Development in Formal Schooling

Chapter 2
Partnership for Sustainability in Science Education: The Japan-Korea Student Science Exchange Program

Kosei Kajiyama, Seong-Jae Yoo, and Hiroki Fujii

2.1 Introduction

Adopting Goal 4, "Quality Education: Ensure inclusive and equitable quality education and promote lifelong learning opportunities for all," of the United Nations Sustainable Development Goals (SDGs) has galvanized efforts to mainstream Education for Sustainable Development (ESD) into each country's school curricula. The new Course of Study (Ministry of Education, Culture, Sports, Science and Technology, 2017) in Japan as the national curriculum guideline emphasizes the importance of equipping students with competencies that enable them to become future builders of a sustainable society. Furthermore, the Republic of Korea's national curriculum has included the high school subjects "Ecology and Environment" and "Environment and Green Growth" since 2007. Both subjects aim to promote students' knowledge and understanding of and attitudes toward environmental issues (Lee, 2014). Further, they have been merged into the subject "Environment" since 2015.

This common efforts form the basis of students' science exchange programs between Japan and Korea which will help address the contentious issues of sustainability on a global level, not just on a national or local level. These exchange programs may lead to new perspectives and insights and, hence, a deeper conceptual understanding of science (Slotta & Jorde, 2010). However, because of the different

K. Kajiyama
Hiroshima University High School, Hiroshima, Japan

S.-J. Yoo
Cheonan Jungan High School, Cheonan, Republic of Korea

H. Fujii (✉)
Okayama University, Okayama, Japan
e-mail: fujii-hi@okayama-u.ac.jp

H. Fujii, S.-K. Lee (eds.), *Science Education for Sustainable Development in Asia*, Education Innovation Series, https://doi.org/10.1007/978-981-99-8711-5_2

socio-cultural background of students from the two countries, the programs must prepare meaningful partnerships that ensure sharing across communities of understanding. In particular, science exchange programs must enhance international cooperation on and access to science, technology, and innovation as well as knowledge-sharing, as indicated by Target 17.6 of the SDG 17: "Partnerships for the goals: Strengthen the means of implementation and revitalize global partnerships for sustainable development."

Regarding the objectives of science exchange programs for sustainability, the programs should develop sustainability competencies in students. These competencies have been presented in various forms since the 2010s (e.g., Wiek et al., 2011), and the latest report by the European Commission (Bianch et al., 2022) listed a total of 12 competences from four areas: embodying sustainability values; embracing complexity in sustainability; envisioning sustainable futures; and acting for sustainability. With these references, the emphasis in exchange programs should be on thinking-related competencies such as critical thinking, systems thinking, and exploratory thinking, as well as value-related competencies, but especially collaborative competency. This is because collaborative competency is considered one of the essential key competencies required to address global sustainability challenges (Cörvers et al., 2016; UNESCO, 2017): Such challenges include wicked problems such as climate change and use of genetically modified organisms, which are far from the certainty of required and available knowledge and far from agreement on norms and values (Block et al., 2019). Encountering others with different ideas, views, values, and perspectives through exchange programs will provide an excellent opportunity for students to gain a realistic experience of how to deal with these problems.

This chapter discusses development and practice of science exchange programs between Japanese and Korean high school students, by answering the following research questions: (1) What are requirements for multicultural science exchange programs aimed on fostering partnership for sustainability? (2) What type of science exchange programs can be developed based on the above requirements? And (3) what effects would the developed science exchange program have on the students who participated?

2.2 Literature Review

Only a few studies have explored the potential of international collaboration for sustainability in science education. Arya and Maul (2016) developed an 8-week online educational program on climate change issues for secondary school students from the United States, China, New Zealand, and Norway in order to clarify through peer-review notions of what counts as scientific evidence. Interviews conducted following the program revealed a significant increase in the use of English terms associated with evidence-based reasoning. Arya and Maul conclude that the most valuable aspect of the students' experience may be the opportunity for students to interpret and evaluate scientific data with their peers, because through open exchanges with peers, students are afforded the freedom to form ideas about the published data, connect it with their own experiences, and learn about the experiences of others.

Similarly, Korsager and Slotta (2015) presented an online module over 6 weeks, designed with inquiry activities for international peer collaboration to teach climate change. Students ages 16–19 participated in this module from Canada, China, Sweden, and Norway, in which they were exposed to peers with other perspectives on climate change, making them better at organizing and connecting relevant scientific concepts to explain climate change issues and at explaining cause and effect patterns in suitable ways.

Focusing on the collaboration between proximate counterparts, Alkaher and Tal (2014) conducted an investigation into collaborative socio-environmental projects that engage eighth-grade Jewish and Arab students in Israel in outdoor learning about environmentally friendly actions. The students gained a better understanding of local socio-environmental issues as a result of completing the projects and also saw a decrease in mutual prejudices toward the environmental awareness and behavior of their counterparts. Further, the differences of the learning outcomes were not related to ethnicity, but rather to students' socioeconomic status and experience in environmental education programs.

Extending international collaboration to the case of university students, we find that Labby (2016) highlighted in place-based sustainable issues with the students' counterparts and conducted an exchange program between university students from Texas, in the United States, and Bucaramanga, Colombia. While immersing themselves in Bucaramanga culture, students of both universities spoke to students in a Bucaramanga local public school about soil erosion, water conservation, and energy production issues and raised environmental, social, and economic awareness of that city's efforts related to these issues. From this outcome, Labby argued that the cultural immersion reinforced providing collaborative educational opportunities to maximize the impact of education on the youth and communities from both cultures.

Furthermore, Finger et al. (2021) analyzed international exchange programs to enhance education for environmental citizenship that representative universities from Iceland, Serbia, Bosnia and Herzegovina, and Romania are developing within funding frameworks such as the European Cooperation in Science and Technology. By providing exchange opportunities for students from countries with obvious differences, that is, Iceland, where renewable energy sustains the country, versus the Balkan countries, which rely on fossil fuels, the exchange program raised the students' environmental awareness and increased the likelihood of them becoming environmental citizens. Further, the majority of the students who are aware of exchange programs believed that it is at least partially the responsibility of the state to provide environmental education to children and students.

Regarding the first research question, the findings of these studies indicate some of the requirements of multicultural science exchange programs that facilitate partnerships for sustainability. First and foremost, participants in the program should be of equal status as well as of heterogeneous backgrounds. The former pertains, for example, to students' age and academic ability to work jointly on the subject matter of the program, and the latter is the culture and society of the country or region to which students belong. Effective learning requires ensuring that students do not feel inferior to their counterparts and creating certain discrepancies between counterparts with different knowledge. Second, the theme of the program is the common concern of participants, and the pedagogical approach is required their collaborative activities based on scientific evidence to act for achieving common goals. The important thing is that they see sustainability issues as their own problem and become decision- and change-makers who connect and interact with others and picture and achieve a desirable future. Third, participants' mother tongues may vary, but smooth communication in one language (in this case English) is a success factor of the program. English skills that allow them to participate in scientific discussions will facilitate partnerships.

2.3 The Japan-Korea Student Science Exchange Program

2.3.1 Program Overview

From this section, the development and implementation of student science exchange programs and their educational effectiveness will be discussed to answer the second and third research questions. The Japan-Korea Student Science Exchange Program (JKSEP) (Fig. 2.1) is a collaborative effort between the Hiroshima University High

Fig. 2.1 Snapshots of the Japan-Korea science exchange program
(Top left, joint science lessons conducted in English; top right, presentation of experimental activity with a team consisting of both Japanese and Korean high school students; bottom left, keynote speech about ESD content; bottom right, group photo after the on-site joint lesson)

School (HUHS) in Japan and Cheonan Jungang High School (CJHS) in the Republic of Korea. Both schools are STEM-focused and have received support from their respective Ministries of Education. The program, which began in 2010, had been held 13 times in total and attracted over 1000 students and 100 teachers in total.

Students from both schools participated in a joint lesson conducted in English in which they learned about various sustainability issues through challenging scientific experiments (Table 2.1). Because of this, they were able to improve their English communication and writing skills in preparation for academic English discussion and presentation. Additionally, they travelled to each other's countries, which helped them become more aware of and appreciative of the other country's cultural traditions.

2.3.2 Evaluation Survey

Data Collection

To assess the impact of the program, teachers from both schools collaborated on the questionnaire for the survey, which was conducted between August and October 2016. In this survey, two questionnaires were administered: (1) a demographic questionnaire and (2) an evaluative questionnaire about the program's effectiveness. Three items made up the former, whereas the later comprised 32 items in 9 different topics. In all, 142 Korean students and graduates and 127 Japanese students and graduates who participated in the JKSEP completed the questionnaire.

Demographics of Respondents

As shown in Table 2.2, Items 1–3 were demographic questions about each respondent's gender, affiliation, and frequency of participation. First, 91% of the participants from Korea were male and 9% female, whereas 55% of the participants from Japan were male and 45% female. The Korean high school is a boys-only institution, so the entire student body at the school is male. Many high schools in Korea are single-gender, and they only accept either boys or girls as students.

However, the Office of Education with jurisdiction over CJHS found female high school students for the program from a nearby girls' high school, so 9% of the Korean students were girls. The gender ratio of Japanese student participants is more balanced. Because HUHS is coeducational, male students outnumber female students by approximately 10 percentage points. This disparity is unavoidable due to the two schools' distinct educational systems.

There is no significant difference between the two countries in terms of the current affiliation of the participants. However, in terms of frequency of participation, approximately 70% of the Japanese students participated twice, while approximately 50% of the Korean students participated once. The reason for this is that HUHS required the students to participate in the program twice during the year, while CJHS asked the students to participate once in order to give more students the opportunity to participate in the program.

Table 2.1 Themes and science experiments of the programs (P: Physics; C: Chemistry; B: Biology)

Year	Program theme	Science experiment	
2010/2011	Eco-friendliness and sustainability	1. Are LED lights eco-friendly!?	(P)
		2. Biodiesel	(C)
		3. Metal resources – Ion	(C)
2011/2012	Renewable energies and sustainability	1. Can we use solar batteries to power our homes?	(P)
		2. Using wind power	(P)
		3. Biodiesel	(C)
		4. Metal resources – Ion	(C)
2012/2013	New perspective on energy development	1. Experiment to measure the Planck constant	(P)
		2. Let us estimate the efficiency of Stirling engine!	(P)
		3. Chemical cell: An invention that has enriched our lives	(C)
		4. Bioethanol production from the fermentation of glucose using yeast	(B)
2013/2014	Disposable food containers	1. Strength and thermal property of food containers	(P)
		2. The properties of plastic	(C)
		To examine the density of plastic and how it burns	
		To examine the volume reduction of polystyrene foam with heat	
		To observe the dissolution of polystyrene foam	
		To reuse polystyrene foam	
2014/2015	Solar energy	1. Energy obtained from the sun – The use of solar cells	(P)
		2. Characteristics of a hydrogen fuel cell	(P)
		3. Electrolysis of water by sunlight – The use of titanium oxide as photocatalyst	(C)
		4. Measurement of rate of photosynthesis by green plants or bacteria	(B)
2015/2016	Hydrogen energy	1. The efficiency of electric motor	(P)
		Measurement of the motor's efficiency as a generator	(P)
		Estimation of the motor's efficiency as a converter of electric energy to kinetic energy	
		2. Study of hydrogen	(C)
2016/2017	Plants factories	1. Electrical properties of LEDs	(P)
		2. Photosynthesis and CO_2 concentration and nutrition in soil and plants	(C)
		3. Effects of different CO_2 concentrations on photosynthesis in plant	(C)
		4. Effect of the light spectrum on the growth of plants	(B)
2017/2018	Solar energy	1. Using solar cells to harness energy from the sun	(P)
		2. Study of solar to chemical energy conversion	(C)
		3. Assembly and evaluation of a "dye-sensitized solar cell"	(C)
		4. Chromatography of a plant's pigment	(B)

Table 2.2 Demographics of the questionnaire respondents

No	Item		Japan		Korea	
1	Gender	Male	70	(55%)	129	(91%)
		Female	57	(45%)	13	(9%)
2	Affiliation	High school second year	32	(25%)	42	(30%)
		High school third year	42	(33%)	37	(26%)
		College students	42	(33%)	59	(42%)
		Others	9	(7%)	4	(3%)
3	Frequency of participation	Once	34	(27%)	73	(51%)
		Twice	93	(73%)	55	(39%)
		Three times	0	(0%)	14	(10%)

Evaluative Questionnaires

The evaluative questionnaire was scored on a 5-point Likert scale: (1) *not at all*; (2) *not really*; (3) *undecided*; (4) *somewhat*; and (5) *very much*. As mentioned above, the questionnaire covered 9 topics and 32 questions (Table 2.3).

Questionnaires Results

The majority of participants from both countries chose "somewhat" or "very much" to respond to the majority of the questionnaire items, indicating that they viewed the program favorably and agreed that it aided in their comprehension of ESD content (Fig. 2.2) and increased their learning output (Fig. 2.3), creativity (Fig. 2.4), and personal growth (Fig. 2.5) while also fostering global talent (Fig. 2.6), language learning (Fig. 2.7), career choice (Fig. 2.8), and Japan-Korea relations (Fig. 2.9). In general, they reacted favorably to the program (Fig. 2.10): It is thought capable of serving as a model for international exchange as well as adding students in understanding sustainability issues. Almost all of the students who took part in this survey stated that they had a general understanding of the concept of sustainable development after completing this joint program (Fig. 2.11).

According to the data, students also improved their creativity and research abilities. Students who participated in this program were afforded a variety of opportunities to broaden their horizons and gain a better understanding of human society and nature on the earth. Because they were all high school students at the time, this joint lesson program piqued their interest in studying abroad and instilled a strong desire in the students to study English. Because it was the first international science exchange program established jointly by a Japanese and a Korean high school, teachers and students communicated in English throughout the program: English is most widely spoken language in global communication.

Furthermore, this program provided excellent opportunities for students from Japan and Korea to change their minds and perspectives on each country. The early data show how much more positive they have become in viewing each other. This program's homestay system also allowed them to immerse themselves in the cultures of the other country.

Students in this program frequently collaborated and held discussions about various issues derived from the agenda, such as sustainable development. They

Table 2.3 Evaluative questionnaires about the program's effectiveness

No.	Topic	Questions
1	ESD content	*Q1. Do you know about the concept of ESD?*
		Q2. Did you understand the keynote speakers' lectures about ESD?
		Q3. Do you think that the course materials were appropriate?
2	Learning output	*Q4. Did this program improve your learning?*
		Q5. Did you obtain any lessons from this program?
		Q6. Did this program improve your problem-solving skills?
		Q7. Did this program motivate you to learn?
3	Creativity	*Q8. Did this program improve your self-management skills?*
		Q9. Did this program improve your information processing knowledge and skills?
		Q10. Did this program improve your creative thinking skills?
		Q11. Did this program improve your aesthetic sensibilities?
		Q12. Did this program improve your communication skills?
		Q13. Did this program improve your sense of community?
4	Personal growth	*Q14. Do you think that this program is an important part of your high school experience?*
		Q15. Do you think that this program affected the formation of your personality?
5	Global talent	*Q16. Do you think that this program is important to foster global talent?*
		Q17. Do you think that this program is important in fostering a sense of internationality?
6	Language learning	*Q18. Did this program affect your perception of the need to learn English?*
		Q19. Did this program affect your desire to learn English?
		Q20. Did this program affect your motivation to learn English?
7	Career choice	*Q21. Do you think that this program will affect your career choice in the future?*
		Q22. Do you think that this program will affect your choice of university major?
8	Japan-Korea relations	*Q23. Did the homestay in this program help you to understand Japanese/Korean culture?*
		Q24. Did this program change your awareness of Japan/Korea?
		Q25. Did this program change your view of Japan/Korea to be more positive?
		Q26. Did this program change your view of Japan/Korea to be more negative?
		Q27. Do you think that this program can improve relations between the two countries?
		Q28. Did this program help you to understand Japan/Korea?

(continued)

Table 2.3 (continued)

No.	Topic	Questions
9	Overall evaluation	*Q29. Did this program help you to understand sustainability issues?*
		Q30. Do you think that this program is a good model for an international exchange?
		Q31. Do you think that this program is a good model to improve teachers' abilities?
		Q32. Do you think that this program can be applied in other classes?

Fig. 2.2 ESD content

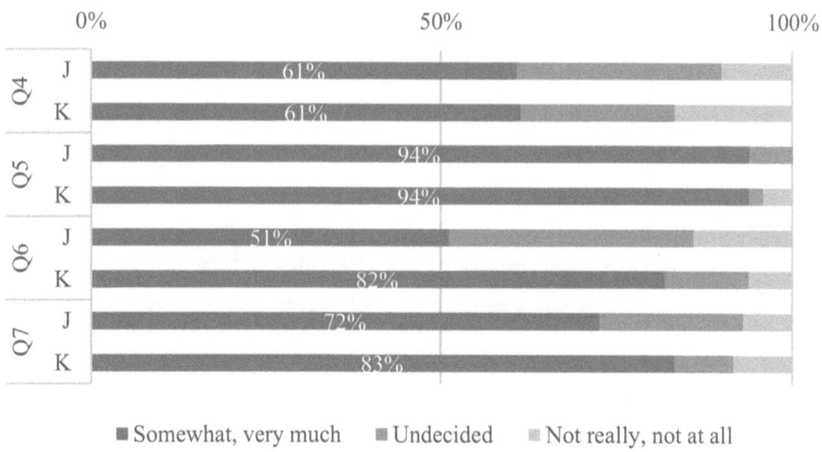

Fig. 2.3 Learning output

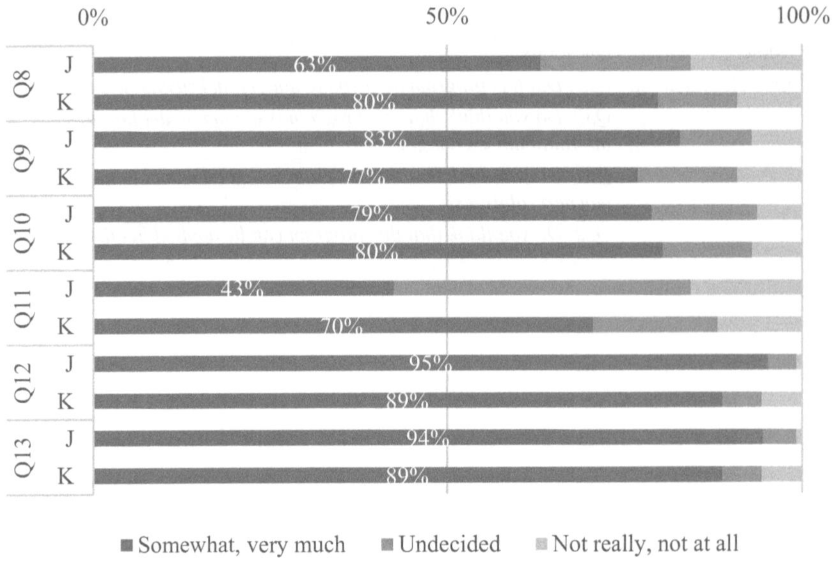

Fig. 2.4 Creativity

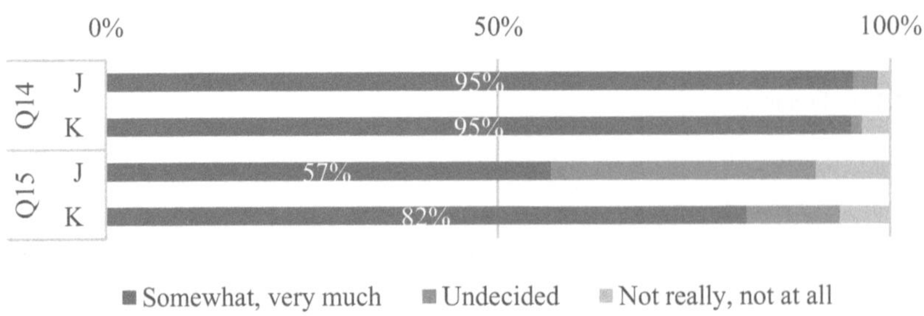

Fig. 2.5 Personal growth

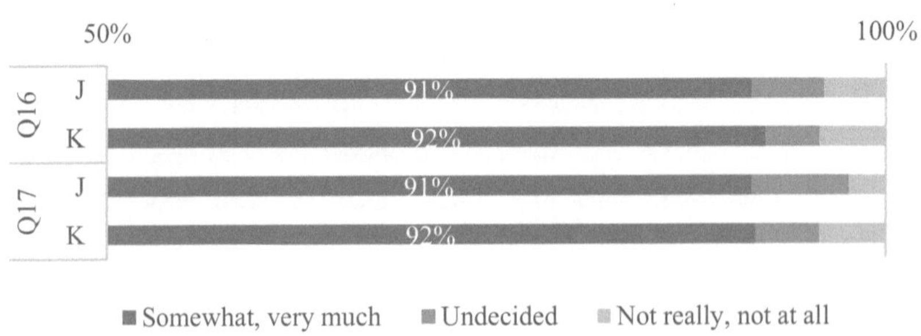

Fig. 2.6 Global talent

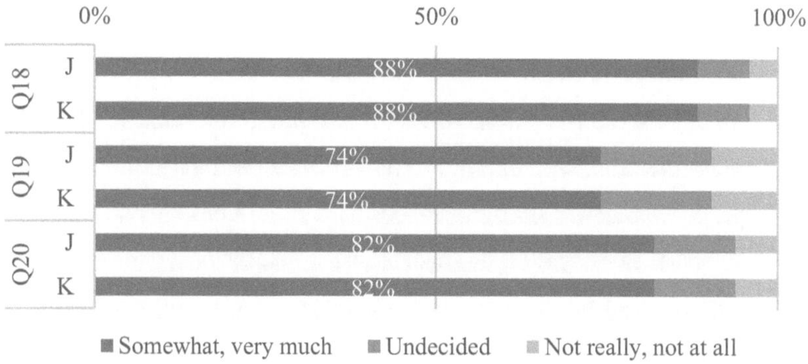

Fig. 2.7 Language learning

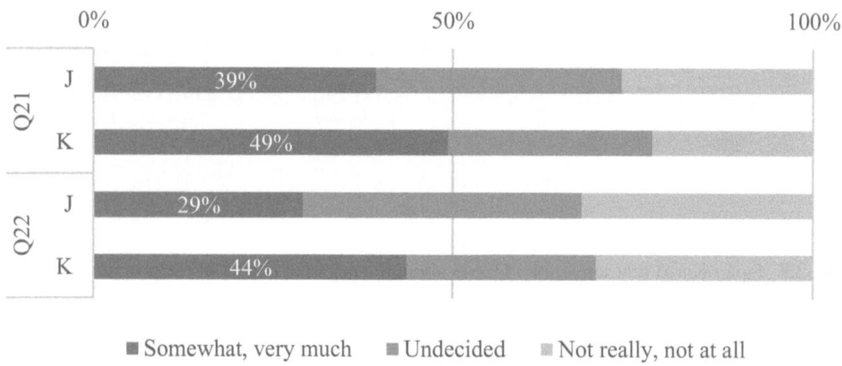

Fig. 2.8 Career choice

collaborated to complete assignments and prepare for English poster presentations. Such a learning environment has been proposed as a model for the science exchange program, in which they can cultivate their creative minds and critical thinking skills.

The collaborative lessons in the program also proved to be extremely beneficial to teachers. They had been involved in this program since its inception, and they finally saw it as a highly appropriate model for developing a diverse range of talented students who are both creative and convergent. To develop a co-teaching syllabus, teachers from both countries frequently exchanged emails, held numerous video conferences, and discussed international sustainability issues. As a result, not only have the students benefited from this program, the teachers have as well.

Many students and teachers traveled to HUHS in Japan as part of this program and to CJHS in Korea. In the long run, this program could serve as a model for a country-to-country exchange program. Furthermore, the teaching and learning contents of this joint lessons in the program can be applied to Japan's and Korea's exchange program policies.

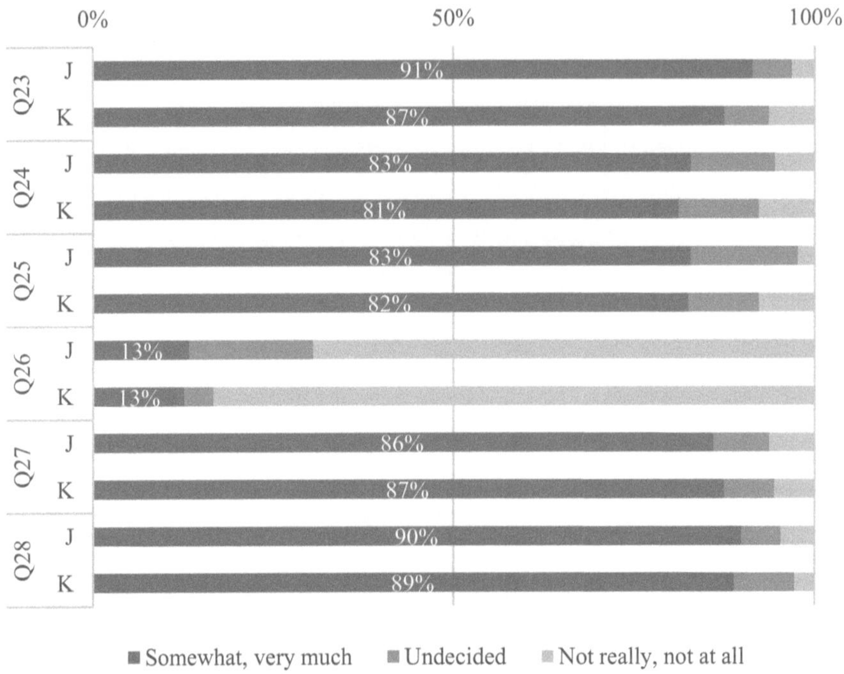

Fig. 2.9 Japan-Korea relations (note: Q26 is about the negative impact of the program)

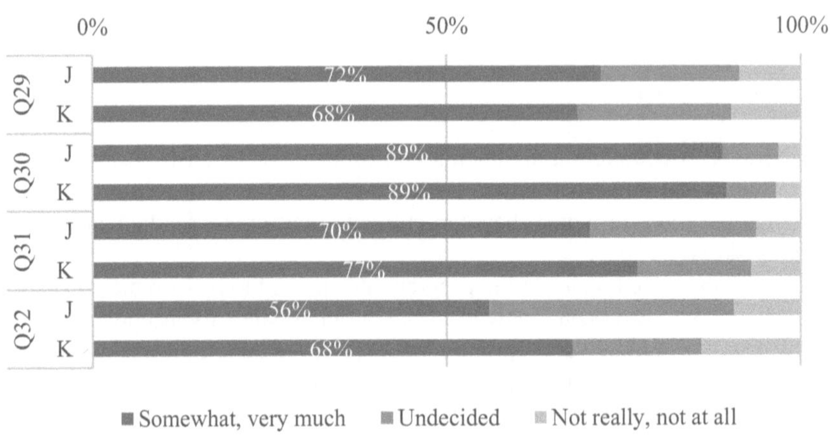

Fig. 2.10 Overall evaluation

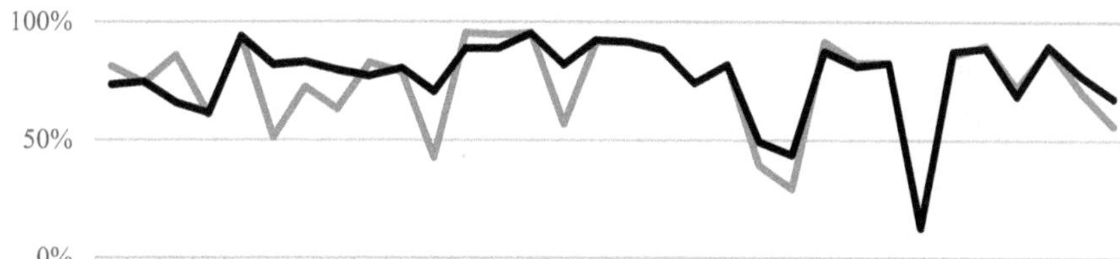

	1	2	3	4	5	6	7	8	9	10	11	12	13	14	15	16	17	18	19	20	21	22	23	24	25	26	27	28	29	30	31	32
J	81	74	86	61	94	51	72	63	83	79	43	95	94	95	57	91	91	88	74	82	39	29	91	83	83	13	86	90	72	89	70	56
K	73	75	65	61	94	82	83	80	77	80	70	89	89	95	82	92	92	88	74	82	49	44	87	81	82	13	87	89	68	89	77	68

Fig. 2.11 Comparison of positive responses by Japanese (J) and Korean (K) students

2.4 The 2014–2015 Student Science Exchange Program on the Theme of Solar Energy

2.4.1 Program Overview and Program Evaluation Method

To provide better understanding about the developed program, this section discusses the 2014–2015 lessons about solar energy from the perspective of HUHS students. In July 2014, prior to the on-site lessons in Korea, 21 male and 19 female students of the HUHS participated in 3 preparation sessions: (1) a 50-min lecture on the physics and chemistry knowledge of the SDGs; (2) a 50-min lecture on Korean language and culture; and (3) a 180-minute lecture and exercise on academic presentations in English. The students then visited the Republic of Korea for a 4-day science exchange program organized from joint science lessons, a cultural tour (visit to the Gyeongbokgung Palace and the Namdaemun Market in Seoul), and a homestay.

In January 2015, students at HUHS prepared to welcome students from CJHS to their school for a 4-day program that included advanced joint science lessons, a cultural tour (visit to the Hiroshima Peace Memorial Museum and the Itsukushima Shrine), and a homestay. Both joint lessons were created to assist participating students in developing their [1] multifaceted understanding, [2] critical and creative thinking, [3] decision-making, and [4] abilities and viewpoints to respond to a global society.

The lesson consists of seven steps (Tables 2.4 and 2.5): First, teachers introduced students to the issue of sustainability through a lecture; second, students made tentative decisions on their own; third, students conducted science experiments; fourth, students collected data through websites and held discussions to support their decision; fifth, students worked in groups to make an evidence-based decision about the use of solar energy; sixth, students made class presentations and had discussions about the result of their experiment and group work; and seventh, students made their final decisions individually.

Table 2.4 On-site lessons in the Republic of Korea (July 2014)

Physics group	Chemistry group
[1] Encountering the issue: Lecture about solar energy (1)	Same as on the left
[2] Individual tentative decision-making (pre-test questionnaire)	Same as on the left
[3.1.a] Question: Will we be able to cover energy consumption nationwide using solar energy?	[3.1.b] Question: What do you think about the future of utilizing titanium oxide for solar energy?
[3.2.a] Science experiment	[3.2.b] Science experiment
Measurement of energy harnessed from the sun using solar cells	Electrolysis of water by sunlight using titanium oxide as a photocatalyst
[4] Data collection and group discussion	Same as on the left
[5] Evidence-based decision-making in group	Same as on the left
[6] Class presentations and discussions	Same as on the left
[7] Individual final decision-making (post-test questionnaire)	Same as on the left

Table 2.5 On-site lessons in Japan (January 2015)

Physics group	Chemistry group	Biology group
[1] Encountering the issue: Lecture about solar energy (2)		
[2] Individual tentative decision-making (pre-test questionnaire)		
[3.1] Question: Will we be able to cover energy consumption nationwide using solar energy?		
[3.2.a] Science experiment Characteristic of hydrogen fuel cells	[3.2.b] Science experiment Assembly and evaluation of dye-sensitized solar cells	[3.2.c] Science experiment Measurement of rate of photo -synthesis by green plants or bacteria
[4] Data collection and group discussion		
[5] Evidence-based decision-making in group		
[6] Class presentations and discussions		
[7] Individual final decision-making (post-test questionnaire)		

The questions in Table 2.6 were used to determine the effectiveness of the developed lessons. Each question was open-ended and was designed to assess mastery of the learning objectives outlined above. Focusing on the on-site lessons in Japan, the following section discusses the theoretical foundations of each indicator in greater detail. Except for Question 4, each question was administered as a pre- and post-test.

Table 2.6 Data collection instrument

No	Indicators	Question
1	Multifaceted understanding	*The use of solar energy has increased recently. Discuss the advantages and disadvantages of using solar energy as a source of energy. Give specific reasons for your ideas.*
2	Critical and creative thinking	*What is the best way to use solar energy in the future? Describe your ideas and provide justifications for them by pointing out their advantages and disadvantages.*
3	Decision-making	*Assume you are the mayor of a city and overseeing the construction of the city's energy supply facility. Determine the ideal facility. Distinguish the relative importance of various energy supply resources (fossil fuel, solar power, wind power, hydraulic power, wave and tidal power, geothermal power, biomass, and nuclear power).*
4	Abilities and viewpoints to respond to global society	*Write a reflection on your participation in the Japan-Korea student science exchange program.*

2.4.2 *Multifaceted Understanding*

The term "multifaceted understanding" of an issue is defined in this study as students' ability to "develop numerous thoughts about a theme." For instance, when students can articulate the advantages and disadvantages of solar energy, it can be inferred that they were able to generate a variety of viewpoints on the solar energy issue. For each student, the number of codes and their rearrangement before and after the lessons were examined. If a student wrote "clean" in both pre- and post-test, it can be inferred that the lessons had no effect on that student. However, if a student who wrote "clean" on the pre-test did not include it on the post-test, it can be inferred that the lessons influenced the student.

Table 2.7 summarizes the number of codes indicating advantages and disadvantages aspects of solar energy use. In the table, the term "total number" refers to the sum of each code that corresponds to at least one pre-post investigation. The analysis revealed no statistically significant differences in the total number of advantages and disadvantages codes. By contrast, nearly 80% of students rearranged the codes.

The term "change in number" is calculated using the following formula. When a student described the same code from pre to post investigation, the number is "0." When a student described a new code, the number is "1" and "-1" when a code was erased from pre- to post-test. As a result, the more significant the absolute change in number, the more frequently students rearranged a code, indicating that the code influenced students' assessment of the benefits and drawbacks of solar energy use.

A two-tailed test of difference for each code showed that the advantage "flexible setting place" increased ($t_0 = 2.111 > t_{0.10}$ (1.734), $df = 18$), whereas "small scale" decreased ($t_0 = 1.860 > t_{0.10}$ (1.753), $df = 15$), indicating that the two codes are marginally significant at the 10% significance level on both sides. The disadvantage "lifetime of equipment" also increased ($t_0 = 2.449 > t_{0.10}$ (2.132), $df = 4$), indicating

Table 2.7 Summary of codes about multiple aspects of solar energy use

	Codes	Total number	Change in number
Advantageous	Clean	30	0
	Unlimited solar energy	29	3
	Flexible setting place	19	7
	Small scale	16	−6
	Usable for long-term	5	−2
	Safety and security	3	0
	Technical innovation	3	1
Disadvantageous	Unstable	26	−5
	High cost	25	1
	Low efficiency	20	3
	Need for a large area	13	−1
	Lifetime of equipment	5	3
	Insufficient energy supply	3	2
	Regional difference	3	2

that this code is marginally significant on both sides at the 10% level of significance.

Before the implementation of the lessons, the students wrote codes such as "clean," "unlimited solar energy," "unstable," and "high cost," which are directly unrelated to the lessons. The fact that 80% of the students rearranged the codes following the lessons indicates that participants appeared to have increased their capacity for multi-perspective thinking. Based on these findings, it is possible to conclude that the lessons provided students with a multifaceted understanding of solar energy use.

2.4.3 *Critical and Creative Thinking*

Ennis (1987 and 1996) defines critical thinking as "reasonable reflective thinking focused on deciding what to believe or do." Additionally, it is strongly associated with creative thinking. As a result, students' responses to the question "describe your ideas" can be interpreted as determining what to believe or do, while the response to "provide justifications for them (ideas) by pointing out their advantages and disadvantages" provides context for examining their critical thinking as a result of reasonable reflection.

Analysis of this indicator involved three steps. First, students' ideas for solar energy use were tallied. After classifying the ideas under five categories (Table 2.8), the frequency of each category was determined. Second, the students' descriptions of the advantages and disadvantages of using solar energy were reviewed, and the frequencies of all text data were counted. Following that, these descriptive data were justified and finally classified into four distinct notions (Table 2.9), and their

Table 2.8 Codes about ideas regarding solar energy use

Category	Sample text
Flexibility of placement and scalability	*Can be placed in home, desert, and space, large-scale*
Technological innovation	*Development other than solar panel, increasing efficiency, wireless transmission of electricity, conversion to hydrogen*
Combination with alternative energy resources	*Combination with geothermal, use of photosynthesis energy, coexistence with existing energy sources*
Government support	*Government subsidy, selling electricity*
Other general idea about usage	*Cost reduction, source of electric and thermal energy*

Table 2.9 Notions about advantages and disadvantages of ideas regarding solar energy use

Notion	Justification	Sample text	
Sustainability	Low environmental burden/scalability/contributes to a recycling-oriented society	[adv.]	*Clean, zero emission, eco-friendly, infinite resource, renewable, easy to start from small scale, durable, stable*
		[dis.]	*Unstable, limited lifetime of the equipment, environmentally unfriendly*
Quality of life	Safety and security/economic aspect/convenience/health aspect	[adv.]	*Flexible installing site, accessible at home, safe, economical, profitable*
		[dis.]	*High cost, increase of taxes, shortages of electric generation*
Challenge to science and technology	Progress of science and technology	[adv.]	*Industrial innovation, high efficiency, maximization of the electric power*
		[dis.]	*Inefficiency, large amount of time required to generate electricity, large area of land required for the equipment installation, the limit of technology*
Regionality	Fairness/no disparity in society/peace/society where human rights are respected	[adv.]	*Complex reasons, self- sufficiency, local production for local consumption*
		[dis.]	*Regional disparities in power generation*

frequency was investigated. Third, correlation analysis was conducted of the data in Tables 2.8 and 2.9.

The Use of Solar Energy

When the total text data for solar energy use before and after the lessons were compared, the number increased by 1.7 times from 53 to 91, and the increases were statistically different ($t_0 = 5.685 > t_{0.01}$ (2.733), $df = 33$). The frequency of solar energy use was calculated using categorized text data, and the result was statistically significant ($\chi^2(4) = 10.68$, $p < 0.05$). It is worth noting that the categories "flexibility of placement and scalability" and "technological innovation" both increased significantly, as shown in Table 2.10.

Table 2.10 Pre- and post-test result about ideas regarding solar energy use ($N = 34$)

Category	Pre-test (%)	Post-test (%)
Flexibility of placement and scalability	26(49)	39(43)
Technological innovation	6(11)	29(32)
Combination with alternative energy resources	5(9)	9(10)
Government support	4(8)	4(4)
Other general ideas and usages	12(23)	9(10)
Total	53(100)	91(100)

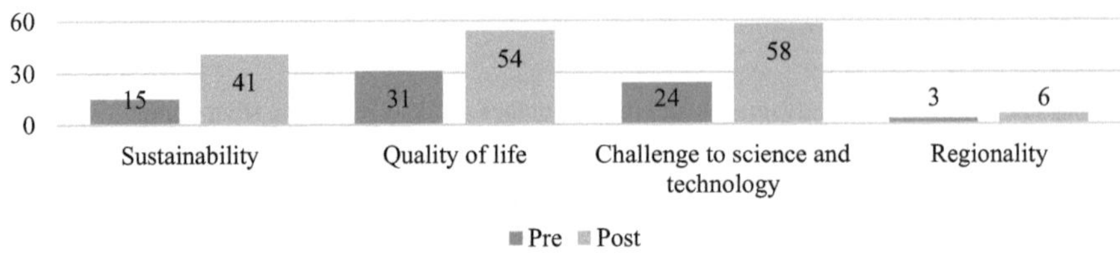

Fig. 2.12 Transformed frequency of four notions ($N = 34$)

Table 2.11 Frequency of the adaptation change of four notions ($N = 34$)

	Pre-test		Post-test		
Notion	%	*SD*	%	*SD*	t_0
Sustainability	0.44	0.87	1.18	1.13	4.019***
Quality of life	0.88	1.01	1.59	0.90	3.630***
Challenge to science and technology	0.68	0.88	1.68	1.86	3.321***
Regionality	0.09	0.29	0.18	0.47	1.547(ns)

$^{**}p < 0.01$, $^{***}p < 0.001$

Advantages and Disadvantages of Ideas Regarding the Use of Solar Energy
Pre- and post-test frequency of all text data related to the advantages and disadvantages of ideas regarding solar energy use increased by 2.2 times, from 73 to 159. A one-tailed *t*-test showed a significant increase in the frequency of the advantages and disadvantages (advantages; $t_0 = 5.398 > t_{0.01}$ (2.733), disadvantages; $t_0 = 5.533 > t_{0.01}$ (2.733)). The frequency of each notion increased 1.7 to 2.7 times for four notions: sustainability, quality of life, challenge to science and technology, and regionality (see Fig. 2.12). The four notions of advantages and disadvantages regarding the use of solar energy were evaluated using a one-tailed *t*-test (different test) of the frequency data for each person before and after the lessons. According to the findings, the students adapted concepts such as sustainability, quality of life, and challenge to science and technology, but not with regionality (Table 2.11).

Table 2.12 Frequencies of the notions of advantages and disadvantages; Categorization: flexibility of placement and scalability ($N = 37$)

	Notion regarding advantages and disadvantages				
	Sustainability	Quality of life	Challenge to science and technology	Regionality	Total
Pre.	6	16	13	1	36
Post.	13	25	14	5	57
Sum.	19	41	27	6	93

Associations Between Idea About the Use of Solar Energy and Its Respective Advantages and Disadvantages

The chi-square test revealed a statistically significant association between the categorization concept "flexibility of placement and scalability" and the concept of advantages and disadvantages ($\chi^2(3) = 12.81$, $p < 0.001$: Table 2.12). The concepts of "sustainability" and "quality of life" increased in frequency significantly more than the concept of "challenge to science and technology." This means that after the lessons, the students were able to apply the various advantages and disadvantages of solar energy to their own lives. The other concept tested for categorization was also statistically significant for "technological innovation" ($\chi^2(3) = 11.75$, $p < 0.001$).

These findings suggested that the students decided about using solar energy and provided a reasonable and reflective explanation for their decision. In comparison, the example adapted the concept of solar energy use to the concept of "regionality," with the result being only 0.18 per person after the lessons (Table 2.12). Furthermore, there was a discrepancy between the categorizations of solar energy concepts such as "combination with alternative energy sources," "government support," and "other general idea and usage," as well as the concepts of advantages and disadvantages.

According to these findings, participants lacked the ability to think critically and creatively about the overall use of solar energy. Therefore, the lessons are insufficiently reflected to "assist them in comprehending the complexities of developing a balanced approach that integrates socioeconomic, ecological, cultural, and ethical factors that all contribute to the sustainability of development" (UNESCO, 2009). The joint lessons were primarily taught through scientific experiments, but there was insufficient emphasis on socioeconomic issues. This is because the challenge remains that while these lessons may benefit the development of the ability to think scientifically about sustainability issues, they have been underutilized as a "balanced approach."

2.4.4 Decision-Making

The question about this indicator was drawn from Bolte et al. (2010). Table 2.13 displays the pre-lesson responses of three students as an example, where the smaller the number, the more critical the decision about energy supply resources. For example, student 1 preferred nuclear energy, while student 2 preferred solar energy.

Table 2.13 The raw data of the order of importance in the energy supply resources

No.	Fossil	Solar	Wind	Hydraulic	Wave	Geo.	Biomass	Nuclear	Total
1	4	5	4	6	5	5	3	2	34
2	8	1	3	7	6	2	5	4	36
3	2	3	3	2	4	3	4	2	23

Table 2.14 The normalized data of the results of Table 2.13

No.	Fossil	Solar	Wind	Hydraulic	Wave	Geo.	Biomass	Nuclear	Total
1	0.12	0.15	0.12	0.18	0.15	0.15	0.09	0.06	1.00
2	0.22	0.03	0.08	0.19	0.17	0.06	0.14	0.11	1.00
3	0.09	0.13	0.13	0.09	0.17	0.13	0.17	0.09	1.00

Table 2.15 Changes in the importance of each energy supply resource ($N = 33$)

	Fossil	Solar	Wind	Hydraulic	Wave	Geo.	Biomass	Nuclear	Total
Pre.	3.65	3.65	4.13	3.68	5.20	4.19	4.85	4.65	33.00
Post.	3.43	3.57	4.41	3.79	4.99	4.21	4.77	4.83	33.00

Although students 2 and 3 both indicated the number "two" (Student No. 2: "Geo.," Student No. 3: "Fossil, Hydraulic, and Nuclear"), their meanings might have differed because each student's total number differed. Furthermore, students 1 and 3 did not express the order from 1 to 8 because the numbers are on an ordinal scale, comparing students is difficult. To avoid this problem and compare the importance of each row of numbers, each row of numbers was converted using the order correlation method (Table 2.14). Before the lessons, the students believed that "Fossil" and "Solar" energy were supply resources of equal importance; however, by the end of the lessons, they decided that "Fossil" energy was the best (Table 2.15). The *F*-distribution equal variances test, which was used to confirm the difference in the importance of energy supply resources, revealed no significant differences in preferences before and after the lessons ($F\ (7,7) = 1.03$, $p > 0.05$).

To investigate shifts in decision-making regarding the significance of each energy supply resource, after the determination of Spearman's correlation coefficients before and after the lessons for each energy supply resource, a *t*-test was performed using a *t*-distribution table ($N = 33 > 30$). The mean ($\overline{X}$), standard deviation (*SD*), correlation coefficient (r), and probability of dominance (π_d) are all shown in Table 2.16. When the correlation coefficient $r = 1$, it means that the importance of energy supply resources remained constant.

As shown in Table 2.16, because the correlation coefficient r of "Fossil" had the smallest numerical value, the program had a significant impact on the decision-making process for the fossil energy supply resource. On the other hand, the greatest numerical value r of "Geo. (geothermal power)" indicates that it did not have a significant influence. The probability of dominance π_d denotes a ratio of change in the importance of energy supply resources; if nothing has changed the average, the numerical value π_d is equal to 0.500; the greater the numeral of the difference in the

Table 2.16 Comparison of pre and post-decision-making about the importance of energy supply resources ($N = 33$)

	Fossil		Solar		Wind		Hydraulic		Wave		Geo.		Biomass		Nuclear	
	Pre-	Post-	Pre-	Post-	Pre-	Post-	Pre-	Post-	Pre-	Post-	Pre-	Post-	Pre-	Post-	Pre-	Post-
$\bar{X}$	0.109	0.102	0.095	0.105	0.122	0.129	0.109	0.111	0.153	0.146	0.123	0.122	0.141	0.145	0.134	0.140
SD	0.043	0.033	0.039	0.044	0.033	0.035	0.042	0.037	0.030	0.035	0.042	0.040	0.047	0.041	0.059	0.062
r	0.361*		0.644***		0.690***		0.789***		0.463***		0.860***		0.510**		0.717***	
π_d	0.562		0.617		0.612		0.519		0.584		0.510		0.528		0.551	

*$p < 0.05$, **$p < 0.01$, ***$p < 0.001$

average increases, the larger the numerical value of π_d. The π_d for "Solar" has the highest numerical value, indicating that many students downplayed the importance of solar energy supply after the lessons.

The numbers in Table 2.16 for the "Solar" and "Biomass" resources reveal a significant difference, indicating that the students changed their decisions regarding these resources. The students discovered the low efficiency of solar and biomass energy supply through the implementation of physics, chemistry, and biology experiments related to solar energy conversion efficiency in the lessons, which is why the change occurred. These findings show that the experience of analyzing and examining science experiments may be a critical factor in their decision-making.

Interestingly, there was a shift in terminology regarding terms such as "Fossil" and "Nuclear" resources, even though these subjects were not directly addressed in the lessons. This change implies that students broadened their own thinking and considered alternative resources through their own research, consideration, or discussion or, as we might say, the expansion and distinction of thought, before deciding. Aside from that, their decision-making for the terms "Wind" or "Geo." resources indicated a conservationist bent, indicating that the students did not always apply the thought expansion and differentiation to every energy supply resource, but instead reacted restrictively.

It can be explained that the students were readily exposed to fossil fuel or nuclear power, but were given little opportunity to engage with wind power or geothermal power, because it would be difficult to arrive at a definite picture of these resources, so they preserve their decision-making as at the start of the lessons. As a result, it could be argued that the disposition to thought expansion and differentiation corresponds to "generate alternative" or "identify consequences of alternatives" in the decision-making process, as stated by Heath et al. (1987).

2.4.5 Abilities and Viewpoints to Respond to Global Society

With reference to UNESCO's (2009) proposal of a "wide range of learning goals" as well as Fisher and Hicks (1985), categories of four abilities and eight viewpoints to respond to global society were used for the analysis (Table 2.17). Students' text data were analyzed, categorized, and tabulated against the following theoretical foundation.

Figure 2.13 depicts the frequency of the eight viewpoints related to the abilities needed to live in a global society, as shown in Table 2.17. The students were more aware of terms like "Abilities to cope flexibly with different cultures"; b-1 Flexibility, b-2 Respect and "Abilities to communicate in English"; d-1 Language, d-2 Positive attitude than "Abilities to solve problems creatively in collaboration with others"; and c-1 Creativity, c-2 Leadership. This change may explain why students perceived the importance of these abilities after participating in corroborative experiments, presentations, and discussions with Korean students. Additionally, these results are consistent with those of the follow-up general survey in Sect. 2.3.2, which showed

Table 2.17 Abilities and viewpoints to respond to global society

Abilities		Viewpoints		Sample text
A	Understand things with wide perspectives	a-1	Intellect	N/A
		a-2	Curiosity	*I felt the importance of the scientific inquiry and curiosity through joining this program.*
B	Cope flexibly with different cultures	b-1	Flexibility	*I began to think that it is more important to cooperate with many foreign countries and find out the solution about this energy problems.*
		b-2	Respect	*Through the discussion with Korean students beyond the differences of language and culture, I felt we were the same high school students and could make good friendships.*
C	Solve problems creatively in collaboration with others	c-1	Creativity	*Without thinking deeply and observing carefully around my usual life style, it must be difficult to make my own claim.*
		c-2	Leadership	*I could persuade others to recognize my conclusion, so this program is very useful one.*
D	Communicate in English	d-1	Language	*This time, I felt that my language skills have grown since the last attendance.*
		d-2	Positive attitude	*I learned that it is important not to be shy but take positive action through participation in this program.*

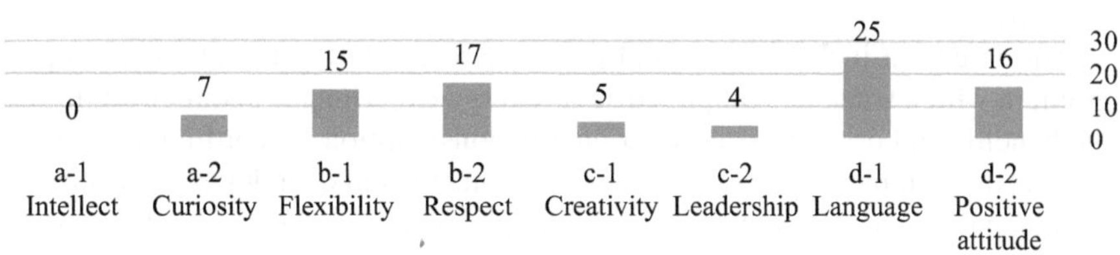

Fig. 2.13 Distribution of viewpoints to respond to global society ($N = 34$)

the positive awareness of the participants in the exchange program, especially on the topics of global talent, language learning, and Japan-Korea relations.

These findings suggest that it is critical to develop and implement an educational exchange program that incorporates experiences such as tackling issues in active communication with others and having the opportunity to share the provided results with others to abilities required for partnership toward sustainability.

2.5 Discussion

The first section examined the requirements for developing a multicultural science exchange program that fosters partnership in the pursuit of long-term sustainability summarized below. To begin, participants should share a common background (e.g., age, academic ability to work jointly on the subject matter of the program). They should do so to avoid feeling inferior to their counterparts. Second, the theme of the exchange program should be a common concern for sustainability that participants can see as their own problem. With such a theme, they will actively participate in the program to a high degree and develop collaborative inquiry activities in order to reach their desired future. Third, the participants in the exchange program should be distinct in some ways, such as their native tongue, which forces them to venture outside their comfort zone. As a result, they will be able to find common ground by communicating in English. As Korsager et al. (2014) point out, in exchange programs involving non-native English-speaking students, English is considered an asset, allowing students to communicate across nationalities, rather than constituting a barrier.

The second section presented the Japan-Korea Science Exchange Program developed by one school each in Japan and the Republic of Korea. Since its inception in 2010, the program has attracted over 1000 students and 100 teachers. Throughout the program, students from both countries took part in combined science lessons that focused on subjects and concerns connected to sustainability. When asked about the program in a general survey conducted in 2016, participants from both countries expressed positive attitudes toward it. Moreover, they praised the program, stating that it aided in their comprehension of ESD content, increased their learning output and creativity, and promoted their personal growth, all while fostering global talent development, language learning, career choice, and Japan-Korea relations.

Third section included pre- and post-test result during the 2014–2015 program, in which students completed solar energy experiments and then addressed the benefits and drawbacks of solar energy consumption. According to the findings of the comparison of the pre- and post-test results, students' multifaceted understanding, critical and creative thinking, decision-making, and abilities and viewpoints to respond to global society all improved. The outcomes of the lessons corroborate prior studies. Participants were given the freedom to develop their own perspectives on the provided data, to connect them to their own experiences, and to learn about the experiences of others. In this way, student learning can be transformed by recognizing the interconnectedness among universe, planet, natural environment, human community, and personal world through critical reflection, holistic approaches, and positive relationships with others (Pavlova, 2013).

To summarize, the developed program demonstrated promise to foster partnership for sustainability. Additionally, one could argue that such a program could successfully bring together people from diverse cultures, including those embroiled in political strife, to discuss common issues. These types of initiatives can assist in bridging the divide between immediate neighboring countries, as attested by one of the participants:

By participating in this Japan-Korea solar energy exchange program, I could gain valuable experience that excited my curiosity and increased my knowledge about solar energy and the energy crisis. I also recognized the critical nature of resolving the energy crisis, which affects all humans. And, even though the two countries' governments are at odds these days,

I'd like to say that collaborating with and exchanging ideas with the Korean people has been the most rewarding experience of my life. (HUHS Student)

These multicultural science exchange programs were made possible because both schools are STEM-focused and have received funding from their respective Ministry of Education. The programs are special extracurricular activities for science in such schools, and it is difficult to apply them to the regular curriculum. However, as this chapter has shown, active participation and collaboration in learning by peers differing in knowledge and opinions to reach a common goal can have significant educational benefits. The acquisition of new knowledge and the acknowledgment of the environmentalism of the other group reinforce a more balanced mindset of education for sustainability toward environmental, human, and social dilemmas (Alkaher & Tal, 2014). Such experiences of students will help to truly achieve cross-border collaboration and solidarity on global issues.

Reflecting on the success factors of this extra curriculum and identifying how to incorporate these factors into the existing science curriculum is the next research question; as Krosager et al. (2014) suggest, this incorporation will help to enrich the local classroom with a global dimension.

Acknowledgments The Japan-Korea student exchange program was modified in 2018 and 2019 to include a presentation of the results of each student's science research project. This modification remains in effect. We would like to express my gratitude to the principals and teachers of Hiroshima University High School in Japan and Cheonan Jungang High School in the Republic of Korea for their commitment and collaboration in developing these programs.

References

Alkaher, I., & Tal, T. (2014). The impact of socio-environmental projects of Jewish and Bedouin youth in Israel on students' local knowledge and views of each other. *International Journal of Science Education, 36*(3), 355–381.

Arya, D., & Maul, A. (2016). The building of knowledge, language, and decision-making about climate change science: A cross-national program for secondary students. *International Journal of Science Education, 38*(6), 885–904.

Bianch, G., Pisiotis, U., & Cabrera, M. (2022). *GreenComp the European sustainability competence framework*. Publications Office of the European Union.

Block, T., Van Poeck, K., & Östmann, L. (2019). Tackling wicked problems in teaching and learning. Sustainability issues as knowledge, ethical and political challenges. In K. Van Poeck, L. Östmann, & J. Öhman (Eds.), *Sustainable development teaching: Ethical and political challenges* (pp. 28–39). Routledge.

Bolte, C., Kirschenmann, B., & Streller, S. (2010). Enhancing pupils' abilities to properly judge and make informed decisions in the field of renewable energy sources – Bioenergy. In G. Çakmakci & M. F. Taşar (Eds.), *Contemporary science education research: Scientific literacy and social aspects of science* (pp. 149–154). Pegem Akademi.

Cörvers, R., Wiek, A., De Kraker, J., Lang, D. J., & Martens, P. (2016). Problem-based and project-based learning for sustainable development. In H. Heinrichs, P. Martens, G. Michelsen, & A. Wiek (Eds.), *Sustainability science: An introduction* (pp. 349–358). Springer Science+Business Media.

Ennis, R. H. (1987). A taxonomy of critical thinking dispositions and abilities. In J. B. Baron & R. J. Sternberg (Eds.), *Teaching thinking skills: Theory and practice* (pp. 9–26). W. H. Freeman and Company.

Ennis, R. H. (1996). Critical thinking dispositions: Their nature and assessability. *Informal Logic, 18*(2 & 3), 165–182.

Finger, D. C., Draghici, C., Perniu, D., Smederevac-Lalic, M., Halbac-Cotoara-Zamfir, R., Sehic, A., & Solomun, M. K. (2021). The importance of international collaboration to enhance education for environmental citizenship. *Sustainability, 13*(18), 10326.

Fisher, S., & Hicks, D. (1985). *World studies 8–13: A teacher's handbook.* Oliver & Boyd.

Heath, P. A., White, A. L., Berlin, D. F., & Park, J. C. (1987). Decision making: Influence of features and presentation mode upon generation of alternatives. *Journal of Research in Science Teaching, 24*(9), 821–833.

Korsager, M., & Slotta, J. D. (2015). International peer collaboration to learn about global climate changes. *International Journal of Environmental & Science Education, 10*(5), 717–736.

Korsager, M., Slotta, J. D., & Jorde, D. (2014). Global climate exchange: Peer collaboration in a "global classroom". *NorDiNa (Nordic Studies in Science Education), 10*(1), 105–120.

Labby, S. A. (2016). Enhancing university STEM students' understanding of sustainability through an international experience in Bucaramanga, Columbia. *Journal of Universality of Global Education, 3*, 175.

Lee, S. (2014). Exploration of the significance of establishing and implementing 'environment-related' subjects in Korea: Based on insider-centered evaluation. *Environmental Education, 24*(1), 34–44. (in Japanese).

Ministry of Education, Culture, Sports, Science and Technology. (2017). *The course of study for high school*. Ministry of Education, Culture, Sports, Science and Technology. (in Japanese).

Pavlova, M. (2013). Education for sustainable development and the transformation of self: How the world cam become a better place to live for all. In H. E. Middleton & L. K. J. Baartman (Eds.), *Transfer, transitions and transformation of learning* (pp. 123–132). Sense Publishers.

Slotta, J. D., & Jorde, D. (2010). Toward a design framework for international peer discussions: Taking advantage of disparate perspectives on socio-scientific issues. *Research and Practice in Technology Enhanced Learning, 5*(3), 161–184.

UNESCO. (2009). *United Nations decade of education for sustainable development (DESD) 2005–2014: Review of contexts and structures for education for sustainable development*. UNESCO.

UNESCO. (2017). *Education for sustainable development goals: Learning objectives*. UNESCO.

Wiek, A., Withycombe, L., & Redman, C. L. (2011). Key competencies in sustainability: A reference framework for academic program development. *Sustainability Science, 6*(2), 203–218.

Chapter 3
Local Nature-Focused Inquiry Learning for Sustainable Development After an Earthquake Disaster in Japan

Shigeyoshi Watanabe and Kei Miyamura

3.1 Introduction

In recent years, large-scale earthquakes have occurred frequently in Japan. In Kumamoto Prefecture, there were two consecutive earthquakes observing seismic intensity 7 in April 2016. Nature has two sides: "benefit" and "disaster." In science education, it is necessary to learn both sides while associating them with the contents of science education. However, it is expected that learners who have experienced natural disasters will have disturbance or negative image, and it is not easy to calmly accept the scientific explanation of the natural phenomenon that caused the disaster. In addition, if information about the disaster experienced is taken carelessly, it may induce PTSD (post-traumatic stress disorder) in the students. Therefore, it is difficult for schools in the disaster area to handle teaching materials about earthquakes, typhoons, volcanoes, and natural disasters. The purpose of this study is to incorporate the geological events of the region into science learning and to guide the students' perspectives that accept earthquakes as part of their natural activities by looking at the region from a multifaceted and bird's-eye view. We developed teaching materials, designed a unit, and practiced lessons, for junior high school science.

S. Watanabe (✉)
Kumamoto University, Kumamoto, Japan
e-mail: shige@educ.kumamoto-u.ac.jp

K. Miyamura
Minamigaoka Elementary School, Koshi, Japan

H. Fujii, S.-K. Lee (eds.), *Science Education for Sustainable Development in Asia*, Education Innovation Series, https://doi.org/10.1007/978-981-99-8711-5_3

3.2 Development of Local Earth Science Teaching Materials

3.2.1 *Field Research and Literature Search*

The subject of our research is a public junior high school (called K School here) in Mashiki, Kumamoto Prefecture, which is one of the schools near the epicenter of the Kumamoto earthquake in 2016. K School suffered great damage to buildings, but no one died among the students. Mashiki is a commuter town adjacent to the east of Kumamoto City. Kumamoto Airport is on the north side of the town, and there is a highway interchange on the west side. The Kiyama River runs through the center of the town, and fields spread around it. In addition, there are low mountains on the east and south sides of the town, and it is a place rich in nature. In Aso City, east of Mashiki, there is a famous volcano called Mt. Aso, and the land of Mashiki is influenced by the volcano.

We conducted a field research and a preliminary literature search in order to discover themes and teaching materials for inquiry learning that make the best use of the area in Mashiki. As a result of the research, we focused on the following geological events that can be observed around K School (Fig. 3.1).

1. An alluvial fan spreads out where the Kiyama River flows out of the mountains.
2. Land use patterns differ between the north and south sides of the Kiyama River.
3. Looking at the geology of Mashiki, igneous rocks are distributed on the east side, and sedimentary rocks are distributed on the west side.

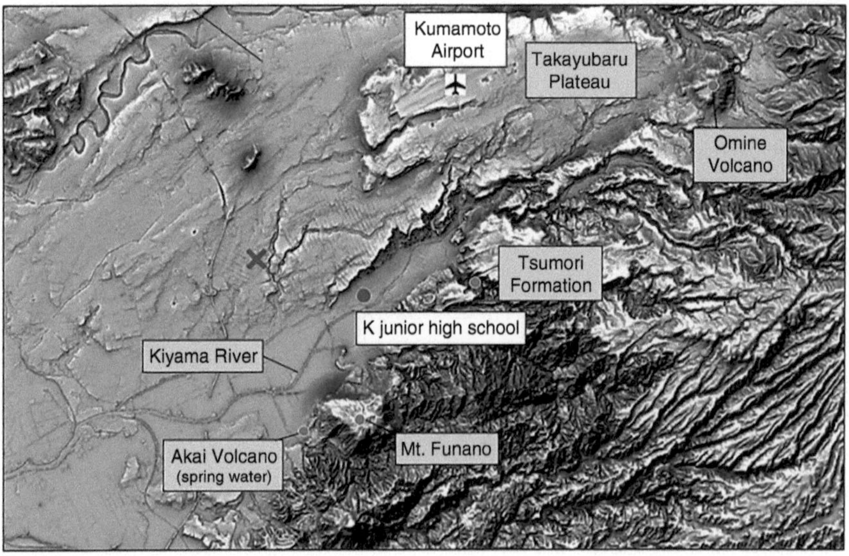

Fig. 3.1 Geological characteristics of the area around K Junior High School. × indicates the epicenter of the Kumamoto earthquake (April 16, 2016). The map was made using KASHIMIR 3D

4. The plateau where Kumamoto Airport is located is a lava plateau derived from Omine volcano, and its characteristics are reflected in the topography.
5. There are volcanoes such as Akai volcano and Omine volcano that erupted in the middle to late Pleistocene (Hase et al., 2016), and water is springing from the place where Akai volcano was.
6. Along the Kanayama River, a tributary of the Kiyama River, there is the Tsumori Formation (lake sedimentary layer) in the middle Pleistocene (Hayashi et al., 2001), where diatom fossils can be collected.
7. On the east side of the Kiyama River, there is an outcrop where Aso pyroclastic flow deposits (called "Aso-4") can be observed.

In the field research, we took pictures showing the topography of Mashiki and videos showing the undulations of the land as teaching materials to spread the viewpoint of the students from the school to the area. In addition, pumice of pyroclastic flow deposits, scoria gravel, tuffaceous mudstone, and outcrop soil were collected for use as teaching materials.

3.2.2 Development of Regional Teaching Materials for Inquiry Learning

Using the results of field research and literature search, we have developed the following three types of teaching materials: (a) teaching materials for directing students' awareness to local geological events in the introduction of the unit; (b) teaching materials used for learning "volcanoes," "earthquakes," and "stratums"; and (c) teaching materials for students' inquiry learning (Miyamura & Watanabe, 2018; Watanabe & Miyamura, 2019). Ideally, students should take the initiative in inquiry if the goal is to change their view of nature in the area. However, Japanese science curriculum is often conducted according to the contents of textbooks that comply with the Course of Study, and it is difficult to secure sufficient time for inquiry activities. Therefore, we set four themes and developed teaching materials for inquiry activities focusing on data analysis.

Theme 1 Mystery of spring water.

Question Why is spring water coming out of the outcrops?

Akai volcano, which erupted about 150,000 years ago, forms a scoria cone, and there is an aquifer of Togawa lava underground. Spring water can be seen at the outcrops where Akai volcano was, making the scenery familiar to the students. Since springs are related to igneous rocks, the knowledge gained from studying volcanoes and geological formations in junior high school science is utilized. We prepared scoria gravel as an observation material and developed a model experimental device for spring water using scoria gravel from Akai volcano, clay, and PET bottles.

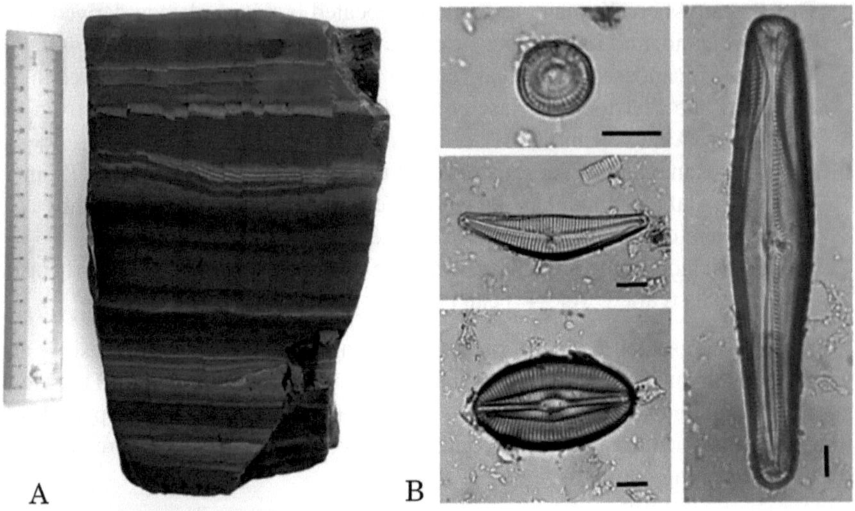

Fig. 3.2 Mudstone (**A**) and diatom fossils (**B**) of the Tsumori Formation. (**A**). Mudstone is polished and flattened. (**B**). Scale is 10 μm

Theme 2 Mystery of Tsumori Formation.

Question What kind of paleoenvironment does the diatom fossil of the Tsumori Formation show?

Since the Tsumori Formation produces diatom fossils in the layer of lake sediments (Tanaka et al., 2005), it is possible to utilize the learned knowledge of water sedimentation and phase-indicating fossils. In order for students to observe diatom fossils under a microscope and infer the paleoenvironment, we polished the mudstone at the outcrops of the Tsumori Formation and made permanent preparations of the collected diatom fossils (Fig. 3.2). In addition, using photographs of diatom fossils observed in the permanent preparation, we made a worksheet that can distinguish whether diatoms inhabited in freshwater or seawater.

Theme 3 Mystery of the plateau where the airport is located.

Question Are volcanoes involved in the formation of plateaus?

The plateau where Kumamoto Airport is located is a lava plateau formed by the outflow of lava from Omine volcano that erupted in the late Pleistocene. Looking at the topographic map, this plateau looks like lava has flowed out. The relationship between the shape of the volcano and the viscosity of magma is what students learned in junior high school science. Students can infer about the formation of plateaus in their area, applying the knowledge they have learned. Therefore, we made topographic maps and cross-sectional views of the plateau using map software (KASIMIR 3D) and web maps (Geospatial Information Authority of Japan) so that students can see the topography from a bird's-eye view (Fig. 3.3A). We also devised a model experiment using starch syrup as a substitute for magma (Fig. 3.3B).

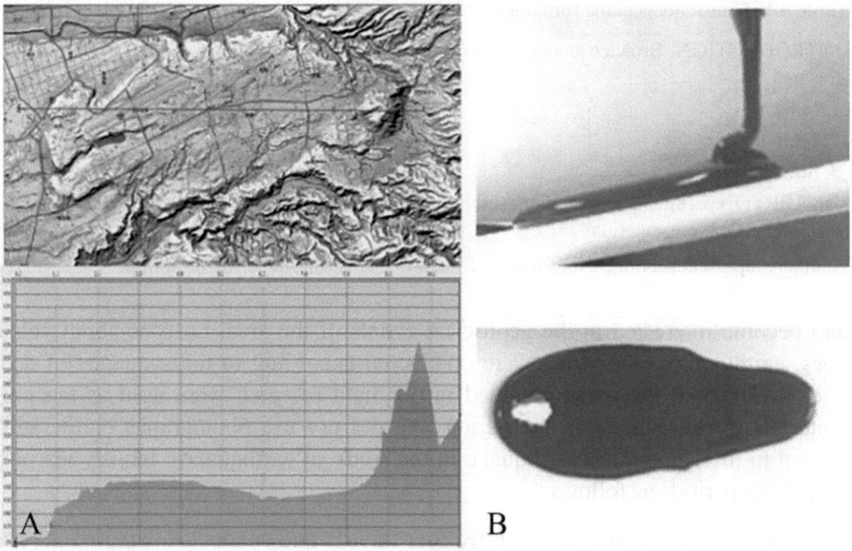

Fig. 3.3 Topographic map and cross section of the plateau (**A**) and model experiment (**B**). (**A**). The cross-sectional view is displayed with the elevation scale enlarged. (**B**). Syrup is colored with red food coloring

Theme 4 Mystery of the topography and land use of Mashiki.

Question Why is the land use different between the north and south sides of the Kiyama River?

Looking at the satellite photographs of Mashiki, students can see that the forests, vegetable fields, rice fields, and residential areas are separated. Also, looking at the geological map, we can see that there is a difference in geology between the upstream side (east side) and the downstream side (west side) of the Kiyama River. The activities to interpret various map information of the area will lead students to a multifaceted and bird's-eye view. We made three distribution maps of igneous rock (red), sedimentary rock (blue), and metamorphic rock (green) in Mashiki and printed each on a transparent sheet. The three types of geological maps are superimposed on a topographic map and used to relate geology and topography.

3.3 Unit Design and Lesson Plan

3.3.1 Unit Design of Inquiry Learning Plan

Table 3.1 shows the sequence of the unit of "The formation and change of the earth" that we planned. In the introduction, teaching materials such as photographs, videos, and maps were used to guide students from K School to the surrounding area

Table 3.1 Unit sequence: the formation and change of the earth

INTRODUCTION: Broaden students' perspectives on nature in the area	1
1. Volcano	6
2. Earthquake	6
3. Strata	5
INQUIRY: Experiments and research activities/Group work and discussion/Oral presentation	3

Numbers represent school periods required for study. One school period is 50 min.

and become interested in the geological events of the area. Lessons about volcanoes, earthquakes, and strata were conducted according to the sequence of textbooks. However, the rocks obtained in our field research were used as teaching materials to relate them to the region. We did not use photographs or literature related to the Kumamoto earthquake in the study of earthquakes. Inquiry learning took three periods as follows.

1. Students selected one of the above four themes and conducted experiments and material analysis in group work (first period).
2. Students exchanged opinions between groups on their results of data analysis and evaluated and reconsidered their inference and conclusion (second period).
3. Students gave an oral presentation on the result and discussion of their inquiry (third period).

3.3.2 Evaluation Method of Learning Outcomes

The following four surveys were conducted to evaluate the effects of the developed teaching materials, unit design, and lesson plan and to investigate the changes in students' thinking about earthquakes.

1. Changes in student's description that introduces the area.
   At the beginning and the end of the unit, we asked the students to write an introduction to Mashiki for carrying on the homepage. We analyzed the changes in students' views of the area based on the differences between the two student descriptions.
2. Description of students' views and ideas about earthquakes.
   At the end of the unit, students were asked to write answers to the following questions. "Did you change your view and ideas about the Kumamoto earthquake through learning about the formation and changes in the earth?"
3. Analysis of student activities using video recordings of lesson practice.
   In order to clarify the learning process of students in inquiry learning, we recorded the activities of students by video and analyzed the utterances and behaviors of students during experiments and discussion activities based on the video recordings.

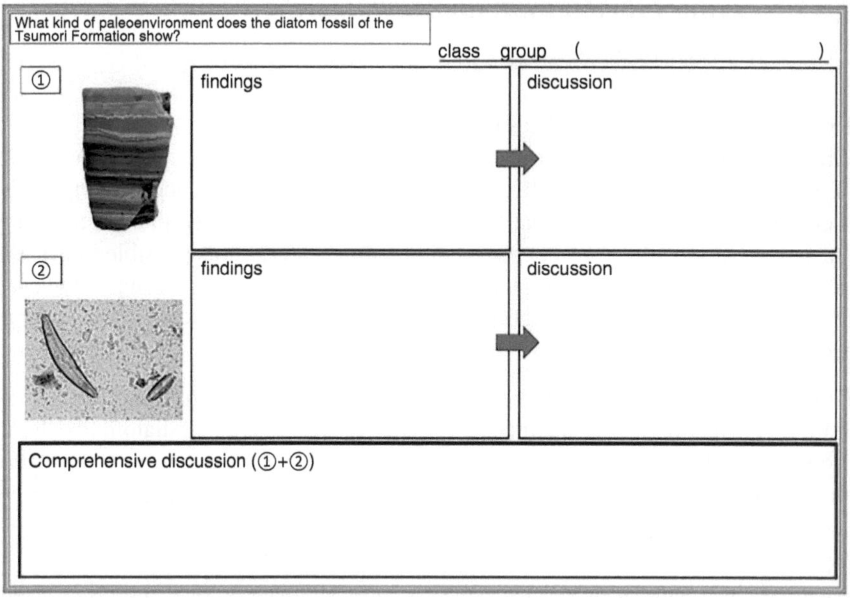

Fig. 3.4 Inquiry learning worksheet

4. Analysis of individual/group worksheets.
   For inquiry learning, we prepared worksheets for individuals and groups (Fig. 3.4). Students first describe their results and considerations on their worksheets and then discuss in groups and describe on group worksheets. We distinguished the space of the worksheet where the students wrote the findings obtained from the experiments and reading data, the result of inference from what they found, and the comprehensive conclusion of the inference. We also prepared another worksheet for rethinking and adding after exchanging ideas with other groups.

3.4 Results and Discussion

3.4.1 Practice of "The Formation and Change of the Earth"

From November 2018 to January 2019, lessons on "the formation and change of the earth" was carried out for three classes (91 students) in the first grade (12–13 year-olds) of K Junior High School. One of the authors (at that time, a K School teacher) was in charge of the introductory and inquiry learning lessons, and another K junior high school teacher was in charge of the other lessons. In inquiry learning, one class of students was divided into eight groups, and two groups worked on each of the four tasks.

3.4.2 Lesson Practice and Students' Reaction in Inquiry Learning

Experiments and Data Analysis (First Period)

Theme 1 Students compared rocks taken from outcrops with springs to other rocks of different type. Then they noticed the difference in the weight and fragility of the rock. Finally, they muttered that it was a pumice stone because there were many holes in the stone. In a model experiment of spring water, it was possible to observe that water comes out on clay soil that does not allow water to pass through. However, few students pointed out that the model experiments were linked to local springs and volcanoes.

Theme 2 Students observed the mudstone collected from the outcrops of the Tsumori Formation with a magnifying glass. In the observation of diatom fossils, all groups were able to find diatom fossils under a microscope. Furthermore, they were able to identify the diatoms of the Tsumori Formation as a freshwater species by using the check sheet.

Theme 3 In a magma model experiment, some students noticed the relationship between stickiness and flow shape as a result of comparing the difference in flow by changing the temperature of starch syrup. In the topographic and cross-sectional data analysis, two out of eight groups could infer as a result of lava flowing through the plateau where the airport is located.

Theme 4 Since the inquiry on this subject did not include experiments, it was worried that the students' activities would be more passive than those who worked on other themes. However, the students repeatedly tried to combine the geological map and topographic map to read the data, and they enthusiastically discussed the results.

Exchange of Opinions (Second Period)

In the second period, two students of the four-member group were responsible for explaining, and two were responsible for listening to the explanations of the other groups. The listener students visited other groups to hear explanations and then exchanged opinions (5 min). After exchanging views with another group, they switched between explaining and listening. The students in the role of explaining used not only worksheets that described their inferences but also materials that used in the experiments or for data analysis. In the 5-min exchange of opinions, the first 3 min were mainly explanations, but during the remaining 2 min, the listener students asked questions and commented, and the interaction between the students could be confirmed. After exchanging opinions, each group reviewed their inferences and conclusions. The students often reconsidered by referring to the explanations of the groups who worked on the same theme as themselves. They also exchanged opinions on exploration activities on a theme different from theirs. However, it did not seem easy for the students to understand what was explained and use it for discussions in their group.

Oral Presentation

The students presented the results of their inquiry activities in each group (3 min for each). In the oral presentation, their worksheets and materials were presented on an overhead camera. There was a group in which four students explained one after another and a group in which the roles were divided between the explainer and the material presenter. Many students were able to distinguish between results and consideration. There was also a presentation explaining that they had taken in the ideas of other groups and reviewed their ideas.

3.4.3 Analysis of Worksheets for Inquiry Learning

Theme 1 Looking at the worksheet for the group, the following was described in the finding column. "There are a lot of holes," "This stone is lighter than other stone" (about scoria gravel observation), "Water came out only A and B holes," and "The water that came out first was muddy, but after that clean water came out" (about a model experiment of spring water). In the column of consideration, there is a description that the observed stone is judged as "pumice of volcanic ejecta" and an inference that "Since water comes out only from the upper pumice stone, it is difficult for the pumice stone to pass water to the bottom." In the comprehensive conclusion, some groups were able to describe the results of stone observations and model experiments in relation to local geological events. The connection of such consideration was confirmed in two out of six groups working on Theme 1. In the review after exchanging opinions, the amount of description on the worksheet increased, the ideas of other groups were added, and their opinions were revised.

Theme 2 The students described the striped pattern and color of the mudstone or the texture of the crushed grains as observation results. In their discussion, there were descriptions such as "It can be seen that there was water in this place in the past" and "Since the grain size is small, it can be seen that the flow of water was gentle downstream." The students probably used their knowledge of water sedimentation for their inference. However, there was also a description that "there was a layer of small grains like mud, so there was abundant flowing water to erode." This group made incorrect inferences because they focused on erosion rather than water deposition. Since the Tsumori Formation is observed in outcrops along the river, the students may have noticed water erosion. Five of the six groups described that diatom fossils are freshwater species as a result of their observations. Regarding the paleoenvironment where the Tsumori Formation was formed, there were multiple inferences in one group, with six groups inferring "ponds," five groups "lakes," and three groups "rivers." The comprehensive conclusion included a description in which only one result and consideration of one observation was transcribed, a description in which both results and consideration were transcribed, and a description in which the paleoenvironment was considered to be a lake with sedimentation.

Theme 3 The students read the topographic map and made the following descriptions of orientation and topography of the plateau: "The north-south area is flat for about 1.5 to 2.8 km." "The eastern side is high, and the west, south, and north areas are flat." On the other hand, there was a description on the premise that the plateau was made of magma and lava, such as "flowing to the west" and "there is a place where lava is agglomerated." The expression of the results of the model experiments differed from group to group. Some groups vaguely expressed how the starch syrup flowed, saying, "It flowed quite a bit / not so much." Another group compared "flowing velocities" and used the term "stickiness." Some groups use the term volcanic ejecta, and it is thought that the knowledge learned in "volcanoes" is reflected in the results and discussions of the students. In the comprehensive conclusion, students were able to explain how plateaus in their area were formed through the process of volcanic eruptions and lava outflows. However, the model experiment showed that when the temperature of the starch syrup was high, the stickiness was weak, and it flowed well and became flat. Therefore, some groups ended up explaining only the description that "the plateau was made of weakly sticky magma."

Theme 4 The students' inquiry activities were limited to the analysis of map information, but the amount of description on the worksheet was larger than that of the other groups. Depending on the group, there was a description that the distribution of igneous rocks and sedimentary rocks was related to the orientation and the height of the land in the results, and the origin and spread of the land were sometimes inferred in the description of the discussion. However, there were few groups who noticed or considered the river map, and there was no description in the comprehensive conclusion column. After exchanging opinions with other groups, there was a description of a review that utilized information on how the plateau was formed and the Tsumori Formation for their own analysis. It is suggested that effective information exchange was carried out with groups that conducted inquiry activities on different themes.

3.4.4 Analysis of Introductory Texts About the Area

In the introductory text ($N = 83$) written by the students at the period of the introduction of the unit, 25% of the students could express it in sentences, 30% of the students wrote only one simple sentence, 19% of the students wrote only word, and 17% of students were blank (Table 3.2). Many students who expressed only words used the terms "mountain," "river," and "rice field." About a quarter of the students ($N = 22$) described specific place names such as "Fine Noodle Waterfall" and "Cluster Amaryllis Road" and their explanations. Only three students took up the earthquake. In the introductory texts at the end of the unit, the amount of description increased significantly, and 99% of the students were able to express them in sentences (Table 3.3). Many of the descriptions were based on the results and discussions of experiments and material analysis in inquiry learning, and the students

Table 3.2 Examples of introductory text at the introduction of the unit

Student A	· There are many rice field and vegetable field · There is a lot of greenery · Surrounded by mountains	
Student B	· Airport · Post photos on homepage · Upload the video	
Student C	Cluster amaryllis road	
Student D	–	

Table 3.3 Examples of introductory text at the end of the unit

Student A	The plateau where the airport is located has arisen by the sticky magma of the volcano. The magma probably flowed toward the west. The distance that the magma flowed is long and the range is wide, so it became flat. Today, there are dams and airports on the plateau. In this way, Mashiki has a lot of land and strata created by nature.
Student B	There are igneous rocks and sedimentary rocks formed by eruptions. There are igneous rocks and sedimentary rocks formed by eruptions. Mashiki is very interesting and you will want to know more if you look it up.
Student C	There are many things related to volcanoes, such as land created by an eruption 90,000 years ago, waterfalls created by an eruption 140,000 years ago, and strata created 500,000 years ago.
Student D	It is thought that the land of Mashiki was created by the eruption of a volcano and the eruption of magma in the past. There is a fault in the Tsumori formation, so I think there was an earthquake. Mashiki was made by such (geological) events.

Students A to D in Tables 3.2 and 3.3 are the same person

explained the origins and characteristics of the land of Mashiki in a style of explaining to others. Terms such as "present" and "past" were used in the introductory texts of 45% of the students. Therefore, it is suggested that some students have a viewpoint to grasp the formation of the land in the area in a temporal context. Five students used the term "earthquake" or "natural disaster." However, most of them were explained in the context that Mashiki was formed by natural disasters such as earthquakes and volcanic eruptions.

3.4.5 Analysis of Changes in Perspectives on Earthquakes

There were 63 students who could answer about their perspective on the earthquake because they did not have enough time in the last lesson of the unit. The change of students' perspective on earthquakes through learning the unit is as follows. There was a description that "the perspective has changed": 21%. There was a description suggesting a change in perspective: 27%. There was a statement that they gained new awareness, interests, and knowledge: 27%. There was a statement that "the perspective did not change": 3%. The change in perspective could not be determined from the description: 21%. Therefore, as a result of lessons including inquiry, about half of the students were influenced by their perspective on earthquakes. Focusing on the content of the description, a quarter of the students used terms that indicate a temporal viewpoint, such as "history" and "tens of thousands of years ago." Some of the students wrote that "the Kumamoto earthquake will be history." In addition, the students answered, "It was serious immediately after the earthquake, but it's been a good experience now." "I felt once again that there are positive aspects (to the earthquake)." "I become interested in the earthquake." These descriptions indicate that a non-negative view of the earthquake has arisen.

Since the descriptions of earthquakes were diverse as shown in the example below, it is possible that the students tried to express their thoughts honestly. One student described, "I was able to deepen my understanding by looking at the wide areas and small parts of Mashiki and noticing and reasoning." This is regarded as a comment related to the development of the "multifaceted and bird's-eye view" aimed at in this research. On the other hand, the expression "scary" was found in the answers of the two students. Therefore, it was reconfirmed that careful handling of earthquakes is necessary for students who have experienced disasters.

Examples of students' description of changes in perspective on the Kumamoto earthquake are as follows:

- *The valuable Kumamoto earthquake that occurs only once every tens of thousands of years. Since I have been studying "The changes of the earth," I thought this was a historical event. Before I learned, I imagined something very unpleasant, but as I learned, I felt it was a very valuable and rare experience.*
- *I thought that I could learn a lot because I experienced the earthquake. If I hadn't experienced an earthquake, I would have studied as someone else's problem. It was serious immediately after the earthquake, but it's been a good experience now. I like Mashiki more.*
- *(My perspective has) changed. Until now, I thought it would be serious because house was broken, and I was looking at the negative side. But, I learned that changes in the strata will bring about positive aspects.*
- *I have never been interested in the topography and history of Mashiki. I was able to deepen my understanding by looking at the wide areas and small parts of Mashiki and noticing and reasoning. Then, I realized that Mashiki was made by a big disaster such as the Kumamoto earthquake, and I became very interested in it.*
- *In the long history of the earth, this quake is only one. But each time, people and animals have survived the earthquake. I think this is amazing, and I think this is proof that Mashiki will overcome this earthquake.*
- *Through the three lessons, I was able to learn about the nature and land of Mashiki, which I had never known before. This earthquake is named "Kumamoto Earthquake," but I learned that there were natural disasters such as similar earthquakes and volcanic eruptions long ago. I felt this earthquake was not the first event.*

- *The perspective of the Kumamoto earthquake has not changed, but it was good to know the ancient times.*
- *It turned out that the Earthquake Early Warning (earthquake alert) is meaningless for earthquakes that take a short time to reach the P wave. But, I will develop a device in the future that can predict any P wave and send a notification to everyone's mobile phone before the S wave arrives.*

3.5 Conclusion

In this study, we incorporated local natural matter and phenomena into inquiry learning as teaching materials and integrated them with ordinary science learning along with the sequence of textbooks. Local natural matter and phenomena may not be easy for students to understand scientifically because they may not be typical cases or may be influenced by special conditions. Therefore, we prepared experiments and materials for inquiring the four themes and guided the students' inquiry using worksheets. As a result of practicing for the students affected by the Kumamoto earthquake, the learning through inquiry influenced the descriptions introducing the area where they live and their perspectives of the earthquakes.

We emphasized that students will be able to see the area from a multifaceted and bird's-eye view in teaching material development and lesson planning. In addition, at the end of inquiry learning, the teacher showed a diagram in which the teaching materials taken up in inquiry were arranged on a cross between the time axis and the space axis as a summary of the learning (Fig. 3.5). The student's picture is

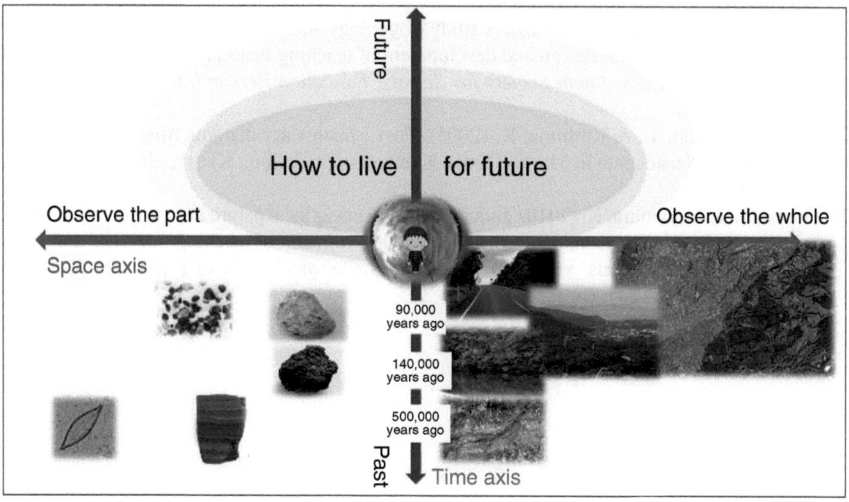

Fig. 3.5 Summary diagram of inquiry learning

positioned in the center of the diagram. The teacher showed the students this diagram and suggested the need to think about how to live for the future. Through studying the entire unit, including inquiry learning, students were able to not only gain new knowledge about the geological events of the region but also explain the formation and change of the region from a temporal perspective. Such learning outcomes may be the result of the effective functioning of the activity of exchanging opinions during the second hour of inquiry learning and the activity of exploring regional characteristics from a macro perspective (Theme 4).

For students who have been affected by an earthquake, there may be an approach that emphasizes the scientific consideration of natural disasters by taking the earthquakes into learning. However, in our practice, we did not take the Kumamoto earthquake directly as a teaching material, but used an approach intended to look at the earthquakes relatively by learning the geological events of the region from the perspective of space and time. As a result, many students changed their perspective on the Kumamoto earthquake and elicited a positive attitude toward learning on the earth science. It must be taken into consideration that it was practiced 2.5 years after the Kumamoto earthquake, but our approach can be proposed as one of the earth science learning strategy for students who have experienced the disaster.

References

Hase, Y., Nakayama, H., Hurusawa, W., & Aramaki, S. (2016). Displacement of sub-alluvial Togawa lava in the southern Kumamoto plain. *Bulletin of Goshoura Cretaceous Museum, 17*, 5–13.

Hayashi, M., Yahiro, K., & Kitabayashi, E. (2001). Middle Pleistocene insects from the Tsumori formation in Masaki-machi, Kumamoto prefecture, Japan. *Bulletin of the Mizunami Fossil Museum, 28*, 239–243. (in Japanese).

Miyamura, K., & Watanabe, S. (2018). A study of geology science to junior high school students in the disaster area: Unit design and development of teaching materials for realizing the living area in multiple ways. *Japan Society for Science Education Research Report, 33*(2), 15–20. (in Japanese).

Tanaka, H., Ngumo, T., & Kashima, K. (2005). Fossil freshwater diatoms from Tsumori formation (middle Pleistocene) in Mashiki town, Kumamoto prefecture, Kyushu, Japan. *Diatom, 21*, 119–130.

Watanabe, S., & Miyamura, K. (2019). Inquiry learning using local nature in an affected junior high school damaged in the Kumamoto earthquake: Development of teaching materials and lesson practices leading students' views of nature. *Bulletin of the Faculty of Education, Kumamoto University, 68*, 239–247. (in Japanese).

Chapter 4
Integrating the Concept of Sustainable Development with Science Education in China: Content Focused on Soil Pollution and Human Health

Xinyi Shen, Nan Liu, and Ying Li

4.1 Introduction

The rapid development of the social economy has led people to gradually focus on seeking sustainable development through regularity, coordination, fairness, high efficiency, and multidimensionality. In 1987, the report "Our Common Future" issued by the World Commission on Environment and Development first put forward the concept of sustainable development. In 1992, Agenda 21 – adopted at the United Nations Conference on Environment and Development – highlighted that education is the key to improving people's ability to solve environmental and development problems, as well as an important medium to promote sustainable development. Subsequently, the relevant international organizations successively convened and issued important documents to guide and promote Education for Sustainable Development (ESD). In 2015, UNESCO announced the "Incheon Declaration and Framework for Action for the implementation of Sustainable Development Goal 4" (UNESCO, 2016) to enrich the connotation of ESD further and emphasized that ESD aims to achieve sustainable development in the aspects of society, economy, culture, and environment.

There is a different understanding of the connotation of ESD in Chinese academic circles. Wang and Li (2014) mentioned that ESD repositions education oriented toward sustainable development to better achieve the sustainable development of educational services. Wang (2015) noted that ESD enables people to establish

X. Shen (✉)
Beijing Academy of Educational Sciences, Beijing, China

N. Liu
Luhe Middle School, Beijing, China

Y. Li
Central University of Finance and Economics, Beijing, China

H. Fujii, S.-K. Lee (eds.), *Science Education for Sustainable Development in Asia*, Education Innovation Series, https://doi.org/10.1007/978-981-99-8711-5_4

their values, scientific knowledge, learning ability, and lifestyle as required by sustainable development and to promote sustainable development in various aspects of society, economy, environment, and culture. Huckle and Sterling (2002) viewed ESD as an example of a new educational theory and practice that emphasizes "the connectedness of education." Regardless of these different perspectives, the concept of ESD includes sustainable development in society, the economy, environment, culture, and human beings. Therefore, we should teach students the scientific knowledge related to sustainable development and cultivate their awareness and learning ability toward sustainable development.

As the primary form of formal education, school-based education plays a key role in implementing ESD. ESD does not consist of a branch of disciplines, e.g., science, but rather should be combined with subject teaching to infiltrate the knowledge, skills, emotional attitude organically, and values related to sustainable development with the taught subject (Wang, 2010). Biology – a discipline that studies the phenomena, laws, and applications of life – is crucial in solving overpopulation, resource crisis, and the deterioration of ecological environments. Through biology, people understand the phenomena and essence of life and devise theories that support the harmonious coexistence between human beings and nature. This subject is therefore closely related to sustainable development. The contents of biology curricula often include the learning objectives of ESD. In other words, biology has a unique value and plays an indispensable role in implementing ESD. Biology teaching can develop the students' core literacy and successfully cultivate the students' life concepts, scientific thinking, scientific inquiry, and social responsibility, which are key abilities, as well as essential values and characteristics. Thus, biology as a taught subject contributes to the students' gradual development, which is necessary to promote the sustainable development of human society. This chapter focuses on teaching biology in schools to explore the deep integration of the concepts of sustainable development and science education.

4.2 Strategy for Incorporating the Concept of Sustainable Development into Classes

4.2.1 Holistic Approach to Promote Content Connectedness

Boeve-de Pauw et al. (2015) used two approaches to integrate the concept of sustainable development with classroom teaching: the "holistic approach to content" and "pluralistic approach to teaching." The holistic approach to content refers to adopting a system that considers the correlation among past, present, and future local and global problems and their correlation with economic, social, and environmental issues. It does not focus on a single teaching content. A pluralistic approach to teaching encourages students to think critically, generate their views, and learn through inquiry. Through this approach, students connect with nature and display emotions that regulate sustainable development – both crucial predictors of sustainable behaviors (Geng et al., 2015; Barbro & Pickett, 2016). From the perspective of scientific knowledge, biology as a subject involves many topics related to the natural environment, such as ecosystems, water and the water metabolism of plants, photosynthesis and respiration, materials and the energy metabolism of animals, as well as genetics and variation. However, this knowledge is spread across different chapters and presented in a scattered manner so as to be easily skimmed over during the teaching process, similar to a dragonfly skimming the surface of the water. As a result, students acquire knowledge related to the natural environment superficially and systematically, without profundity. The class developed in this study combines the relevant knowledge organically and comprehensively, which is more conducive to helping students build knowledge networks and develop a higher level of cognitive learning. As a highly adaptable teaching method, the "Case Study Teaching Method" is particularly suited to science classes (Bonney, 2015). This method also conforms to cultivating the students' abilities to solve the issues in sustainable development.

The class highlighted events that occurred in the surrounding environment, created a real-life situation for students, and guided them to observe, ask questions, discuss, and exchange experimental results regarding specific biological phenomena. Backed by biological facts and evidence, the class applied methods such as induction, generalization, deduction, and reasoning to explore and explain the phenomenon of life and examine social issues related to biology. During the process of solving practical problems, students can gradually develop their core literacy and cultivate any learning abilities related to sustainable development while acquiring knowledge of ecological environment protection.

The class did not simply help students sift through and consolidate the knowledge they had acquired. More importantly, while solving problems in real-life situations, students boosted their scientific thinking to understand concepts, classify and summarize data based on biological facts and evidence, interpret life phenomena, devise rational explanations and opinions, as well as develop ecological consciousness. While mastering such knowledge and skills, students understood that production and environmental protection are two wheels of human social development, both indispensable.

4.2.2 *Pluralistic Teaching Approach to Promote the Transfer of Knowledge*

The pluralistic teaching approach mainly adopts problem-based learning and inquiry learning, both of which are widely applied. With students at the center, this approach reduces their cognitive burden, enables them to develop higher-order thinking in complex situations, and helps them acquire knowledge, practical skills, cooperation, and self-access learning (Hmelo-Silver et al., 2007). Therefore, this approach is beneficial when teaching biology to achieve sustainable development.

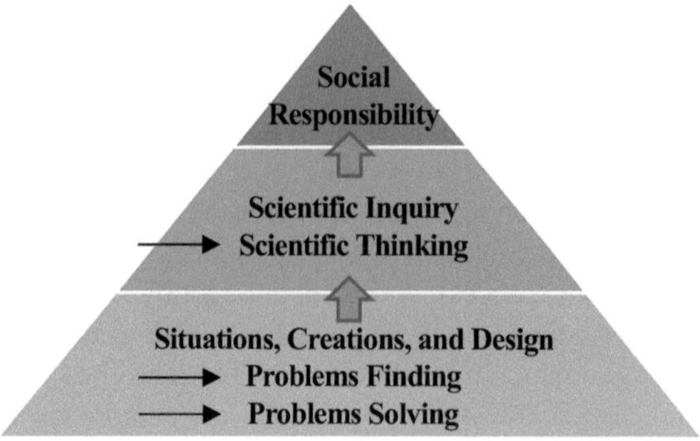

Fig. 4.1 Learning process of the pluralistic teaching approach in science

Considering the harm and remediation resulting from soil pollution, teaching the developed class focuses on analyzing actual case studies of soil pollution and integrating the scattered biology contents organically. Through problem-oriented, self-access learning, and cooperative participation, the class helps students use existing knowledge, evidence, and logic of biology to explore and solve practical problems. With a particular emphasis on inquiry learning, this approach to teaching develops the students' core literacy in aspects of the concept of life, as well as their scientific thinking, scientific inquiry, and social responsibility. Consequently, students not only acquire knowledge about the environment but also enhance their awareness of environmental responsibility and actively participate in the process of sustainable development.

Using inquiry learning as the leading approach throughout the class inspires students to discover soil pollution cases around them, consider situations related to their life experiences, and solve problems by themselves. Under the design and guidance of teachers, students can investigate by themselves, conduct data analyses, draw scientific conclusions, and offer reasonable suggestions and ideas. By participating in activities on generating and producing as many environmental protection strategies as possible, students can exhaustively understand their value and function. Through this learning process, students can master the method of scientific inquiry, cultivate scientific thinking, and enhance their sense of social responsibility and life concepts related to environmental protection (Fig. 4.1).

4.3 Teaching Design and Learning Process

4.3.1 Analysis of the Students' Learning Situation

In this study, senior high school and first-grade students in the class acquired knowledge of the structure and basic components of cells, the movement (in and out) of materials, as well as canceration, heredity, and variation in cells. Moreover, they gradually developed abstract thinking to think comprehensively and develop their dialectical logical thinking to a certain extent. During the class, the students could obtain the current situation of environmental pollution and the results of environmental governance. They can assess problems comprehensively and dialectically, recognize the concept of sustainable development, and implement practical strategies to achieve this effect.

4.3.2 *Teaching Design*

We focused on the real-life case of soil pollution and the redevelopment of the Dongfang Chemical Plant in Tongzhou District, Beijing. The students' cognitive mainline in the class was composed of: (1) What is soil pollution? (2) Why should we focus on soil safety? (3) How to resolve soil pollution? The teaching steps are illustrated in Fig. 4.2.

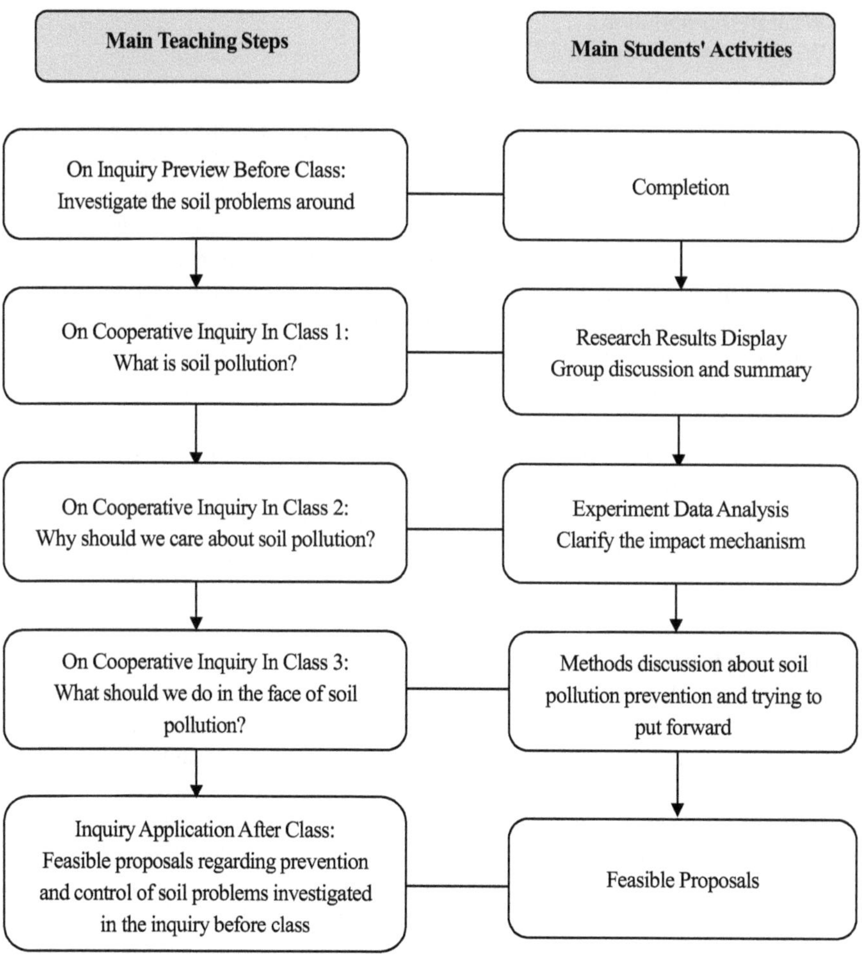

Fig. 4.2 Teaching steps of the developed class

Inquiry Preview Before Class

Before class, the teacher assigned students an inquiry task to investigate soil problems from a local ecological environment. Through a short research report or sides, the students answered the following three questions:

1. Where are the obvious soil problems around us?
2. What are the problems and consequences of soil problems in local areas?
3. What causes soil problems in this local area?

They focused on changes in the surrounding ecological environment and understood the causes of soil pollution in that area.

Cooperative Inquiry During Class

Using the Dongfang Chemical Plant as an example, the teacher analyzed the potential safety problems in the chemical plant and guided the students to focus on soil pollution. The teacher then asked three questions to the student:

1. What is soil pollution?
2. Why should we focus on soil pollution?
3. What can we do?

Regarding Question 1, soil pollution was defined based on the existing forms and components of the pollutants. The students received explanations from the teacher and interacted with one another by asking: "What elements are in the polluted soil?" and "What is the difference between the elements in the human body and those contained in polluted soil?" The teacher guided the students to review their existing knowledge and construct a knowledge framework on soil pollution and human health.

Regarding Question 2, the student further advanced their research based on the first question. To answer the subsequent questions: "How do excess elements in the soil enter cells?" and "What effects do these elements have on the human body?", students analyzed documents and data from some examples of soil pollution and interpreted the impact of soil pollution on the life system at the individual, cellular, and molecular levels.

Regarding Question 3, the students reported the results of their research by group and received informative suggestions from the teacher. Furthermore, they assessed the prevention and treatment of soil pollution by considering the interrelationship between nature, the economy, and society.

Inquiry Application After Class

After the class, the students conducted follow-up research on the soil problems listed in their inquiry activities before class. They analyzed the leading causes of pollution and offered suggestions to promote the prevention and control of soil pollution.

Consequently, the students developed the inquiry activities from the local industrial and ecological conditions, as well as their daily lives, explored the overall problems cooperatively, and devised solutions to these issues. They understood the importance of appropriate behaviors for the protection and scientific utilization of resources and, finally, established awareness that "ecological security starts with me."

4.4 Effect of the Class and Reflection

The best way to promote ESD is to identify facts. Science disciplines determine facts, opportunities, and the time required to educate students. Through the inquiry activities illustrated in the previous section, the developed class offered relevant learning content for students and enabled them to research, conduct data analysis, draw scientific conclusions, present appropriate suggestions and ideas, participate in creating knowledge of the environment, and thoroughly understand the significance and function of environmental protection. Through learning in and out of the classroom, the students assimilated knowledge of environmental protection by integrating theory with practice, cultivated their ability to communicate and cooperate with others, and enhanced their sense of responsibility and obligation to protect their local environment. Scientific disciplines deal with natural and ecological environments. By combing the concept of sustainable development with the science class, we created an effective way to improve environmental awareness and cultivate the future builders of ecological civilization.

4.4.1 Extracting the Contents of ESD from Real Life

In the developed class, the teacher introduced that the condition of the polluted land around the industrial plant was improving. Starting from this basic information, the students analyzed and solved the problems from specific situations and understood the merging of theory and practice. They discussed the influence of soil pollution on life activities and cell functions, considered the relationship between the concept of life and sustainable development, and recognized that human beings must respect the laws of nature. Ultimately, they reconstructed the concept of life from the perspective of sustainable development.

ESD seeks knowledge of and actions in nature. To develop science classes that incorporate ESD, we should find more pertinent subject matters from events relevant to the students' real-life settings and lead the students to understand the ecological environment and natural world through the immersed experience and discovery of ecological laws (Hu, 2019). UNESCO and other countries have successively constructed the core themes of ESD, such as poverty eradication, gender equality, rural reform, and sustainable production and consumption, which focus on heated issues among countries and within the international community, in order to respond to the problems of sustainable development (Wang, 2006). By learning the core themes of ESD from real-life situations, students can better understand the actual state of their communities, cities, and countries, connect what they learn in and out of the classroom, and encourage themselves to take measures toward sustainable development.

4.4.2 *Paying Attention to the Comprehensiveness and Connectivity of the Class Content*

In terms of teaching content, the developed class emphasized the comprehensiveness within and among disciplines, providing students with multidimensional and multi-level thinking spaces. During the teaching process, the core content organized by problem orientation led the students to critically assess and solve the problem as individual researchers. Their cognitive development in the class was based on an empirical study and subject exploration, which required the students to conduct the scientific research process, grasp the research methods, and improve their understanding of sustainable development.

ESD is comprehensive and interdisciplinary. It should not be limited to the classroom but should be expanded from local, national, and global perspectives to integrate multiple disciplines. Currently, primary and secondary education in China shows a good convergence; however, there is a lack of continuity between the disciplines and the different types of knowledge within each discipline. Only by paying attention to the vertical progression in education and the horizontal integration between disciplines can we better adapt to the students' ability for sustainable learning, as required in the new era (Huang et al., 2019).

4.4.3 *Considering the Students' Core Literacy and Knowledge Application and Transfer*

Although several literacy guidelines exist for students, the concept of sustainable development should be a standard to define core literacy. For example, UNESCO (2017) mentions key competencies for sustainability, such as systems thinking, anticipatory, normative, strategic, collaborative, and critical thinking, as well as self-awareness and integrated problem-solving competencies. In the developed class, the teacher adopted a pluralistic teaching approach and provided the students with opportunities to directly contact society, deepen their understanding of soil safety in practice, and create an open space as an active learning environment. The students could discuss the problems from real-life situations and make cooperative inquiries so that they significantly fostered their ability to engage in systems thinking, problem-solving, and cooperative learning.

In the class, the students' consciousness of environmental protection and transfer to action were designed synchronously. Based on the idea of "problems finding – principle analysis – consciousness touching – behavior implementation," the students acquired knowledge and awareness of sustainable development. Furthermore, they discovered minor issues in their immediate surroundings, considered the importance of civilized behaviors, and advertised them with practical actions. This learning process is consistent with that of ESD.

In essence, ESD was created for the sustainable development of education, and its goal is practical. Humans are required to have knowledge, awareness, attitude, and values toward sustainable development. We should not only actively participate in the practices of sustainable development, but also master life skills to support this practice (Song & Tang, 2009). Therefore, we should design classes for students to correctly grasp and flexibly use knowledge. Through this design, students would break the barriers of classroom learning, expand their learning environment and contents, and master the internal logic of knowledge transfer based on necessary knowledge and skills (Liu et al., 2019), rather than simply replacing and applying knowledge. Thus, students would effectively solve the practical problems of sustainable development.

4.5 Conclusion

Sustainable development is an important concept in social development, with school-based education primed to implement this idea. Biology is a subject that studies the phenomena, laws, and applications of life. It is possible to generate a solution to overpopulation, resource crisis, and deterioration of the ecological environment through sustainable development. The holistic approach to content concerns local and global issues and their interrelationships with environmental, economic, and social aspects. The pluralistic approach to teaching encourages students to carry out inquiry learning, foster critical thinking, and put forward their views. The study described in this chapter suggests that these are two fundamental approaches to integrating the concept of sustainable development with science education. To further explore the effectiveness of these approaches, we will expand the integration with other branches of biology, as well as those of other subjects closely related to sustainable development, such as chemistry, geology, and geography.

References

Barbaro, N., & Pickett, S. M. (2016). Mindfully green: Examining the effect of connectedness to nature on the relationship between mindfulness and engagement in pro-environmental behavior. *Personality and Individual Differences.*, *93*, 137–142.

Boeve-de Pauw, J., Gericke, N., Olsson, D., & Berglund, T. (2015). The effectiveness of education for sustainable development. *Sustainability, 7*, 15693–15717.

Bonney, K. M. (2015). Case study teaching method improves student performance and perceptions of leaning gains. *Journal of Microbiology and Biology Education, 16*(1), 21–28.

Geng, L., Xu, J., Ye, L., Zhou, W., & Zhou, K. (2015). Connections with nature and environmental behaviors. *PLoS One, 10*(5), e0127247. https://doi.org/10.1371/journal.pone.0127247

Hmelo-Silver, C. E., Duncan, R. G., & Chinn, C. (2007). Scaffolding and achievement in problem-based and inquiry learning: A response to Kirschner, Sweller, and Clark (2006). *Educational Psychologist, 42*(2), 99–107.

Hu, J. (2019). The value vision and goal construction of ecological civilization education. *Journal of the Chinese Society of Education, 2019*(4), 34–38. (in Chinese).

Huang, Z., Wu, J., & Tang, L. (2019). How far is China from the goal of sustainable development education in 2030: The analysis based on compulsory education curriculum standards. *Educational Research, 40*(2), 140–148. (in Chinese).

Huckle, J., & Sterling, S. (2002). *Education for sustainability*. Translated into Chinese by M. Wang et al. China Light Industry Press.

Liu, W., Qi, W., & Song, S. (2019). Research on deep learning dedicated to knowledge transfer. *Modern Educational Technology, 2019*(3), 25–31. (in Chinese).

Song, N., & Tang, Z. (2009). The aim of sustainable developing education. *Journal of Southwest China Normal University. Social Sciences Edition, 35*(2), 123–127. (in Chinese).

UNESCO. (2016). *Education 2030: Incheon declaration and framework for action for the implementation of sustainable development goal 4: Ensure inclusive and equitable quality education and promote lifelong learning*. UNESCO. https://unesdoc.unesco.org/ark:/48223/pf0000245656

UNESCO. (2017). *Education for sustainable development goals: Learning objectives*. UNESCO.

Wang, M. (2006). The core theme of sustainable development education. *Environmental Education, 27*(1), 27–30. (in Chinese).

Wang, P. (2010). The strategy of implementing sustainable development education in subject teaching. *Educational Science Research, 2010*(7), 52–55. (in Chinese).

Wang, X. (2015). Commentary on the meaning and content of international sustainable development education. *Journal of World Education, 28*(5), 54–57. (in Chinese).

Wang, H., & Li, H. (2014). *Introduction to ecological sustainable development*. Northeast Normal University Press. (in Chinese).

Chapter 5
Developing Competencies for Sustainability Through STEM Education in Mongolia

Uuriintuya Dembereldorj, Burmaa Tsoodol, Namjildagva Raash, Nyamjargal Batmunkh, and Dulguun Jargaldsaikhan

5.1 Introduction

5.1.1 Education for Sustainable Development and STEM Education in Mongolia

The Mongolian education sector has always aimed at pursuing policies that are concurrent with the stages of global sustainable development. Various policies and strategies have been adopted and implemented over the past 20 years in Mongolia to promote sustainable development while preserving the country's traditions. These include the "Mongolian Action Programme for the 21st Century (MAP 21)" (Mongolian Government, 1998), the "Millennium Development Goals-based Comprehensive National Development Strategy of Mongolia" (Mongolian Government, 2008), the "Mongolia Sustainable Development Vision 2030" (State Great Khural of Mongolia, 2016), and the "Guidelines and Requirements to Implement the Concept of Sustainable Development Education in Secondary Schools" (approved by the A/458 order of the MECSS (Ministry of Education, Culture, Science, and Sports, 2015)).

These guidelines and requirements by the MECSS point to a full alignment of quality changes in the content, methods, and assessment in schools' core curricula with the government's educational reforms. They further emphasize the importance

U. Dembereldorj (✉) · B. Tsoodol · N. Raash · N. Batmunkh
Mongolian National University of Education, Ulaanbaatar, Mongolia
e-mail: uuriintuya.d@msue.edu.mn

D. Jargaldsaikhan
National University of Mongolia, Ulaanbaatar, Mongolia

H. Fujii, S.-K. Lee (eds.), *Science Education for Sustainable Development in Asia*, Education Innovation Series, https://doi.org/10.1007/978-981-99-8711-5_5

of fostering teaching and learning processes and school management in developing students' knowledge, practical skills, attitudes, and values in line with Education for Sustainable Development (ESD). Mongolian education for sustainability emphasizes the need for further development and intensification of the changes suggested by these guidelines and requirements.

ESD assists students in acquiring knowledge, practical skills, and attitudes to maintain and protect natural ecosystems entrenched in the Mongolian traditional way of living, customs, and beliefs while aligning with global trends. Since 2000, the concepts, contents, and methods of ESD have been adequately reflected in the general education standards, national core curricula, and textbooks.

In accordance with Goal 4 – "Quality Education: Ensure inclusive and equitable quality education and promote lifelong learning opportunities for all" – of the SDGs, STEM education has been introduced to develop the quality of education at all levels while recognizing the importance of science, technology, engineering, and mathematics. The approach in STEM education highlights new teaching and learning methods aimed at practical and open exploration (Honey et al., 2014; Thomas & Watters, 2015; Cameron & Craig, 2016). This approach supports students' interests, skills, and experiences to enhance their competencies, such as problem-solving, creative thinking, collaborative teamwork, and technology literacy, required for the twenty-first century. However, STEM education has been indirectly integrated into the objectives, content, and methods of general education in Mongolia. The MECSS first began introducing STEM tools (kits) at the primary school level.

The goal of the "Core Curricula for Primary Education" (Ministry of Education, Culture, Science, and Sports, 2019) is to organize activities that enable students to develop their own learning strategies, discover their talents, be confident and creative, live a healthy life, seek lifelong learning, and be viable citizens contributing to the community and to work. In a broader sense, the core curricula value the Mongolian language, national culture, and traditions. As per this goal, in 2017, a three-phase national training was organized to provide primary education teachers training in a theoretical and methodological approach using STEM tools. This training provides basic concepts of STEM education at primary school level as well as teaching strategies to increase students' interest and creativity in mathematics and natural sciences. In the first phase, 120 national trainers were trained in Ulaanbaatar in April 2017, who then trained 30% of all primary school teachers as local trainers in the second phase. These teachers became trainers in their schools of their communities and organized STEM training with experts from the local Department of Education, Culture, and Arts. Additionally, STEM tools were delivered to all primary schools to increase support for all primary school students. Since 2017, all primary school teachers have started teaching STEM lessons using the tools in their everyday instruction, including two sets of ArTeC STEM tools produced in Japan for grades 1 to 4 and a set of Fischertechnik STEM tools produced in Germany for grade 5.

5.1.2 Relationship Between Education for Sustainable Development and STEM Education

The STEM educational practices in primary schools developed through these tools are closely related to education as per the SDGs, with ESD at its core. The 17 goals include overcoming the environmental, economic, and social challenges of the twenty-first century that have a local impact alongside global relevance. The topics covered by the goals focus on the complex relationships among the environment, economy, and society and therefore provide an appropriate thematic structure for STEM lessons that are intended to be interdisciplinary in approach. For example, topics such as purification of drinking water and sustainable electricity supply can be themes for STEM lessons.

STEM education has a student-centered approach that ensures close links between theory and practice, as well as the integration and coherence of all subjects, and develops students' creative thinking, teamwork skills, and problem-solving skills. Thus, it is considered an appropriate educational process that assists students to acquire these twenty-first-century skills. Further, these competencies that STEM education focuses on overlap with the competencies that students should gain in ESD.

The eight key competencies for sustainability identified by the United Nations Educational, Scientific and Cultural Organization (UNESCO) in 2017, based on previous competency research (Rieckmann, 2011; Wiek et al., 2011; Pachner, 2018), are generally required of "sustainability citizens" and are the output of ESD research and practice (Box 5.1). These competencies are divided into three domains: (1) cognitive domain (knowledge and thinking abilities); (2) socio-emotional domain (social abilities); and (3) behavioral domain (action competencies) (Taimur & Sattar, 2019). If these competencies are arranged into a three-domain framework (see Fig. 5.1), for effective ESD, the competencies in the cognitive domain must be deployed first, followed by those in the socio-emotional and behavioral domains.

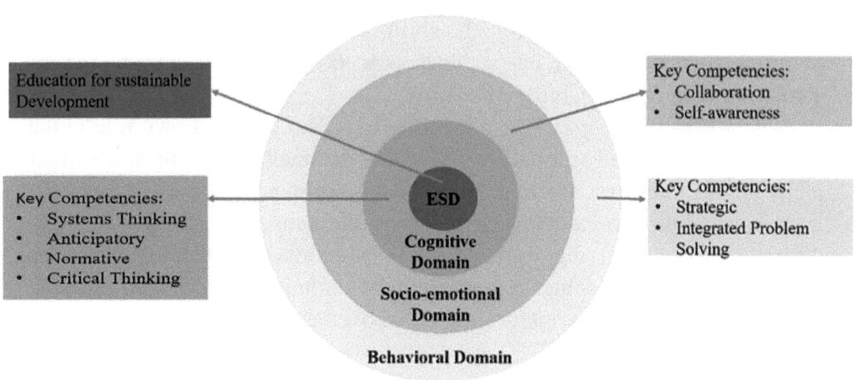

Fig. 5.1 Key competencies for sustainability arranged a three-domain framework

Systems thinking, anticipatory, normative, and critical thinking competencies in the cognitive domain are the foundation for developing collaboration and self-awareness competencies in the socio-emotional domain. Furthermore, knowledge and thinking competencies in the cognitive domain and awareness of oneself and others in the socio-emotional domain leads to the development of strategic and integrated problem-solving competencies in the behavioral domain, which are necessary for individuals to act as sustainable citizens.

Box 5.1: Key Competencies for Sustainability (UNESCO, 2017)

Systems thinking competency: the abilities to recognize and understand relationships; to analyze complex systems; to think of how systems are embedded within different domains and different scales; and to deal with uncertainty.

Anticipatory competency: the abilities to understand and evaluate multiple futures – possible, probable, and desirable; to create one's own visions for the future; to apply the precautionary principle; to assess the consequences of actions; and to deal with risks and changes.

Normative competency: the abilities to understand and reflect on the norms and values that underlie one's actions and to negotiate sustainability values, principles, goals, and targets, in a context of conflicts of interests and trade-offs, uncertain knowledge and contradictions.

Strategic competency: the abilities to collectively develop and implement innovative actions that further sustainability at the local level and further afield.

Collaboration competency: the abilities to learn from others; to understand and respect the needs, perspectives and actions of others (empathy); to understand, relate to, and be sensitive to others (empathic leadership); to deal with conflicts in a group; and to facilitate collaborative and participatory problem-solving.

Critical thinking competency: the ability to question norms, practices, and opinions; to reflect on own one's values, perceptions, and actions; and to take a position in the sustainability discourse.

Self-awareness competency: the ability to reflect on one's own role in the local community and (global) society; to continually evaluate and further motivate one's actions; and to deal with one's feelings and desires.

Integrated problem-solving competency: the overarching ability to apply different problem-solving frameworks to complex sustainability problems and develop viable, inclusive, and equitable solution options that promote sustainable development, integrating the abovementioned competences.

5.2 Competencies for Sustainability Gained Through STEM Lessons

5.2.1 *Study Scope*

In 2019, 20 national trainers from the Institute of Teacher's Professional Development conducted a field monitoring consisting of lesson observations and discussion and focus group interviews with teachers and students in primary schools where STEM tools (kits) were implemented. They collected research data from a total of 1260 primary school teachers and 1680 students (Institute of Teacher's Professional Development, 2019).

5.2.2 *Methodology*

Lesson Observation and Discussion

The national trainers observed one STEM lesson in each primary school in Ulaanbaatar, the provinces, and *soums* collaboratively with local trainers and experts from the local Department of Education, Culture, and Arts. The observation usually lasted 80 min, focusing on teaching and learning activities. During and after the observation, the national trainers recorded in their notebooks the teachers' STEM lesson planning, instruction, and diagnostic assessments of the students. They recorded, for example, how teachers were promoting creative thinking, design thinking, and participation among students. The notebooks were used to discuss with and advise teachers on how to motivate students to learn through STEM lessons and how to support students' learning with STEM tools.

Focus Group Interviews with Teachers

The teachers and local trainers were interviewed on the implementation of STEM lessons in the following three areas:

- What impact do STEM tools have in improving students' knowledge, skills, and abilities?
- What educational methods are embedded in STEM lesson plans? For example, cabinet principles, club activities, projects, after-school activities, integration with other lessons, etc.
- What additional support is needed to properly organize and apply STEM lessons?

Focus Group Interviews with Students

Twenty-five students from grades 1 to 5 were selected randomly from one school and interviewed on the following questions:

- What is most interesting in STEM lessons?
- What are the difficulties faced in STEM lessons? How do you solve them?
- How many STEM lessons do you attend per week?

- What models do you assemble in STEM classes?
- How many students are working with one STEM tool?

5.2.3 Implementation of STEM Lessons

The field monitoring revealed that the implementation of STEM lessons varied from school to school. 1.8% of the schools included STEM lessons in their curricula for more than 2 hours per week; 26.3% for 2 hours per week; and 66.7% for only 1 hour per week; and 5.3% did not include STEM lessons in their curricula at all. By location, schools in Ulaanbaatar are the most advanced in incorporating STEM lessons into their curricula, followed closely by schools in the provinces; however, schools in soums are less advanced (Fig. 5.2).

The number of students per teacher is about 39 in the country as a whole; 42 in Ulaanbaatar; 35 in the provinces; and 28 in soums (National Statistical Office of Mongolia, 2019). This leads to problems in implementing STEM lessons, especially in Ulaanbaatar, as a large number of students in a class negatively affect their teamwork. Some teachers face issues in reaching out to each student and in facilitating group work owing to overcrowded classrooms and limited possibilities for the use of tools by each student.

Table 5.2 shows teachers' cooperation status, indicating that 26.3% of all primary school teachers did not collaborate with anyone while preparing and conducting their STEM lessons; 35.1% prepared lessons in collaboration with other primary school teachers; 7% with physics teachers in an attached secondary school; 14% with art teachers in an attached secondary school; 3.5% with both physics and art teachers; and 14% with the students' parents. STEM lessons are largely being implemented in collaboration with other teachers and the students' parents.

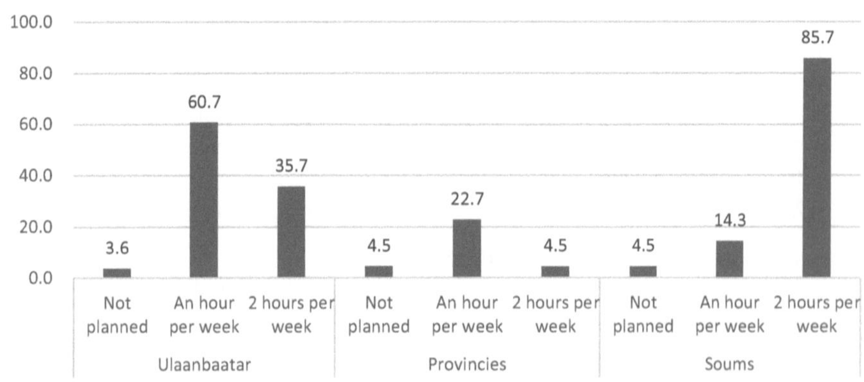

Fig. 5.2 Incorporation of STEM lesson in curriculum

Table 5.2 Teacher cooperation in STEM lessons

Status	%
No cooperation	26.3
Cooperation with other primary school teachers	35.1
Cooperation with physics teachers	7.0
Cooperation with art teachers	14.0
Cooperation with physics and art teachers	3.5
Cooperation with the students' parents	14.0

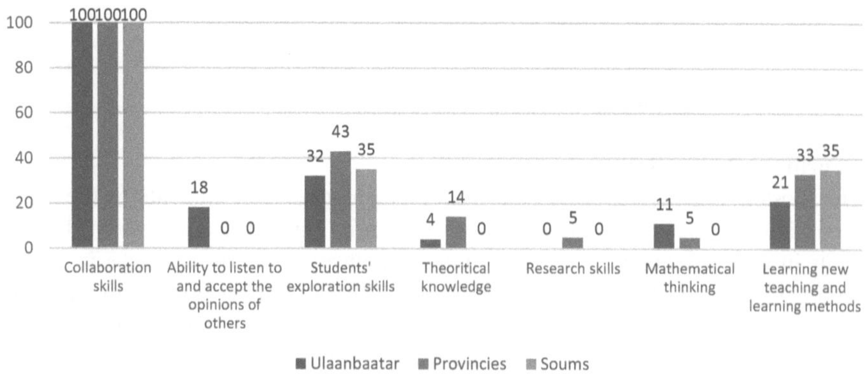

Fig. 5.3 Teachers' gains through STEM lessons ($N = 1260$)

5.2.4 *Outcomes of STEM Lessons*

Improvement in Teachers

The most frequently reported data by elementary school teachers in the focus group interviews were tabulated and are shown in Fig. 5.3. All the teachers that were interviewed report that their team work and collaboration skills improved, and 18% of the teachers in Ulaanbaatar report that they became more willing to listen to and accept the opinions of others. In addition, 32% of the teachers in Ulaanbaatar, 43% of those in the provinces, and 35% of those in soums were convinced that their students' exploration skills were enhanced, while 21% of the teachers in Ulaanbaatar, 33% in the provinces, and 35% in soums reported that they learned new teaching and learning methods; this data for teachers in soums was relatively high compared to the others, which might be due to a lower teaching load.

Teachers in Ulaanbaatar, the provinces, and soums uniformly enhanced their skills while working with others, learning more about other teachers and their students, and learning new methods and techniques for teaching STEM. In addition, some teachers noted that they gained theoretical knowledge and technical and mathematical thinking skills through STEM lessons.

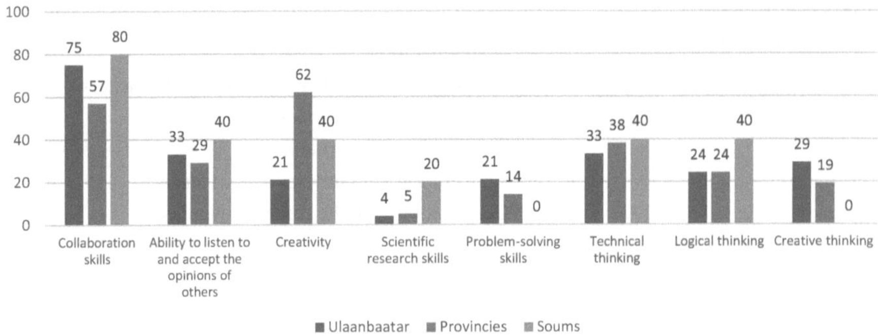

Fig. 5.4 Students' gain through STEM lessons ($N = 1680$)

Improvement in Students

The most frequently reported data from elementary school students in the focus group interviews were tabulated and are shown in Fig. 5.4. Here, 75% of the students in Ulaanbaatar, 57% of those in the provinces, and 80% of those in soums reported that they developed collaboration skills. Following this, many students indicated that they developed creativity, the ability to listen to and accept the opinions of others, technical thinking, and logical thinking. However, the values for creativity varied widely by location, with the students in Ulaanbaatar noting it less than those in other locations.

Overlapping specifically with the competencies for sustainability are collaboration skills, problem-solving skills, and creative thinking. Not many students indicated that they acquired the latter two; more students in Ulaanbaatar acquired them than those in the provinces, while students in soums did not indicate their acquisition at all.

5.2.5 Summary

STEM lessons have been taught in primary schools in Mongolia since 2017. The field monitoring conducted by the Institute of Teacher's Professional Development indicated that, depending on the location, schools allocated varying amounts of instructional time to STEM lessons. The majority of schools in Ulaanbaatar (61%) and the provinces (68%) scheduled STEM lessons for 1 hour per week, with some province schools extending them to more than 2 hours per week. However, in soums, the majority (86%) of schools included STEM lessons for only 1 hour per week, while the remaining 14% did not have them at all.

Statistics show that the number of students per teacher in Ulaanbaatar and provincial schools is 25–35% higher than the national average, which challenges the quality of STEM instruction. In fact, a lower percentage of students in Ulaanbaatar and the provinces responded that they gained collaboration skills, ability to listen to and accept the opinions of others, technical thinking, logical thinking, and scientific research skills, compared to students in soums, where there are fewer students per teacher. However, the students in Ulaanbaatar and the provinces are more likely than those from soums to indicate gains in problem-solving skills and creative thinking that overlap with the competencies for sustainability. The fact that more students per teacher led to the attainment of these competencies may indicate that teachers in Ulaanbaatar and the provinces are relatively more proficient in STEM lesson instructions.

As for teachers, preparing and teaching STEM lessons gave them more opportunities to learn about their students and tailor their instruction considering the unique characteristics of each student. This, in turn, improves teachers' creativity in organizing and applying STEM lessons. In the focus interviews with teachers, they noted that they learned about teachers' collaboration and new teaching and learning methods through STEM lessons. They also pointed out that teachers' teamwork in implementing new teaching and learning methods in their STEM lessons probably led to the improvement of their students' exploration skills. It can therefore be inferred that these positive outcomes are linked to their student gains mentioned earlier.

However, the number of teachers reporting improvement in students' exploration skills is less than the majority in all locations, and the number of students reporting gains of problem-solving skills and creative thinking is less than 30%, even in Ulaanbaatar and the provinces. Teachers need to pay more attention to teaching and learning strategies for developing these skills and ways of thinking.

5.3 Conclusion

The field monitoring results presented in this chapter demonstrated the potential for primary school students to develop competencies for sustainability through STEM lessons. In particular, they suggested that such development builds on students' competencies in the cognitive domain, such as creative, logical, and technical thinking, extending to the social-emotional domain, which includes collaboration skills and the ability to listen to and accept the opinions of others, and the behavioral domain, comprising problem-solving skills. Therefore, educational policies to improve the quality of STEM education are critical to the development of competencies for sustainability of primary school students and should address the following issues:

1. The official curriculum does not include the content for STEM education per se. Therefore, schools incorporate STEM lessons in the content of Human Environment in grades 1 through 3, in that of Human and Nature in grade 5, and in that of Human or Society in grades 4 through 5.
2. The objectives of the supplied STEM tools are not always properly conveyed during the teacher training. Therefore, teachers lack a basic understanding of pedagogy to create better STEM lessons.
3. Large class sizes have a negative impact on students' teamwork skills, ability to accept others' opinions, and creative thinking.

Furthermore, a successful STEM education necessitates the dissemination of its lessons not only in primary schools but also in secondary schools, which requires ongoing training aimed at improving the expertise of natural sciences teachers, including physics and technology teachers, in secondary education. Some of the primary school teachers interviewed worked closely with physics teachers in order to implement STEM lessons. Thus, future training should link primary and secondary education to address specific topics related to students' development and learning while continuing to advance teachers' professional development in curriculum, pedagogy, and assessment. At the same time, the professional development should be based on the relationship between ESD and STEM education as well as on the promotion of ESD through STEM education.

References

Cameron, S., & Craig, C. (2016). *STEM lab for middle grades*. Mark Twain Media Publishing.

Honey, M., Pearson, G., & Schweingruber, H. (2014). *STEM integration in K-12 education: Status, prospects, and an agenda for research*. The National Academies Press.

Institute of Teachers Professional Development. (2019). *Report on the review and evaluation of the STEM training process and implementation in primary education*. Institute of Teachers Professional Development. (in Mongolian).

Ministry Education, Culture, Science, and Sports. (2015). *A/458 order. Guideline and requirements to implement the concept of sustainable development education in secondary schools*. Ministry of Education, Culture, Science, and Sports. (in Mongolian).

Ministry of Education, Culture, Science, and Sports. (2019). *Core curricula for primary education* (2nd ed.). Ministry of Education, Culture, Science, and Sports. (in Mongolian).

Mongolian Government. (1998). *Mongolian action programme for the 21st century (MAP 21)*. Mongolian Government. (in Mongolian).

Mongolian Government. (2008). *Millennium development goals-based comprehensive national development strategy of Mongolia*. Mongolian Government. (in Mongolian).

National Statistical Office of Mongolia. (2019). *Education sector statistics for 2018–2019*. National Statistical Office of Mongolia. (in Mongolian).

Pachner, A. (2018). *Reflexive Kompetenzen – Bedeutung und Anerkennung im Kontext erwachsenenpädagogischer Professionalisierung und Professionalität*. Berhard Karls Universität Tübingen.

Rieckmann, M. (2011). Future-oriented higher education: Which key competencies should be fostered through university teaching and learning? *Futures, 44*(2), 127–135.

State Great Khural of Mongolia. (2016). *Mongolia sustainable development vision 2030*. State Great Hural of Mongolia. (in Mongolian).

Taimur, S., & Sattar, H. (2019). Education for sustainable development and critical thinking competency. In W. L. Filho, A. M. Azul, L. Brandli, P. G. Özuyar, & T. Wall (Eds.), *Quality education* (pp. 1–11). Springer.

Thomas, B., & Watters, J. (2015). Perspectives on Australian, Indian and Malaysian approaches to STEM education. *International Journal of Education, 45*, 42–53.

UNESCO. (2017). *Education for sustainable development goals: Learning objectives*. UNESCO.

Wiek, A., Withycombe, L., & Redman, C. L. (2011). Key competencies in sustainability: A reference framework for academic program development. *Sustainability Science, 6*, 203–218.

Chapter 6
Value-Centered Sustainability for Environmental Literacy: A Perspective from the Philippines

Jestoni P. Babia and Jennifer D. Paño

6.1 Introduction

The World Commission concludes that the most relevant and essential outcome is that developmental issues and the environment are interrelated. These challenges are interrelated as development cannot subsist on a deteriorating environmental resource base; the environment is in harm when growth leave out of account the costs of destruction. These problems cannot be treated separately by fragmented institutions and policies. They link to a complex system of cause and effect (UNESCO, 2018).

For a couple of years, the notion of "VUCA" is gaining popularity as a term to cover the various dimensions of this "uncontrollable" environment (Kraaijenbrink, 2018). Harvard Business Review determines VUCA as volatile, uncertain, complex, and ambiguous. The world has many "unknown unknowns." The more complex and volatile the world is, the more it becomes uncertain (Kraaijenbrink, 2018).

The environmental issue is a major element that disturbs the complex and uncertain realities of this volatile and ambiguous world. This compels humans to take concrete actions to protect and preserve the environment to fight against major social catastrophes caused by climate change and global degradation. Such phenomena affect health and sanitation, economic growth, and many others.

J. P. Babia (✉)
University of San Jose-Recoletos, Cebu, Philippines
e-mail: president@citycollegecdo.edu.ph

J. D. Paño
Cebu Normal University, Cebu, Philippines

H. Fujii, S.-K. Lee (eds.), *Science Education for Sustainable Development in Asia*, Education Innovation Series, https://doi.org/10.1007/978-981-99-8711-5_6

Therefore, various sectors must investigate these challenges in everyday life, instilling the need for reorientation in society to address and understand sustainability in a local and global context. Thus, educators, at the heart of their teaching, should know the most appropriate and essential competencies in making education at present more sustainable. UNESCO's 17 Sustainable Development Goals (SDGs) are a blueprint that will help society to redirect its goals toward achieving a better and more sustainable future. Contextualizing and translating these goals in the educational setting will create responsible students and citizens, making them aware of the right actions for sustainable development.

Science education in its systematic and complex nature can have a positive impact on environmental sustainability. Integrating Education for Sustainable Development (ESD) concepts into the science curriculum is essential to capture the social, political, environmental, and economic aspects of society. This can be reflected in classroom studies, developing a more comprehensive perspective on the science processes, methods, and approaches that students will have to experience. In general, social transformation can be affirmed by understanding the competencies, frameworks, and theories for mainstreaming and integrating science education and sustainable development into a more transdisciplinary context. Hence, this chapter will utilize the "head," "heart," and "hand" approach to ESD, where the sets of competencies are seen in the *head*, signifying the knowledge of ESD; the *heart* reflects the values toward sustainability; the *hand* centers on the skills for ESD transformation. Specifically, this section will discuss the "head," while the heart and hand approach will be discussed in the subsequent sections.

6.2 Sustainability Competencies in Addressing Social Concerns

6.2.1 UNESCO's ESD Integration Framework

ESD integration defined in UNESCO (2018) is the all-embracing and systemic concept and objective. ESD mainstreaming requires more activity focused on specific areas of the curriculum and institution. The integration framework recognizes that the interrelatedness of three important dimensions, social, economic, and environmental is viewed as a holistic and integrated vision of education (Fig. 6.1). Further, the interplay of these dimensions is embedded in culture, which significantly influences the prevailing political context.

The dimensions contain the elements divided between *ESD practice* and *ESD context*. ESD practice contains the *content*, defined as the knowledge element that serves as the "entry point" for ESD learning; *methods* is described as a teaching and learning approach element, typically identified as student-centered and participatory. Curriculum mainstreaming is the element that determines the teaching and learning practices to be utilized in institutions to integrate ESD for curriculum change.

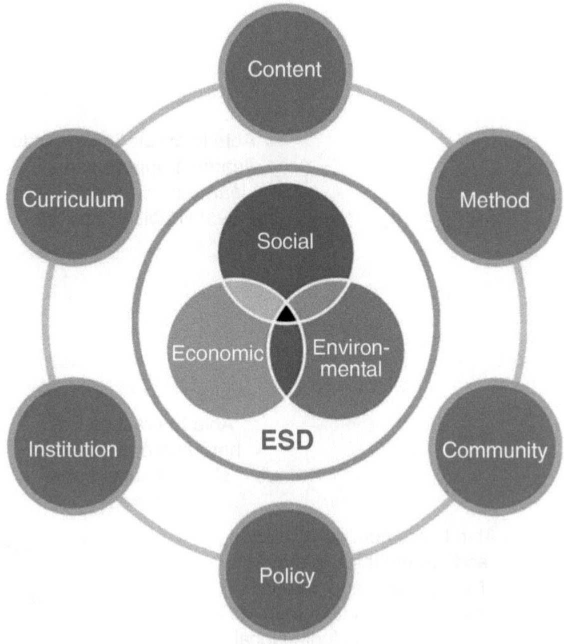

Fig. 6.1 UNESCO's ESD integration framework. (UNESCO, 2018)

ESD context contains the *policy*, *community*, and *institutional mainstreaming*. The policy is determined at varying levels: global, national, ministry of education, and school levels; *community* includes stakeholders and other local and community realities working to shape education and sustainability. *Institutional mainstreaming* results in transformative leadership, which is reflective of the policies developed and implemented using resources that support ESD integration toward institutional transformation.

6.2.2 The Asia-Pacific ESD Teacher Competency Framework

The guide for the effective dissemination of the Asia-Pacific ESD teacher competency framework, the project of Okayama University and UNESCO Japan and UNESCO Bangkok, presented the framework for ESD, highlighting the three competencies and domains in *Facilitating Learning*, *Continue to Learn and Create*, and *Connect, Collaborate, and Engage* to shape sustainable futures (Okayama University ESD Promotion Center, 2020) (Fig. 6.2).

The framework discusses key issues related to sustainable development from multidisciplinary perspectives and viewpoints. Additionally, it talks about the essential elements of sustainable development for one's educational practices in assessing, monitoring, and evaluating the sustainability competencies of the learner, as reflected in these practices.

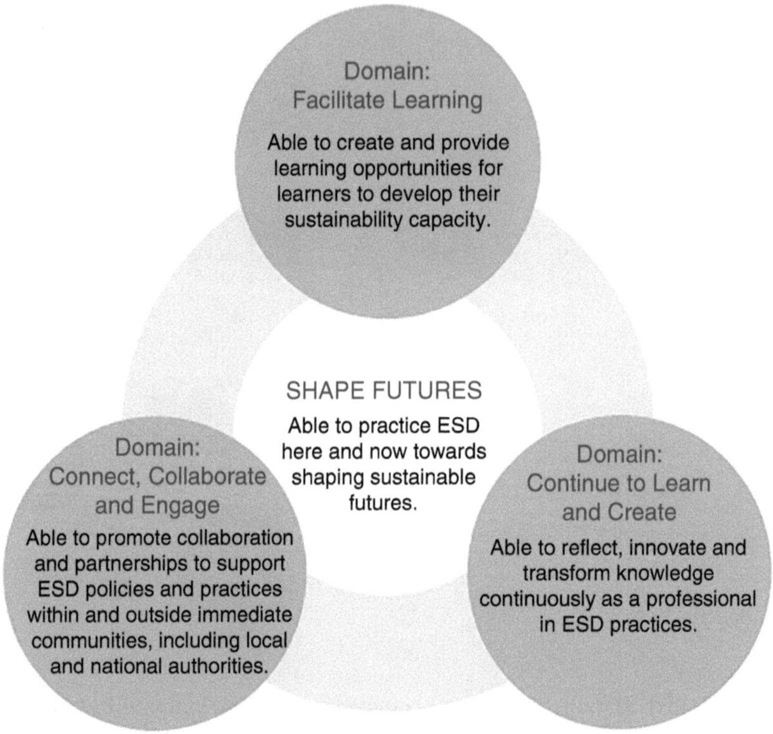

Fig. 6.2 Overview of the Asia-Pacific ESD teacher competency framework

Further, it presents cooperation and action to disseminate and inform the ESD practices to other schools and communities, thereby expanding ESD engagements to other stakeholders. Hence, the framework expands ESD by making students aware of how ESD can be lived out and how collaboration can be used to strengthen ESD in communities and societies. These domains are essential to disseminate the best ESD practices in teaching and education in Asia and the Pacific.

Domain I: Facilitate Learning

This domain enables the creation and provision of learning opportunities to develop the sustainability capacity of learners. This highlights four major competencies, such as culture, pedagogy, technology, and content.

Culture describes the way of practicing knowledge, values, and skills for individual families and their well-being and intercultural cooperation and practices; and it is the evidence of transforming attitudes and lifestyles based on these practices.

Pedagogy is to the process of teaching, focusing on the pedagogies of the teaching and learning approaches and methods for ESD; planning, implementing, and evaluating these pedagogies based on the requirements of the learners; and collaboration with stakeholders, both internal and external, to assess these pedagogies.

Content is defined as the "scope" or the "what" of the ESD teaching and learning process. Specifically, this component identifies the goal set toward sustainability, the identification of issues and problems related to sustainability, and the encouragement of students to raise questions about real-life situation context.

Technology is the instructional tool employed to deliver the ESD pedagogies, and this will be measured using the basic understanding of the principles of the appropriate technology and its utilization to facilitate the learning of students.

Domain II: Continue to Learn and Create

This domain enables the learner to continuously reflect, innovate, and transform knowledge as a professional in ESD practices. The components, *reflect*, *innovate*, and *transform*, are cyclical, indicating the manifestation of behavioral changes for learners, enabling them to think of many ways to innovate as a product of continuous reflection and eventually create transformational change within the self and society. As the transformation process continues, reflection for continuous growth is embedded to further create other innovations and transformational changes.

Reflect is a process that enables learners to engage in critical self-reflection based on what they have learned, relearned, and unlearned, which includes the historical, socio-cultural, and emerging trends of ESD integration. Additionally, it designs the plan of action for the continuous improvement of the ESD integration into the teaching and learning process, and eventually, the progress is monitored and evaluated toward lifelong learning.

Innovate is a process that enables learners to engage in transdisciplinary and trans-sectoral professional learning and development programs to develop sustainable thinking quality. Such engagement can be best achieved through extensive resource mobilization mechanisms and strategies and by co-designing effective and creative solutions to problems to achieve sustainability.

Transform is a process that enables learners, after reflecting and innovating, to promote and enhance campaigns on sustainability. The "transformation" process is manifested by practicing a sustainable life by modeling the ESD practice and by stimulating collective changes to advocating a sustainable way of living and lifelong learning.

Domain III: Connect, Collaborate, and Engage

This domain enables learners to promote collaboration and partnerships to support the ESD policies and practices within and outside immediate communities, including local and national authorities. This domain emphasizes the importance of connecting individuals to schools and communities, thereby developing leadership capacities through these three important components: *participate and engage*, *cooperate*, and *collaborate* toward meaningful partnerships for ESD.

Participate and Engage, done at the individual level, capacitates the self to find key stakeholders and engage with the community for meaningful partnerships and engagements.

Cooperate, done at the school level, enables the understanding and demonstration of sociocultural, political, historical, and environmental concerns. This is manifested through active engagement and response to the needs of the community and society. This is an agency for advocating change by being a role model for policies and practices at the community, national, and international levels.

Collaborate, done at the community level, facilitates and/or leads formal and informal collaboration by generating ethical and safe working conditions. Such manifestation can be seen through the trust that is built, as well as in maintaining reciprocal caring relationships and negotiating conflict situations.

The interconnectedness of the individual, school, and community for participating and engaging, cooperating, and collaborating creates leadership competence to promote and advocate ESD on local, national, and international levels toward transformative change for the present and future generations.

6.2.3 Cell Resilience Theory Toward Understanding Challenges in Achieving Sustainability Competencies

The cell's resilience theory by Szanton and Gill (2010) provides an understanding of the delineating factors of society vis-à-vis personal qualities. As transformation for sustainable development also creates challenges and opportunities, this theory can help individuals and society to achieve the goal of transformation by becoming resilient and dynamic in the process.

Resilience entails surpassing the trying times by accepting the changes as part of a dynamic process and embracing positivity to provide solutions to existing problems (Bonanno, 2004; Luthar, 2006). Schools at present are experiencing changes because of the pandemic, and therefore, they require resiliency as the world is experiencing its complexities resulting from many unexpected events. Without becoming resilient, the sustainability efforts might not be continued, and the goal for collaboration at the local, national, and international levels may be a huge challenge instead of an opportunity.

The framework of the theory emboldens the constructive impact of environmental and societal factors to shape the internal variables of each person, such as psychological, physiological, and cellular coping abilities to achieve resilience and sustainability (Fig. 6.3). Additionally, this framework is an important tool for reflection, innovation, and transformation as to how schools, communities, and the nation should understand their functions toward environmental sustainability.

In the article "the history of twins, as a criterion of the Relative powers of nature and nurture," Francis Galton concludes that nature has more effect than nurture on development. In relation to the individual, community, and society for the cells resilience theory, the nature and nurture aspects are harmonized without any conflict since both are beneficial to the living. It suggests unity and the acceptance of the roles of people in the environment and society. The consideration of the external and internal variables is substantial, as one factor can affect the other. To achieve resilience in the overall structure of the society, the adaptability to the changing world phenomenon, awareness and understanding of the cause of the natural disasters, and

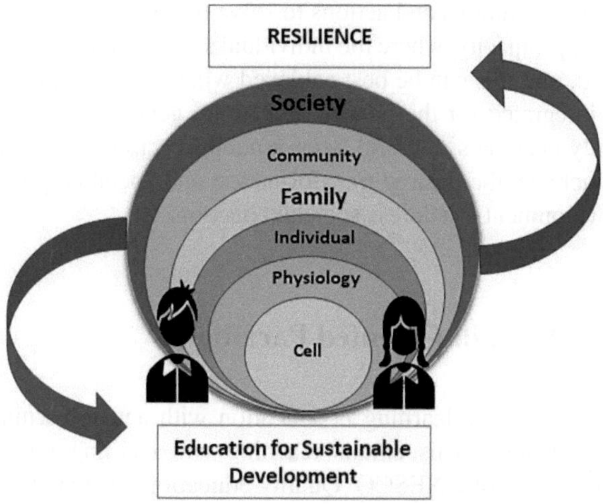

Fig. 6.3 Society-to-cells resilience theory anchored with the three pillars of ESD

fortifications and solutions to avoid potential risks to human lives, animals, and the ecosystem are highly relevant for sustainability (UNISDR, 2005).

Schools should understand that there is an interrelationship between and among the cellular functions of living organisms in society. From cells as the basic unit of society to maintaining and creating good physiology, an individual becomes successful in advocating and involving in capacity-building initiatives toward sustainability. Without looking at its relationship, the initiatives and innovations to achieve the competencies may be hard to actualize. For instance, understanding that the pandemic is a threat to sustainability, collaborative efforts must center around the utilization of appropriate resources for environmental preservation, for example, the presence of medical experts and scientists to determine the cause and origin of the pandemic can be a good opportunity to achieve sustainability capacity.

This capacity through societal resilience can be best achieved through the preservation and conservation of environmental resources for future generations (Mayena, 2006). Adequate funding, environmental programs, disaster preparedness, new buildings and infrastructure, numerous job opportunities, support from the local and international government, unity among people, a just and human society, and communication and information to all people are the identified factors to help achieve resilience in the society (Malalgoda et al., 2014).

Laying all these sustainability competencies for teachers, including how these competencies are utilized in the present world of many "unknown unknowns," will enable schools to realize that they play a major role in society. Thus, schools should critically determine what knowledge, skills, values, and attitudes should be integrated into the school curriculum and the entire institutional practice.

In summary, the use of competencies for sustainable development and the framework that guides society toward a sustainable future can bring educators together to create synergistic actions on issues and challenges that affect the entire system and provide alternative solutions and actions to solve these problems. This synergy is a form of leadership capacity, where the individual self, school, community, and society are involved, and this can be best achieved when such competencies are mainstreamed and integrated for the present and future generations. Most importantly, schools can only become successful when students "act out" and value these competencies and achieve the desired transformation and social empowerment, all for sustainable development to address societal concerns.

6.3 ESD as the Skill-Oriented Paradigm

The key to transformative learning is education with active learning engagement and dynamic pedagogy. This entails teaching and learning with competencies anchored on the SDGs of UNESCO. Quality education necessitates the incorporation of a skills-oriented teaching model and active approaches to direct students in solving real-life issues in society.

To achieve skills orientation for ESD integration into the curriculum, innovative educational approaches and pedagogies are required for transformative and lifelong learning. Significant efforts have been made to achieve the target of SDG 4 for quality education. New platforms were utilized, a plethora of pedagogies were applied, and various learning opportunities were realized through collaboration, partnership, and various support systems to ensure lifelong learning and sustainability. However, with the rapid technological changes, the resource constraints and the impact of the health pandemic presented various opportunities and challenges for educators to hurdle to keep pace with the changing paradigm of learning in the COVID world.

The transformative potential of innovative teaching and learning must direct learners to problem-solving and critical thinking through action-oriented and learner-directed pedagogy. Thus, the careful planning of educators on the pedagogical strategies to use should lead learning to a strong sense of ownership and responsibility over the decisions made on the most pressing concerns of society. ESD teaching is holistic; this indicates that various pedagogies should use the *head, heart, and hand* approach. Typically translated, the cognitive, psychomotor, and affective domains of a learner should altogether be tapped to ensure profound learning and the ownership of responsibilities. In this section, the *hand*s approach will be supported to accommodate sustainable development.

Action Competence for Shaping the Teaching and Learning Process

Knowledge acquisition should be perforated with well-thought actions and decisions from local and global perspectives. Education should set a high premium for teachers to improve their teaching skills for sustainability to facilitate their personal development in that field. Various educators integrate sustainable development into the curriculum via subject infusion, extracurricular and cocurricular activities, or community engagement. Using various modalities, the principle of sustainable development and the social, economic, environmental, and cultural lenses should be discussed using multidisciplinary, interdisciplinary, and transdisciplinary perspectives. In 2014, UNESCO stated that ESD will be integrated at all levels and should be taught as early as possible (Paño, 2019).

The big picture of the 17 SDGs of the world raises the cognizance of "global citizenship" and the need for active partnerships to resolve social, economic, and environmental problems.

Five I's of the Skill-Oriented Paradigm of ESD
In science education, building skills are vital to solve the current economic, environmental, and social problems. Below are the skills that educators should inculcate and nurture among students to become proactive learners.

Inclusiveness "Inclusive education" or "inclusive learning" refers to the inclusion and teaching of all children in formal or non-formal learning environments without regard to gender, physique, intellect, society, emotion, language, culture, religion, or other characteristics.

The twenty-first-century learners are digital citizens or digital natives; hence, they can smoothly navigate the information highway; however, the availability of Internet connection poses a challenge to some economically challenged families and some places. Additionally, other societal issues, such as poverty, malnutrition, and gender, and cultural issues may lead to isolation. The reality of the digital divide and the societal impediments are critical in ensuring inclusive education, considering the skills, understanding, and knowledge required to work effectively in an inclusive setting. These sources of apprehension may limit learning, stigmatize, and socially exclude some learners, making it challenging for educators to facilitate learning. Therefore, it is important to understand the principles of inclusive education.

Principle 1 Diversity in the classroom. Students in every classroom have diverse backgrounds of cultures, experiences, values, and beliefs. Educators should be skilled to handle diversity to draw knowledge and wisdom from the various experiences of the students.

Principle 2 Strength-based. The strength-based approach in education enhances the strengths and capabilities of every learner. Educators should understand the dynamism of learners by teaching them how to learn best and supporting them in their challenges and difficulties. Considering this, the educator can provide the required educational support and motivation for the proximal and potential zones of development where teaching and learning can best take place (Hopps-Wallis et al., 2016).

Principle 3 Student engagement. Allowing students to speak out inculcates confidence, freedom, and trust in their immediate environment. The active engagement of students boosts their motivation to learn and eventually increases their learning satisfaction (Martin & Bolliger, 2018).

Principle 4 Feedback. Feedback improves the learning experiences of the students and is considered a vital approach to encourage their development as independent learners as they track, assess, and control the phasing of their learning. This enables them to gauge their performance, assess their learning, set their goal, self-motivate for further learning, and shape the future of their learning (Al-Bashir et al., 2016).

Principle 5 Commitment. An educator should be fully committed in terms of the "head" (i.e., knowledge), "heart" (ethical and moral standards), and "hand" (skills). Commitment is a hackneyed word; however, it entails dedication and responsibility for education to be sustainable (Sharma & Pace, 2019).

Integrativeness Subjects in the curriculum should not be taught as a separate field of knowledge but should be linked to other fields of specialization to have a fresh approach. Integrative teaching connects the principles and theories of a book to a real-life scenario, particularly societal issues, which encourages educators to expose learners to problem-solving and critical thinking to achieve the most efficient solution.

Principle 1 Contextualized. This is useful in drawing the interest of students, thereby facilitating a relatively deep understanding and concept retention (Cabansag, 2014). Contextualized lessons are integrated with topics that are meaningful, useful, and germane to the experience of students using driving questions, events or experiences, artifact development, and culminating activities (Picardal & Paño, 2018). Genetic lessons can use a zipper model; for example, popsicle sticks can be used to better understand the central dogma of molecular biology and a twin phenomenon.

Principle 2 Relevant. Learning may be relevant if students can connect their prior knowledge with the new lesson. Integrative learning should be culturally relevant to the lives of students. Byrd (2016) has proposed three approaches to making relevant culture-based lessons, including having a positive outlook, promoting cultural competence, and fostering critical consciousness. In having a positive outlook, teachers should offer a dynamic curriculum built on the strength of students. In promoting cultural competence, the way of life of the student is given a high value; the individual and community values and the existing community problems may be integrated into the classroom lessons.

Principle 3 Evidence-based teaching and learning (EBTL). This is an approach to provide information to educators about scientifically effective pedagogies and assessment; it provides empirically supported tools for a variety of courses and proposes the most efficient technique for studying. EBTL is a teaching modality that leads students to make decisions based on objective evidence derived from empirical literature, experiments, and other data sources.

Principle 4 Attainable goals. Setting specific targets guides both teachers and learners to better realize the targets. Further, these attained targets are positive signs of progress for the academic achievement of the learners. It is not simply stating the lesson goals or targets, rather goal-setting sets the direction for learning, identifies the parameter for monitoring progress, and confirms teachers' expectations of their students.

Principle 5 Educational space availability. Learning space is a conducive atmosphere created in a classroom if students are encouraged to learn independently or collaboratively. Allowing students to share their thoughts and think through, giving them time to nurture metacognition and assess their learning through class activities, and providing them with spaced practice results in the long-term retention of information and fewer anxieties.

Intuitiveness Pedagogical tact is the manifestation of the intuitive action of the teachers in the classroom; it is defined as the capacity of the teachers to considerately, instantaneously, yet appropriately act on the complexities of classroom circumstances. Several researchers agree that intuitiveness is useful to strengthen the teaching process, considering that it can overcome the limitations of analytical thinking, it is instrumental to effective classroom management, and it gives justice to differentiation in the classroom and flexibility for a relatively safe atmosphere (Sipman et al., 2019). Intuition may be categorized into two: inferential and holistic intuition. The former is reliant on automatic responses based on the rapid recognition of memory patterns learned from experience; the latter arises from the integration of fragments of previously unconnected memory into a new information system (Sipman et al., 2019).

One skill required by an educator is intuitiveness, and Winch et al. (2015) believed that technical know-how and colloquial wisdom can develop pedagogical tact. Technical know-how entails the use of principles, methodologies, or materials to ensure productivity. This technical know-how enables teachers to foresee or meditate on the effects of the interventions. Conversely, colloquial wisdom involves virtue to rightly judge the current circumstance. Biesta (2014) proposed that virtue is required to judge appropriately and can negotiate complexity in decision-making. Pedagogical tact cannot be learned from textbooks alone; however, it is taught by experiences. Considering that unavoidable circumstances, whether good or bad, may occur in a classroom setting at any time, therefore, an educator should possess pedagogical tact in speech and action to avoid damage to the feelings of the students.

Innovativeness (Future-Oriented) Human potential is unlimited and hedonistic in principle, i.e., we aim at pedagogical success wherein our students will simultaneously enjoy and learn. Every learner expects learning to be interesting, engaging, and motivating, and this classroom ambiance should integrate innovative teaching methods to promote in-depth learning. To foster innovative teaching, curricula should be skillfully and thoroughly developed, where the teacher can adopt different innovative teaching strategies according to the diverse needs of the students. The innovative strategies should be implemented and modified according to the requirements of the students (Naz & Murad, 2017). Mynbayeba et al. (2017) proposed the following features of innovative teaching:

1. Anticipation of development. Anticipation requires us to know the future but not necessarily what will happen. In science, for example, we knew that in the future, there can be changes in the environment. Thus, we can anticipate future developments out of the existing technology but with the utmost consideration of techno-moral or socioeconomic and environmental scenarios to make the earth sustainable for the next generation. Anticipating development is a skill to train our minds to anticipate the future or discern what is to come.
2. Flexibility. The trends of diversification and flexibility are seemingly interrelated. Therefore, higher education programs should adapt to have increased access to education (Teichler, 2004). Similarly, in the classroom setting, this provides a paradigm shift from having a student-centered to becoming a student-led classroom to accommodate a flexible curriculum (Paño, 2019); therefore, the flexibility and variety of teaching methodologies are necessary.
3. Creativity. Monotonous teaching decreases the effectiveness of learning among students and restricts their creativity, which leads them to become bored or suffer academic losses (Mynbayeba et al., 2017). Creative teaching is as an intelligence and a personal trait, and the geniuses of Newton, Fleming, Roentgen, and other scientists, which led to some foundational truths in science, may be considered creativity in action (Ala-Mutka et al., 2008).
4. Collaboration. Teacher collaboration reinforces peer support and the sharing of teaching practices.

Inquisitiveness (Didactic Orientation) Inquisitiveness is the didactic reorientation of the mind, which could lead to scientific discovery or exploration. This may entail curiosity about the things with favorable traits, i.e., beauty, aroma, softness, and other biological characteristics. For example, the aroma of the volatile compound released by the Venus fly trap is a possible reason for the attraction of insects to it, causing its death. Anton von Leeuwenhoek's curiosity led him to observe anything with his microscope. He was able to vividly describe bacteria, green alga, protists, plants and animal tissues, mineral crystals, fossils, and so on (ucmp.berkeley.edu). Curiosity was defined by Kashdan and Fincham (2004) as the volitional recognition of a novel, uncertain, complicated, and conflict-oriented opportunity to pursue an active search to settle discrepancies between the existing knowledge and present transaction.

ESD is a skill-oriented paradigm that can characterize a relatively deep understanding of the complex society. Incorporating the skill-oriented teaching model in the integration of the concepts of sustainability will propel the reorientation of the mind toward sustainability. Thus, the pursuit of environmental objectives and knowledge, skills, and values can be conducive for the students to adopt and be adept at a sustainable way of life.

6.4 Value-Centered Approach in ESD

Having the right knowledge, skill, and attitude will manifest positive values for sustainability. ESD guides every human being to acquire not just knowledge and skills but also values to acquire a healthy and livable future. Sustainable behavior is the positive manifestation of the acquired knowledge, skills, and attitude. The previous sections of this book already discussed the required building block for a sustainable future: knowledge and skills. This section presents pedagogical strategies for integrating values through ESD and examines how implementing these pedagogical practices in the integration of holistic knowledge develops environmental literacy and positive attitudes toward the environment.

Sustainability education involves the activation of the principled mechanisms of behavioral and lifestyle transformation through value integration. Science education must contribute to the formation of a *whole person* who is openly engaged in harmonious coexistence with the world they live in.

Definition of Sustainability Value

Values are the manifestation of one's appreciation, recognition, and degree of importance to the natural system and the objective existence of both tangible and intangible material. It may be demonstrated in the behavior and attitude of individuals (Keong, 2020) and eventually shape one's attitude and actions. For sustainable development to flourish, the three pillars, economic, social, and environmental dimensions, should be balanced. The anthropocentric view of nature, wherein the ecosystem is a means to meet the needs of humans, should be overcome to achieve environmental sustainability (Keong, 2020). One method to achieve this is by inculcating the right sustainability values.

Different Value Orientations

Steg et al. (2014) purported four value orientations, namely, hedonistic, egoistic, social-altruistic, and biospheric orientations. A hedonic individual tends to sacrifice the environment for their pleasure and comfort; egoistic environmental orientation believes that people protect the environment because of the benefit we derive from them, such as fresh air and clean water. Social-altruistic environmental attitude takes care of the environment because of human goals or benefits with some amount of sacrifices. These people are open to change and are willing to implement pro-environmental practices despite some challenges. Examples of this include the use of solar panels to save electricity despite some financial consequences. Biospheric attitude is concerned about the environment because of the inherent value of its existence, i.e., all species have a right to live (Sadovska, 2019).

Reports reveal that the human population has been on exponential growth in the last 50 years. This has led to increased pollution and destruction of the ecosystem, having devastated half the species of plants, destroyed 83% of the wild animals, and consumed 30% of the known resources (Arora, 2019). The ecosystem services are at stake, and the accompanying economic benefits humans derive from it have become relatively scarce rather than an abundant gift to enjoy.

Values: The Cornerstone of Environmental Management

The surge of technological advancement, modernization, and rapid industrialization demands the harmonization of knowledge, skills, and attitude to align one's values to the natural system. Science educators play a pivotal role in inculcating these values to produce active, productive, and transformed citizens who are advocates of nature. Effective values should draw out the necessary skills to solve societal issues. By having a values-centered education, the teachers are guiding the young people toward a sustainable future.

Values and Environmental Education

It is the responsibility of every science teacher to pass on the invaluable treasure of values to their student so that tomorrow they will pass these to the next generation. Vijayalakshmi (2019) categorized the values of human activity connected to environmental education into the following dimensions:

1. Aesthetic. The appreciation of beauty relative to other environmental affordances.
2. Cultural. The attitudes, customs, traditions, and rituals of the community.
3. Ecological. The conservation of the natural systems in a state of interaction with other living things.
4. Economic. The trade between people and ecosystem services, resource ownership, and the accessibility of resources, calculating the costs and benefits derived from the system.
5. Ethical. Human responsibility as consumers of resources guided by conscience.
6. Existential. The management of one's life in relation to the environment, making people's lives more comfortable.
7. Intellectual. The reasoning capacity of students through the various understanding of science concepts toward better utilization of resources.
8. Political. The management of the interaction between individuals and society according to some guiding policies.
9. Psychological. The application of scientific attitudes to an enhanced way of thinking and reasoning.
10. Transactional potentiality. The established relationship or significant control or influence on others to share awareness and meaningful insights about nature.
11. Recreational. The creative and effective use of the natural system through leisure.
12. Vocational. The application of values in any vocation for professional development to become productive members of the society.

Pedagogical Strategies for Values Formation

Values in science are best taught in an integrated manner rather than as a separate subject. The work of Pierre acknowledges that science teaching will be improved if values are taught in relation to social issues. There is a wide plethora of pedagogical strategies for inculcating values in science to further guide the most appropriate environmental decision based on normative commitment.

The "whole-school approach" (WSA) was part of the Global Action Program promoted by UNESCO in 2014. The WSA for values formation is herein proposed to establish a culture of sustainability in the entire academia, along with its partners and other stakeholders. In this section, the in-school pedagogies will teach values in the classroom setting; the out-of-school pedagogies will emphasize and extend values to the community, while lifelong learning will instill values among educators, parents, organizations, and learning materials.

A truism of teaching holds that there is not one best strategy in teaching; however, good teachers are flexible, creative, and resourceful in using these strategies. Mahmud (2017) and APEID (1993) identified several pedagogical strategies, which may be used to emphasize values formation.

In-School Pedagogies

1. Experiential and participatory approach. Carefully planned and structured activities may be given to students to cull out their introspection and reaction to a specific scenario. Experiential learning will provide profoundly enriching experiences to students, which can eventually shape future feelings or thoughts that will forecast consequent action. In this case, values are neither taught nor imposed; rather educators are guiding the learning experiences.

   (a) Field-based coursework. Outdoor learning coupled with some reflections will provide direct hands-on learning for the students.
   (b) Internship program. This approach will intensify the skill required by the students to learn.
   (c) Invitation of guest speakers. Guest speakers provide social capital for students to expand their networks and expose them to a variety of experiences.

2. Place-based approach. This approach intends to contextualize learning using the immediate environment of the students, identifying the issues and concerns, and guiding them toward a practical, doable, and sustainable decision.
3. Integrated approach. The infusion of values in the lesson content is done purposively and systematically to develop both the cognitive and affective dimensions of the lesson. This may be done through the following:

   (a) Value Clarification Strategies. This may be done using individual or group dynamics, wherein the students are led to understand their thoughts and feelings about specific situations using the systematic steps of problem-solving.
   (b) Moral Development Strategies. This is the strategy appropriated to the age of maturity, wherein the pros and cons of specific actions have corresponding consequences.
   (c) Value Analysis. This strategy follows the steps in problem-solving wherein students identify the problem, make inferences, gather and weigh evidence, and create conclusions to represent their stand on the presented situation.
   (d) Values Inculcation. This is the most commonly used strategy, which uses direct teaching through moralizing.

4. Questioning techniques. The effective processing of questions helps students better understand their thoughts and feelings.
5. Role modeling. The attitude of the teachers is a silent epitome of the values a student may learn, and school regulations or policies are the "silent" teachers of the values that the students must emulate.

Out-of-School Strategies

1. Community outreach. The class may be brought to less privileged and depressed areas to expose students to the realities of life.
2. Volunteer work. A teacher may encourage students to work for a cause and take an active part in solving a problem.
3. Research work. This provides students with an in-depth feel of the scientific truths through research.
4. Field trips. This exposes students to realize human interventions in relation to the actual scenarios.
5. Environmental campaigns. Students conduct lectures or symposia and posters or parades to raise the awareness of the community and enjoin them in such advocacies.
6. School projects. This affords opportunities for students to experience a tangible output of the lesson through a school project, which promotes ecological concerns.
7. Environmental education camping activities. This will provide exposure to young people on the existing environmental problems of the society for them to take an active role in solving the concern.

Lifelong Learning Activities

1. Teacher Training. Teachers should be updated on values training and formation, teaching/learning strategies, materials development, training strategies, and assessment strategies.
2. Parent-Teacher Activities. This is a meaningful way of encouraging partnership between the school, family, and community to take care of the environment.
3. Supplementary Resource Materials. This refers to the vast array of learning materials, which may be available in print and online forms, that may be prescribed by the teachers to give emphasis and further depth to the lesson discussed.
4. School Clubbing. This is a creative expression of the interests of the students as they participate in value-laden activities through organizations.

These strategies are useful in imparting values through authentic and practical lessons. A WSA toward a value-centered ESD necessitates the agreement between the school administrators, educators, parents, community, and students. Education must direct people and society to think scientifically and analytically with utmost consideration of its environment. Further, science educators must hone students with affective behaviors to become productive members of society. The value-centered ESD is a unifying pedagogical strategy that leads to the development of a "whole person" (cognitive, affective, and psychomotor dimensions) in solving societal issues and concerns, offering a greater opportunity for learning for sustainability.

6.5 Emerging Pedagogical Teaching Approaches in ESD

The manifold changes in society, the rapid development of knowledge, the shifting technology infrastructure, and the demands of the twenty-first century resulted in major changes in teaching and learning, thereby necessitating changes to the educational landscape. A meta-synthesis of 20 research journals, as to the emerging pedagogical teaching and learning in this digital age, resulted in the framework presented in Fig. 6.4.

There is multidimensionality in the emerging pedagogy, wherein four major components are visible: learning attitude, learning environment, learning process, and learning content. These components have a significant impact on how teaching and learning processes are structured. Learners are more expressive, and learning is more experience-based and significantly focuses on metacognition. The learning environment has become more open and accessible anytime and anywhere; it nurtures creativity, hybridization, and collaboration and aims for positive learning for knowledge construction. The learning process is considerably shaped by technology, where learners prefer more authentic solutions and activity-oriented and multisensory heutagogy. Learning content maximizes the contextual environment of the students, making learning more meaningful and relevant. The emerging pedagogy empowers both educators and students to leverage learning in a rich and complex environment; thus, there is an anticipated urge to reconsider roles, authority, and learning evaluation.

ESD teaching is holistic; this indicates that various pedagogies must use the *head, heart, and hand* approach. Typically translated, the cognitive, psychomotor, and affective domains of a learner must be used together to ensure profound learning and the ownership of responsibilities. Additionally, ESD teaching is transformational; students learn to positively approach societal concerns and evaluate decisions to arrive at an innovative solution. Conversely, educators should teach using the lens of critical perspectives, above the status quo, and level up from a stereotype to a more transformative approach in teaching, which requires flexibility and dynamism in the use of teaching strategies. The varying practices, concepts, and contexts of teachers as to their social, economic, environmental, and cultural backgrounds affect the perspectives and performances of teachers in teaching. This provides the leeway for the freedom to use a plethora of teaching strategies that they may perceive best.

Table 6.1 shows a sample of a lesson plan in the field of science, which highlights the *head, heart, and hand* teaching approach using the 5E's of lesson planning, guided by the framework shown in Fig. 6.4. The 5E model of instruction stands for engagement, exploration, explanation, elaboration, and evaluation.

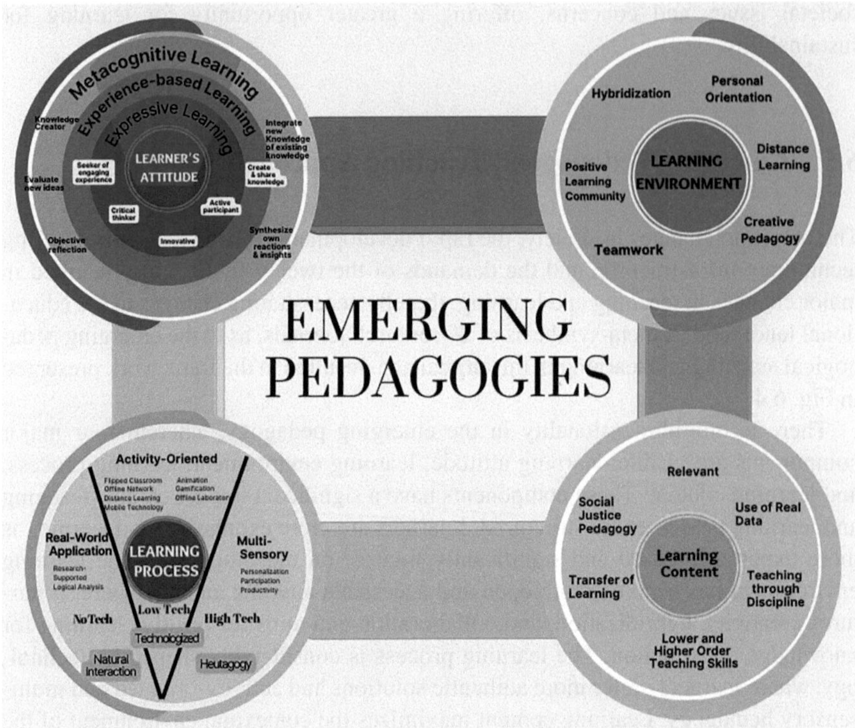

Fig. 6.4 Emerging pedagogies and teaching approaches in ESD

From the lesson plan, the *head* is the information input necessary for students to know. The "heart" is the sustainability value of ensuring the safety of lives and properties. The "hand" strategy is the expected skills of the students during a calamity based on what they have learned. This strategy exposes the students to mitigation and prevention, preparedness, response, and recovery in case of an earthquake. Moreover, the actual drill and training of the teacher will set the minds of the students on the most ideal thing to do before, during, and after an earthquake. This lesson plan has encompassed the four thematic areas of disaster risk management and will realistically orient and prepare students for an actual calamity. This lesson plan is responsive to accomplishing the SDG, focused on sustainable cities (11), good health and well-being (3), climate action (13), and quality education (4).

The ESD-based teaching pedagogies were implemented years ago before it has a name; however, not all educators are fully aware of the advantages and opportunities they can create. More often than not, science education focuses on the creation of new knowledge and the invention of new technologies, thereby neglecting the ways and methods to restore or preserve the environment.

Table 6.1 5E's lesson plan in science, highlighting the "head, heart, and hand" teaching approach. (Paño, 2019)

Subject/grade level: Science STEM Grade 10
Lesson objectives: At the end of this lesson, students will be able to: 1. Describe the mechanism of earthquake generation. 2. Explain how epicenters are determined. 3. Discuss how tsunamis are formed. 4. Determine ways to mitigate the impact of earthquakes.
Materials: Pictures of the effect of the 9.0-magnitude earthquake and tsunami that hit Tohoku, Japan, in 2011; the 7.9-magnitude earthquake in Sichuan, China, in 2008; a 7.8-magnitude quake in Lamjung, Nepal, in 2015; and 7.1-magnitude quake in the Philippines in 2013. Each picture provides information on casualties and damage to lives and properties. World Map Video clip on "Plate Tectonics Mechanism" Slinky Compass, pencil, paper Worksheets on determining the earthquake epicenter Checklist on building safety Four stations with information sheets on the following: 1. Three types of seismic waves and their characteristic motion. 2. Effects of the seismic waves on buildings. Factors leading to tsunami formation. 3. Principle of the seismograph machine. Determining epicenters. 4. Building safety standards and checklist. 5. Demonstration area on seismic waves using a slinky.
Engagement Class activity: 1. Unlocking word difficulty. 2. Ask students about their experiences with the recent earthquake. 3. Students will do a think-pair-share to discuss the topics in the four stations. 4. Watch the video clip (5-minute video) and answer the following guide questions: A. Why are Japan, China, Nepal, and the Philippines susceptible to earthquakes? B. Why are earthquakes destructive? C. How does earthquake affect the economy, society, and environment?
Exploration Students, in triads, will take turns to go to each of the four learning stations, pictures, and the world map. Each station will input information regarding the topic. 1. They will explore and relate the movement of the slinky, seismograph, and earthquake waves. 2. They will perform the activity worksheet on determining the earthquake epicenter.
Explanation Students will discuss their answers to the guide questions in each station to their group. Explain the relationship and principle of the slinky, seismograph, and earthquake waves. They will explain how they were able to obtain the epicenter of the quake.
Elaboration/extend Each triad will be provided a checklist on building safety; they will be assigned one building in the campus to evaluate.
Evaluation Students will propose ways the impact of earthquakes can be mitigated. Teacher will have a drill/training on what to do before, during, and after an earthquake. Give summative test.

In the context of science education, ESD-based teaching pedagogies have emerged to redirect its goals, mission, and objectives to conserve, preserve, and make the earth sustainable for the next generation. These initiatives should lead teachers to incorporate those participatory approaches that will enable them to develop their critical perspectives about a sustainable world. Organizing and implementing these projects results in the sustainability capacity to continue. The ESD-based lesson plans are feasible guides for the implementation of these plans.

However, there is a need to evaluate the implementation of these emerging pedagogical approaches to check whether these concepts are practiced or not. Further, it is a priority of the education sectors to monitor and evaluate sustainability practices so that the teachers can involve and engage private and public sectors, such as the community, to join hands in protecting and restoring the environment.

Teaching sustainable education is more than just a concept. Rather, it is a technique to instill critical, creative, and ingenuous ways to be mindful of what we do to our environment. The goal of ESD is to provide solutions to the plethora of problems that gradually exterminate the beauty of nature.

With these specific explanations and illustrations, the teaching pedagogies in ESD should be placed in the context of the curriculum so that the learners will be ready to respond and be resilient to the volatile and complex world. For the *head, heart, and hand* goal to be achieved, the teachers should "walk" the sustainability talk.

6.6 Transformative and Transformational Learning Experiences

6.6.1 Transformational Education Through ESD

ESD as a transformational education addresses the learning outcomes, pedagogy, and learning environment. Transformational learning is a process of deep, constructive, and meaningful learning that goes beyond simple knowledge acquisition and supports critical ways of making a learner's life meaningful. The learners have become aware of tacit assumptions or expectations and assess their relevance for interpreting (Mezirow, 2000).

As transformation is seen in the different aspects of ESD components, schools must create important milestones and learning tools to ensure that knowledge, skills, and values are added to the overall learning experience. As Manzini puts it, the transition toward sustainability is a "socially complex learning process," that is, behavior, patterns, values, and aspirations dramatically change.

Connecting the knowledge, skills, and values of the students to achieve sustainability is seen as a gap by Manzini. Hence, schools, as an agent of transformation, must lead in facilitating dynamic movements from theory to practice by involving various sectors and disciplines in the society toward transdisciplinary engagements.

As defined by Mezirow (1994), transformational learning theory is a constructivist approach, focusing on learner-centered approaches, where learners are exposed to hands-on activities. Moreover, it is an orientation that determines how learners interpret and reinterpret their sense experience, which is central to making meaning and learning.

The role of the learner is actively participating in the discourse for validation. This dialogue provides the opportunity to critically examine the evidence, arguments, and alternate points of view that foster collaborative learning. It highlights the importance of the open-mindedness of learners, their courage to take risks and change their readiness to unlearn and self-motivate, and their self-confidence to make decisions and take responsibility.

To foster informative learning, the educator should assist learners to become aware of critical assumptions and provide learners with practice in recognizing frames of reference and practice in redefining problems from different perspectives. The frame of reference helps the learners to understand the kind of environment they live in. The goal is to create a community of learners who are "united in a shared experience of trying to make meaning of their life experience."

Box 6.1: Case Study on Transformative Learning Using Collaboration and Participatory Approaches: USJ'R's Bloom Project on Developing Decodable and Leveled Books for Children

The University of San Jose Recoletos (USJR)-College of Education (COE) has reflected on improving the implementation of Mother Tongue Multilingual Education in the respective communities by integrating and mainstreaming ESD to revitalize and strengthen the early grades reading using the child's mother tongue. The USJR-COE team has responded to the call for improving literacy instruction by acting as the organizers of the enabling writers' (EWT) training program to engage the community as a major reference in contextualizing and indigenizing materials and resources. It has made use of the award-winning book creation software, Bloom, for writers to create small decodable books for the elementary school (Babia & Cotejo, 2021).

6.6.2 Transformational to Transformative Learning Experiences

The target of the SDGs can serve as a foundation for socioeconomic pathways; however, they should be operationalized to reduce complexity and account for long-term sustainability concerns beyond 2030 (Zim et al., 2018). The operation has to be constructively and deeply processed to achieve effective outcomes for sustainable transformation.

Transformative learning can transcend the notorious value/action gap that divides the awareness of environmental threats from our capacity to take appropriate action (Boehnert, 2011). Further, the environment has a tremendous impact on cultivating the holistic aspects of the development of learners, including physical, emotional, spiritual, and psychological health. The environment provides natural resources for consumption, education, and habitat. ESD pillars must be understood and valued.

Box 6.2: Case Study on Curriculum Integration to ESD Action-Based Approach: USJ-R Senior High School Experience

The paradigm of Education for Sustainable Development (ESD) has pervaded discourses on enhancing education for national development. The University of San Jose Recoletos Senior High School Department recognizes this commitment to quality education and uses it to inculcate the principles of sustainable development into the learning experiences of its learners. With this in mind, the USJR-SHS curriculum has been utilized as an entry point.

Senior high school students are provided with opportunities to be more aware and knowledgeable of ESD through the inclusion of the ESD concepts and values integration toward environmental literacy into the SHS subjects, research, and other department initiatives. Strand organizations were created and more than 30 clubs were formed. These organizations can be categorized according to their focus on the 5P's that shape the SDGs: people, planet, prosperity, peace, and partnership. Student-centered teaching and learning methods were established through the implementation of the Peer Learning Program's #REACHing4You Learning Sessions. Environmental education initiatives were also started through the Mega Cebu and Green Thumbs' Environmental Education and Awareness Camps and Workshops, Semestral Mangrove-planting Initiatives, and various Conserve Energy Challenges.

These initiatives are done to influence the learners' values, interactions, and general ways of living to contribute to sustainable development. By slowly translating these into the curriculum and transmitting them to the students, USJR SHS is proud to have started to utilize education as an armor to make sustainable development an attainable goal.

Mezirow's transformative learning theory (2000) in a more simplified process begins with *critical reflection* through identifying problems and dilemmas, examining the self, and critically assessing the assumptions. It follows with *reflective discourse* where discontent and the process of transformation are shared, including the exploration of new roles, relationships, and actions, and next, *action*, where an individual utilizes the acquired knowledge and skills in planning, implementing one plan, trying out new roles, building competence and confidence in such roles and relationships established, and making this process to be reintegrated into his or her perspectives.

Box 6.3: Case Study on Transformative Learning Theory (Critical Reflection, Reflective Discourse, and Action) at Philippine Normal University-Visayas

"Sa katunggan Damo Kita Mahibaluan" ("We can learn a lot in a mangrove area" is an ESD community-based project borne as a result of critical reflections and reflective discourse with the school and other stakeholders on the success of the other extension projects of the Philippine Normal University-Visayas conducted in other areas of Cadiz entitled "Bungalon, Padamuon, and Tatapon Mangrove Rehabilitation Among Fisher Folks." This is in response to the team's finding on the need to reverse the degradation and loss of mangrove forest wetlands and their associated coastal ecosystems as a consequence of the severe typhoon "Yolanda" in Purok Kakahuyan, Barangay Daga. PNU-V successfully restored these wetlands and has been in close contact with the fisherfolks in planning, designing, implementing, and evaluating the status of the coastal areas.

The fishermen have shared their stories, and these stories are made into Filipino-level decodable books and are distributed and used by Daga Elementary School, Cadiz. This transformation became the continuous advocacy of the Philippine Normal University-Visayas to protect its environment through rehabilitating its mangrove ecosystem.

6.6.3 *Schools as Agents of Change and Transformation*

How do we gauge transformation in academia and society through ESD? Sterling and Orr (2001) presented and described the four-stage level of sustainability in education. As emphasized in his model, transformative learning is not an easy process either for the teacher or the learner. Further, he stressed that learning for sustainability must transcend the traditional learning approach and methods because acquiring information alone does not automatically lead to change. There must be a relatively broad learning process that should occur for the "change" to work. Sterling's four stages include no change to transformation (Fig. 6.5).

The basic step after determining the "no-learning" or "ignorance" context of the schools is to emphasize content and skills as the entry point for implementing ESD. The process of accommodation happens when an individual learns about change. This process is known as the accommodative response. Once the competencies are maintained, schools will add values as an emphasis for the reformation of purpose and practice. Thus, institutions in this context will learn to change by adapting to change. This is called the reformative response. Lastly, when ESD is

Fig. 6.5 Levels of learning in ESD. (Adapted from Sterling and Orr 2001)

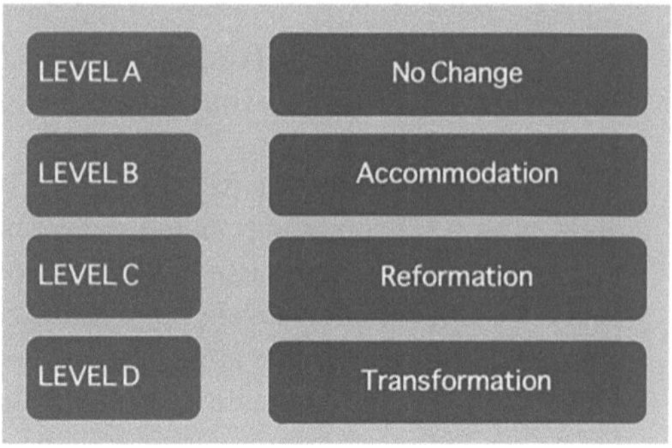

accommodated and used to reform the self and society, schools should be ready to apply capacity-building initiatives toward sustainability, integrating it into the entire school structure (curriculum, methods, projects, plans, etc.) This process is called a transformative response, where enactment engages stakeholders and partners working toward individual transformation and societal change.

The year 2030 is fast approaching, and this will etch another milestone for countries worldwide to embed the SDGs into the universal quest for peace, prosperity, the planet, and people. All educational levels have worked hard to ensure that the present generation contributes toward having a sustainable planet and that each one will be a responsible citizen.

ESD, being an integrative concept of economic, environmental, and social dimensions, should be viewed as a requirement for the future. The head, heart, and hand approach of ESD should be considered in the overall school practices. An extensive support system is required to mobilize technological resources and innovations to be developed and disseminated, requiring broad-based scientific support (German Advisory Council on Global Change, 2011; Dannberg & Grapentin, 2016).

There is no doubt about the strong relationship between science education and the concept of sustainability, which led to the advancement of the SDG. This transformation in schools can only be seen depending on the extent of ESD and integration into the curriculum and institutional practices. This reinforcement concept of sustainability in the school curriculum can be manifested by presenting science education transformation in the SDG era and by presenting exemplars of sustainability concepts using the lens of pedagogical teaching and learning pedagogies and approaches through collaboration, connection, engagement, and creation among and between educators and students in ways that are meaningful and relevant.

Box 6.4: Case Study on ESD Transformation in Schools Toward Family and Community Development and Partnership: CTU'S Hablon sa Cebu

"Hablon sa Cebu, as a community extension project of Cebu Technological University-Argao clearly addresses the United Nations' Sustainable Development Goals, particularly in addressing poverty, hunger, quality education, decent work and economic growth, responsible production and consumption, and partnerships for the goals. Through Hablon of CTU-Argao, the dying cultural heritage of Handloom weaving was reborn and revived. It involved the community women to go back to handloom weaving, raising economic livelihood or Argao, Cebu, Philippines, and uplifted the economic status of the entire community.

The CTU-Hablon has manifested the transformation from a very academically focused university into a social entrepreneur university. The school was able to transform each individual to be more committed and passionate in weaving as they worked hand in hand to mobilize the weavers to link with the government institutions, foundations, and professional organizations, which gave them consultations, skills training, infrastructure, and the like, which, in return, gave them opportunities for employment. It expanded, transformed, and provided economic security to sustain the livelihood of the beneficiaries and would continue to provide good income to the future beneficiaries in the next 5 years. Since handloom weaving was a predominantly female industry, wives became the weavers, while their husbands and children transformed as assistants in the very tedious pre-production processes of weaving. Their families took pride in the livelihood of their mothers as they were helping them to weave. The pride was seen among children coming to center and helping their mothers to weave. The fathers and the children eliminated gender bias in the family structure. The project transformed each member of the family with a great sense of responsibility."

References

Ala-Mutka, K., Punie, Y., & Redecker, C. (2008). Digital competence for lifelong learning. *JRC Technical Note,* JRC48708, 271–282.

Al-Bashir, M., Kabir, R., & Ismat, R. (2016). The value and effectiveness of feedback in improving students' learning and professionalizing teaching in higher education. *Journal of Education and Practice, 7*(16), 38–41.

APEID. (1993). *Strategies and methods for teaching values in the context of science and technology*. UNESCO Principal Regional Office for Asia & the Pacific.

Arora, N. (2019). Earth: 50 years challenge. *Environmental Sustainability, 2*, 1–3.

Babia, J. P., & Cotejo, D. G. I. (2021). Education sustainable development community-based projects of the Philippines. *Psychology and Education, 58*(3), 3998–4007.

Biesta, G. (2014). *The beautiful risk of education*. Paradigm Publishers.

Boehnert, J. (2011). *Transformative learning in sustainable education*. https://www.slideshare.net/ecolabs/skin-deep-ecolabs-june2011-boehnert.

Bonanno, G. A. (2004). Loss, trauma, and human resilience: Have we underestimated the human capacity to thrive after extremely aversive events? *American Psychologist, 59*(1), 20–28.

Byrd, C. M. (2016). Does culturally relevant teaching work? An examination from student perspectives. *SAGE Open, 6*(3), 1–10.

Cabansag, M. G. S. (2014). Impact statements on the K-12 science program in the enhanced basic education curriculum in provincial schools. *Journal of Arts, Science and Commerce, 5*(2), 29–39.

Dannenberg, S., & Grapentin, T. (2016). Education for sustainable development-learning for transformation: The example of Germany. *Journal of Futures Studies, 20*(3), 7–20.

German Advisory Council on Global Change (WBGU). (2011). *World in transition – A social contract for sustainability*. WBGU.

Hopps-Wallis, K., Fenton, A., & Dockett, S. (2016). Focusing on strengths as children start school: What does it mean in practice? *Australasian Journal of Early Childhood, 41*(2), 103–111.

Kashdan, T. B., & Fincham, F. D. (2004). Facilitating curiosity: A social and self-regulatory perspective for scientifically based interventions. In P. A. Linley & S. Joseph (Eds.), *Positive psychology in practice* (pp. 482–503). Wiley.

Keong, C. Y. (2020). *Global environmental sustainability: Case studies and analysis of the United Nations' journey toward sustainable development*. Elsevier.

Kraaijenbrink, J. (2018). *What does VUCA really mean?* https://www.forbes.com/sites/jeroenkraaijenbrink/2018/12/19/what-does-vuca-really-mean/?sh=408a545d17d6&fbclid=IwAR1W1THysxAMQRtfYqPKaGoS-BQXOKVw7tWKussXMErwOuoFOSq6WhLVjVw

Luthar, S. S. (2006). Resilience in development: A synthesis of research across five decades. In D. Cicchetti & D. J. Cohen (Eds.), *Developmental psychopathology: Volume three: Risk, disorder, and adaptation* (2nd ed., pp. 739–795). Wiley.

Mahmud, S. N. D. (2017). Engaging head, heart and hands: Holistic learning approach for education for sustainable development. *International Journal of Learning and Teaching., 9*(2), 298–304.

Malalgoda, C., Amaratunga, D., & Haigh, R. (2014). Challenges in creating a disaster resilient built environment. *Procedia Economics and Finance, 18*, 736–744.

Manyena, S. B. (2006). The concept of resilience revisited. *Disasters, 30*(4), 433–450.

Martin, F., & Bollinger, C. D. (2018). Engagement matters: Student perceptions on the importance of engagement strategies in the online learning environment. *Online Learning Journal, 22*(1), 205–222.

Mezirow, J. (1994). Understanding transformation theory. *Adult Education Quarterly, 44*(4), 222. https://doi.org/10.1177/074171369404400403

Mezirow, J. (2000). *Learning as transformation: Critical perspectives on a theory in progress*. Jossey-Bass.

Mynbayeba, A., Sadvakassova, Z., & Akshalova, B. (2017). Pedagogy of the twenty-first century: Innovative teaching methods. In O. Bernad-Cavero & N. Llevot-Calvet (Eds.), *New pedagogical challenges in the 21st century – Contributions of research in education*. IntechOpen.

Naz, F., & Murad, H. S. (2017). Innovative teaching has a positive impact on the performance of diverse students. *SAGE Open, 7*(4), 1–8.

Okayama University ESD Promotion Center. (2020). *Guide for the effective dissemination of the Asia-Pacific ESD teacher competency framework*. Okayama University. http://ceteesd.ed.okayama-u.ac.jp/pdf/200511.pdf.

Paño, J. (2019). Sustainable development goals integration in lesson planning among pre-service teachers. *CNU Journal of Higher Education., 13*, 1–10.

Picardal, M. T., & Paño, J. D. (2018). Facilitating instruction of central dogma of molecular biology through contextualization. *Journal of Teacher Education and Research, 13*(2), 118–132.

Sadovska, A. (2019). What do youth really care for? Towards ethical education, education for sustainable development and youth values. *Forum Pedagogiczne, 9*, 205–218.

Sharma, U., & Pace, E. M. (2019). Teachers' commitment to teach in inclusive schools, preparation of. In M. A. Peters (Ed.), *Encyclopedia of Teacher Education* (pp. 1–6). Springer Singapole.

Sipman, G., Martens, R., & McKenney, S. (2019). The role of intuition in pedagogical tact: Educator views. *British Educational Research Journal, 45*(6), 1186–1202.

Steg, L., Perlaviciute, G., van der Werff, E., & Lurvink, J. (2014). The significance of hedonic values for environmentally relevant attitudes, preferences, and actions. *Environment and Behavior, 46*(2), 163–192.

Sterling, S. R., & Orr, D. (2001). *Sustainable education: Re-visioning learning and change* (Vol. 6). Green Books for the Schumacher Society.

Szanton, S. L., & Gill, J. M. (2010). Facilitating resilience using a society-to-cells framework: A theory of nursing essentials applied to research and practice. *Advances in Nursing Science, 33*(4), 329–343.

Teichler, U. (2004). The changing debate on internationalization of higher education. *Higher Education, 48*, 5–26.

UNESCO. (2018). *Integrating Education for Sustainable Development (ESD) in teacher education in Southeast Asia: A guide for teacher educators*. UNESCO.

UNISDR. (2005). *Hyogo Framework for Action 2005–2015: Building the resilience of nations and communities to disasters*. UNISDR.

Vijayalakshmi, M. (2019). *Role of biology teacher in inculcating biological science values*. Conference paper of Value Education in Teacher Training – Innovations and Challenges at PPG College of Education for Women, Coimbatore.

Winch, C., Oancea, A., & Orchard, J. (2015). The contribution of educational research to teachers' professional learning: Philosophical understandings. *Oxford Review of Education, 41*(2), 202–216.

Zimm, C., Sperling, F., & Busch, S. (2018). Identifying sustainability and knowledge gaps in socio-economic pathways vis-à-vis the sustainable development goals. *Economies, 6*(2), 20.

Chapter 7
SEAMEO's Journey to Promote Education for Sustainable Development in Southeast Asia

Ethel A. P. Valenzuela, Orawan Sriboonruang, Piyapa Su-angavatin, Wahyudi, and Weerawan Theeroungchaisri

7.1 Background

7.1.1 Brief History of ESD

Education for Sustainable Development (ESD) has a long history starting in the 1980s. In 1992, UNESCO was asked by the United Nations Secretary-General to act as the lead agency of ESD. The UNESCO Chair launched in 1999 was the first of its kind in the field of ESD. In 2002, the concept of creating a UN Decade of Education for Sustainable Development (UNDESD) was discussed and endorsed by many nations during the World Conference on Sustainable Development in Johannesburg, South Africa. Many nations agreed that more ESD progress was needed. Consequently, the UN Decade (2005–2014) was declared through a resolution by the UN General Assembly (57/254).

As the decade was ending, nations called for a continuation of the work begun during the UNDESD and requested that UNESCO develop a continuing strategy. In 2014, UNESCO launched the Global Action Programme (GAP) on ESD highlighting five priority areas for action. The GAP has proven useful in maintaining the momentum of ESD that has now emerged as a crucial implementation element in the Sustainable Development Goals (SDGs).

Ministers of education adopted a global education strategy to implement the SDG 4, entitled Education 2030, at the World Education Forum 2015 in Incheon, Korea (UNESCO, 2015). The ministers also approved the yearly publication of the

E. A. P. Valenzuela · O. Sriboonruang (✉) · P. Su-angavatin · Wahyudi · W. Theeroungchaisri
The Southeast Asian Ministers of Education Organization (SEAMEO) Secretariat, Bangkok, Thailand
e-mail: orawans@seameo-stemed.org

H. Fujii, S.-K. Lee (eds.), *Science Education for Sustainable Development in Asia*, Education Innovation Series, https://doi.org/10.1007/978-981-99-8711-5_7

Global Education Monitoring Report, which requires nations to make yearly reports on their SDG 4 progress. In 2016, the 2030 Agenda for Sustainable Development was put into effect. ESD is arguably at the core of these 17 SDGs that aim for a sustainable future for our planet (Leicht et al., 2018). In 2019, the 40th session of the UNESCO General Conference adopted a new global framework on ESD called "Education for Sustainable Development: Towards Achieving the SDGs" or "ESD for 2030" (UNESCO, 2019).

The history of ESD continued to unfold when the global pandemic hit the world and affected more than a billion children and youth who had to temporarily leave school, and more than 24 million are in danger of not returning to their education. It is time for us to strengthen ESD as an important driver of change, empowering learners to take the decisions and actions needed to build a just and economically viable society that is respectful of both the environment and cultural diversity in this challenging time.

7.1.2 Overview of the ESD Landscape in Southeast Asia

The development and progress on ESD in the Southeast Asian Ministers of Education Organization (SEAMEO) member countries vary. ESD has been implemented by education ministries via either curriculum enhancement or mainstreaming in educational activities and policies at different levels of implementation; some countries are at an advanced level, while others are just starting (UNESCO & SEAMEO Secretariat, 2017).

SEAMEO spearheaded a situation analysis of ESD in 2017 and assessed how ESD was being implemented in Southeast Asian countries via selected teacher training institutions. The findings are summarized in the following paragraphs.

In Cambodia, two universities that are involved in ESD integration are the Royal University of Phnom Penh (RUPP) and the National Institute of Education (NIE). The RUPP integrated ESD in 2013, ensuring inclusive and equitable quality education and promoting lifelong learning opportunities for all. ESD in Cambodian universities is included by introducing the Association of Southeast Asian Nations and global studies in the curriculum. The NIE formed a committee on ESD integration in collaboration with experts from UNESCO, the Ministry of Environment, and the Royal Academy of Cambodia as well as internal experts to design training for subject area heads to train ESD trainers and other NIE trainers in ESD. It reported that the awareness of ESD increased significantly among teacher educators at teacher education institutions in Cambodia. ESD concepts became important and were mobilized from cities to provinces. Major teacher education institutions started working together and closely collaborating. There is a possibility of investing in ESD further (SEAMEO Secretariat, 2018).

As for Indonesia, one of the universities that integrated ESD is Gadjah Mada University. The University Strategic Plan (2012–2022) focuses on developing learning systems that can solve local, national, and global challenges. The Indonesian

River Restoration Movement taught communities along rivers how to properly interact with their environment. This movement sought the communities' and schools' cooperation in restoring the rivers, which involved clean-up drives and harvesting rainwater for reuse. This resulted in a better society and improved the communities' economic and environmental quality, at the same time as developing their culture (SEAMEO Secretariat, 2018).

For Lao PDR, one of the universities involved in ESD integration is the National University of Laos. ESD has been integrated into the entire education system, including the concepts of climate change, global warming, and environmental education supported by the College of Education (SEAMEO Secretariat, 2018).

In the Philippines, two of the universities that integrated ESD are Philippines Normal University (PNU) and Cebu Normal University (CNU). PNU Visayas was designated as the Environment and Green Technology Education (EGTE) Hub by the Board of Regents. This hub was created to attend the non-mainstreaming courses for internationalization. The PNU, as an EGTE hub, is committed to producing teachers that are distinctively knowledgeable on EGTE, as well as research scholars who are proficient in groundbreaking and cross-disciplinary inquiries (SEAMEO Secretariat, 2018).

Additionally, CNU aims to develop high-performing professionals and intellectuals proficient in generating new knowledge to create a progressive and peaceful pluralistic society. Its mission involves the transformation of education that nurtures thinking individuals who are valued members and leaders of the society, high-impact research that pushes the boundaries of knowledge in education and contributes to improving communities, and strong partnerships that collectively and creatively address development gaps in the community. The university has around three ESD modules, which are Introduction to Disaster Risk Reduction in Education, Climate Change Action for Sustainable Development, and Youth Exchange Climate Change and Lifestyle Guidebook (SEAMEO Secretariat, 2018).

As for Thailand, two universities that integrated ESD and Global Citizenship Education (GCED) are Chulalongkorn University and Khon Kaen University. Chulalongkorn University integrated ESD and GCED by involving thematic training modules and networking with practitioners, academia, supporters, and policymakers. It builds bonds between school teachers and educators in teacher education institutes, which involves the usage of a learning platform that is constantly updated. Moreover, this university has a capacity-building program to equip regional mobile trainer teams in 18 areas and implements MIDL classrooms in schools and universities, DCE schools and camps, and GCED pilot schools and training centers aided by OBEC, the Ministry of Education, and UNESCO Bangkok. On the other hand, Khon Kaen University established an ESD integration demonstration school based on the findings of research and professional development workshops and since then has been integrating ESD into pre-service and in-service teacher training via curriculum design and instructional management (SEAMEO Secretariat, 2018).

7.2 SEAMEO Program Activities Related to ESD

The SEAMEO Regional Centre for Education in Science and Mathematics (SEAMEO RECSAM) based in Penang, Malaysia, contributed to inclusive and STEM education development, to develop science and mathematics education in Southeast Asia (Khor, 2019). As a training institution, the SEAMEO RECSAM needs to participate in the delivery and achievement of the 17 SDGs that aim to end poverty, protect the planet, and ensure prosperity for all (UN, 2015). One key program conducted by the SEAMEO RECSAM is professional development programs that educate in-service teachers from Southeast Asia on the SDGs. This program aims to underline ideas and practices that support teachers who include students with diverse needs in the classroom; apply appropriate teaching and learning strategies to develop individual strengths with high and appropriate expectations for each child; and assist students through good inclusive practices such as collaboration, teamwork, innovative instructional practices, and peer strategies.

The SEAMEO Regional Centre for Community Education Development (SEAMEO CED), based in Lao PDR, introduced ESD and GCED through its border school and rural education projects. The SEAMEO CED also trains lifelong learners, particularly in educating rural communities and providing learners with knowledge on sustainable livelihoods (SEAMEO CED, 2019).

The SEAMEO Regional Centre for Graduate Study and Research in Agriculture (SEAMEO SEARCA) based in the Philippines introduced ESD by training schools on the school-plus-home gardens project (S + HGP). This initiative was co-implemented with the Philippine Department of Education-Laguna and the University of the Philippines Los Baños to improve the nutritional condition and dietary habits of school-age children by increasing the production of locally grown vegetables through school and home gardening, which then contributed to the security of the community's food and nutrition (SEAMEO SEARCA, 2019). Moreover, the S + HGP contributed to the achievement of the SDGs to achieve a better and more sustainable future for global citizens.

Global challenges that humanity has encountered, especially those related to poverty, inequality, climate, environmental degradation, prosperity, and peace and justice, are acknowledged as key components of GCED (UNESCO, 2016). SEAMEO centers aim to contribute to the goals that are interrelated to achieve the aim of leaving no one behind by the year 2030.

The SEAMEO Inter-Centre collaboration project of the SEAMEO Regional Centre for Archaeology and Fine Arts (SPAFA) and the Ministries of Education in Southeast Asia includes mangrove conservation and education. SEAMEO SPAFA has been involved with the Khlong Phitthaya Longkorn School and Bangpakong Bovorn Witthayayon School to promote best practices in mangrove conservation education in Thai schools (SEAMEO SPAFA, 2017). Consequently, the four SEAMEO centers under this program are the SPAFA, SEAMEO Regional Centre for Tropical Biology, SEAMEO Regional Open Learning Centre, and SEAMEO Regional Centre for Quality Improvement of Teachers and Education Personnel.

The four SEAMEO centers, along with other participating institutions, cooperated in helping schools in Indonesia to learn and share knowledge with schools in Thailand in the areas of mangrove conservation and restoration, education and awareness-raising, and socioeconomic and cultural development. This is made possible as the project involves facilitating the exchange of teaching-learning materials and experiences in mangrove education and sharing appropriate ways to conserve and explore the sustainable use of mangrove ecosystems. The project benefits other schools and their communities in SEAMEO member countries.

The SEAMEO Secretariat and UNESCO Bangkok jointly established the Southeast Asia ESD Teacher Educators' Network (SEA-ESD Network) during the workshop held from April 24 to 26, 2018, in Bangkok, Thailand. The network serves as a collaboration platform for teacher education institutions and universities in Southeast Asia to promote ESD in their communities. A total of 26 universities and teacher education institutions in Southeast Asia joined the network (UNESCO, 2018). The network keeps in close contact through online channels, particularly the WhatsApp messaging service, continuing to share the successful ESD projects that are taking place in each of its institutions and communities. It can be said that the SEA-ESD Network led to the successful networking of Southeast Asian teacher educators who are committed to promoting ESD in their communities and countries across Southeast Asia. The members have expressed their desire to expand and grow their ESD networks and carry out ESD actions in each of their communities through a holistic approach.

7.3 SEAMEO-Japan ESD Awards

In 2012, the SEAMEO Secretariat and the Japanese Ministry of Education, Culture, Sports, Science and Technology, in cooperation with the UNESCO Asia and Pacific Regional Bureau for Education, made a collective commitment to promote best practices in ESD in schools across Southeast Asia, through the SEAMEO-Japan Education for Sustainable Development (ESD) Awards. The awards aim to raise awareness of ESD in schools and communities across Southeast Asia, promote ESD best practices in schools and communities across Southeast Asia, share and exchange knowledge and best practices on ESD in schools across Southeast Asia and Japan, encourage networking among schools and communities that implement ESD practices in Southeast Asian countries and Japan, and support the SDGs. Since then, the SEAMEO-Japan ESD Awards have become one of the SEAMEO flagships in promoting ESD systematically.

Referring to the SDGs and UNESCO's roadmap for implementing the GAP on ESD, the SEAMEO-Japan ESD Awards have set a theme related to the key themes of ESD including climate change, biodiversity, sustainable production and consumption, global justice, and the reduction of poverty. The number of participating schools and countries varies from year to year. From 2012 to 2020, SEAMEO

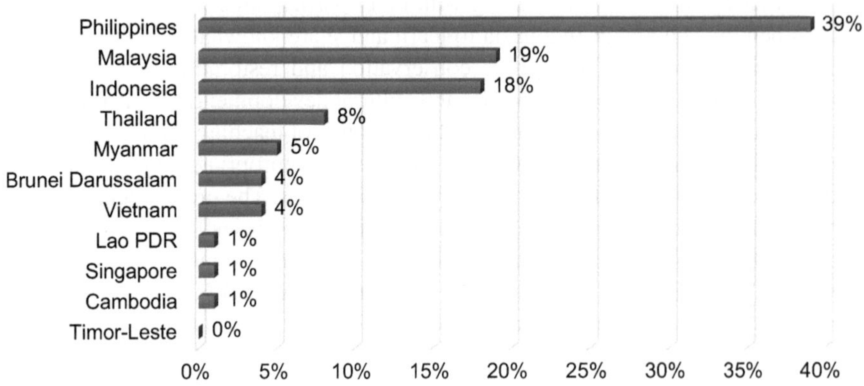

Fig. 7.1 Percentage of entries from SEAMEO member countries

member countries actively participated in the SEAMEO-Japan ESD Awards, as shown in Fig. 7.1.

The themes of the SEAMEO-Japan ESD Awards from 2012 to 2020, the number of participating schools and countries, and the winners are shown in Table 7.1.

7.4 Voices from the SEAMEO-Japan ESD Awards Participating Schools on Their ESD Program

SEAMEO conducted a study to investigate the sustainability of the ESD program carried out by the participating schools, including the winners and non-winners. A questionnaire was developed and sent to the participating schools to gather insights on the participating school program's progress and sustainability. The survey questions mainly focused on the achievements and challenges experienced by participating schools in sustaining the program/project and its impacts on the students, school faculties, and greater communities. A total of 56 participating schools, consisting of 9 winner schools and 47 non-winner schools representing the entries from 2012 to 2019, provided responses. A summary of the findings is as follows.

First, the majority of participating schools, both winners and non-winners, confirmed that their ESD program was still active (Table 7.2).

Interesting feedback from the participating schools in their responses explained the reasons for their ESD programs ending. While the schools that had ended their program plainly explained that they had met their goals and completed the program, the majority of participating schools confirmed that their programs will surely continue for the sake of community development and were still ongoing amid the current COVID-19 pandemic:

> *Peaceful school activities will always be carried out in order to teach students how to respect others properly and naturally. In addition, it makes students able to live in a better future.* (School X)

Table 7.1 Themes, related the SDGs and SEAMEO Education Agenda/7 priority areas, participating schools, countries, the first place winner of the SEAMEO-Japan ESD Awards

Year	Themes	Number of Schools	Number of Countries	The first place winner	SDGs	SEAMEO Education Agenda/7 priority areas
2012	Education for Disaster Risk Reduction	69	7	State High School No.1 Kungyangon, Myanmar	4/11/13	3
2013	Values Education	126	8	Sekolah Menengah Sayyidina Hasan, Malaysia	4	7
2014	Fostering global citizenship for sustainable future	63	9	Udonpittayanukoon School, Thailand	3/11	7
2015	Safeguarding our cultural traditions	90	10	Sekolah Kebangsaan (Perempuan) Methodist, Kuantan, Pahang, Malaysia	11/16	7
2016	Saving energy	43	8	SMK Negeri 1 Bandung, Indonesia	7	7
2017	Improving health and nutrition	56	8	SJK(C) WOON HWA, Johor Bahru, Malaysia	2/3	7
2018	Applying local wisdom for environmental conservation	94	7	Widya Mukti Vocational School, Indonesia	13/14/15	7
2019	Building peace in schools and communities	79	9	South Hill School, incorporated, Philippines	16	7
2020	Addressing plastic problems for transforming communities,	181	8	SM St. Patrick, Tawau Malaysia	12/13	7
	Total	801				

Table 7.2 Current status of the ESD programs in participating schools (as of November 2020)

Status of the program	Percentage: Winner school ($N = 9$)	Percentage: Non-winner school ($N = 47$)
Active	78%	83%
Postponed	11%	9%
Ended	11%	6%
Transformed into a small scale	–	2%

Second, the majority of participating schools, both winners and non-winners, confirmed that their ESD program involved a significant number of students, as shown in Table 7.3.

Table 7.3 Number of students participating in the ESD program

	Percentage	
Status of the program	Winner school ($N = 9$)	Non-winner school ($N = 47$)
More than 500	38%	56%
101–500	36%	33%
51–100	15%	–
Less than 50	11%	11%

Third, the majority of participating schools, both winner (89%) and non-winner schools (98%), confirmed that their ESD program involved the communities in their area. The top four community stakeholders involved included community leaders, parents/relatives, youth activists, and youth clubs.

Fourth, regarding the possible achievements of their ESD program, the majority of participating schools, both winners and non-winners, identified the following: (1) Students acquired soft skills/teamwork skills, (2) improved students' attitude toward ESD, (3) improved teachers' competencies in project management, (4) strengthened community engagement, and (5) improved the environment/atmosphere of the school and community.

Fifth, the top five challenges faced by participating schools in implementing their ESD programs include (1) funding limitations, (2) time constraints, (3) disasters/emergency situations, (4) coordination constraints, and (5) curriculum and cultural issues.

In response to COVID-19, we also asked the schools about the impacts of COVID-19 on the program operation. They identified that the lockdown due to the COVID-19 triggered students' disengagement in the program. Postponement of the program until the end of the pandemic was another option chosen by some schools. The pandemic also led to them having very limited resources and time for program operation. Nevertheless, they still saw the possibility of the program being carried out using alternative solutions.

Furthermore, the schools also shared their plans for program implementation after the pandemic, which included (1) re-designing the program for the new normal, (2) resuming the program as planned, (3) seeking external funding and resources, and (4) postponing the program until the school has sufficient resources.

7.5 Lessons Learned from the SEAMEO-Japan ESD Awards

What did our participating schools learn from the ESD programs? We classified their feedback into the micro level (student level) and macro level (school and community level). At the micro level, it was undeniable that the students' engagement in the ESD programs enabled them to learn skills that could not be learned using paper and pencil in the classroom:

> *The ESD program makes us more aware and encourages us to change the way we gain and deliver knowledge, skills, values, and attitudes that enable us, students, stakeholders, and communities to have more sustainability and a just society for all.* (School A)

> *The students improve not only in terms of living towards a healthy lifestyle but also improve in terms of other important soft skills such as communication skills, teamwork, and sustaining the best learning to other students in the school and the community.* (School Y)

> *Our graduate students [gained] human faith, deepness of religious knowledge, obedience, mastering science and technology, possessing 21st century skills (creativity, critical thinking and problem solving, communication, collaboration), [and became] independent and environmentally friendly.* (School Z)

At the macro level, ESD programs enabled schools to give benefits to and establish strong collaborations with their communities. Interestingly, the ESD programs triggered communities' engagement in other school programs as well:

> *This award is treasured by our institution because we have been given the chance to showcase in our own little way the project intended for everyone living in the community, particularly the people around the school. They just walk-in to get the necessary vegetables and plants available in our garden as their needs arise since the vegetables in our schools are located within the vicinity of the community. Many have benefited from this project. Even in the pandemic time since lockdown they were able to have vegetables that were available in our garden. It really helped them a lot in their day-to-day living. Thank you SEAMEO-JAPAN ESD Award for having this kind of involvement in a small school like us, considered the second smallest and poorest school located in the remote area of our town. Because if you did not challenge us with this, maybe without this garden we would not be able to help people during this time of the pandemic. We are truly blessed to have and have maintained this garden from this time. [We are] looking…to make this accessible to everyone around the school.* (School P)

> *In school, we learnt that the implementation of the ESD program has a great impact on learners in the sense that it really encourages 100% attendance at school and it also strengthened the engagement of our stakeholders for they were tapped and acknowledged for their voluntary services rendered in school aligned with this ESD implementation.* (School H)

> *This ESD program [caused] a tremendous cultural reboot journey at my school. After the introduction of the SDGs at my school, 'sustainability' became the ultimate goal in every school activity. My school community gives more attention to sustaining any activity for students' wholesome developmental process. Indeed, this ESD made us travel the extra mile to meet more people from different walks of life. That's how my school was able to incorporate 17 SDG goals into learning activities and make the school work towards the SDGs. Apparently, I made people in Malaysia realize that the SDGs benefit the school, and every school must practice a minimum of one goal in their school teaching and learning activities. My school became a hub to share the ESD and SDG goals. I believe ESD is needed for children's sustainable development.* (School S)

> *We can spread our message and spirit in addressing local environmental degradation in our community.* (School E)

> *We are able to maintain the priceless cultural tradition [for the] young generation.* (School F)

The 9-year journey of the SEAMEO-Japan ESD Awards has paved the way to achieve the SDGs at the school level. Lessons learned and shared by most of our participating schools showed the possible sustainability of this program. Indeed, the

SEAMEO-Japan ESD Awards 2020, which were run during the COVID-19 pandemic, received more than 180 entries, the highest number in its history. This indicates that many schools in the region are well aware of the importance of ESD for the sake of our future learners and ensuring that they can live sustainably in the future world.

7.6 The Way Forward

The past experiences of the SEAMEO-Japan ESD Awards in promoting ESD indicated positive impacts on the participating schools. The current SEAMEO-Japan ESD Awards 2020 with more than 200 entries, a very significant number, designates the future direction to transform the ESD program into a more solid and resilient program in response to the current challenges posed by the COVID-19 pandemic. Therefore, it is timely for SEAMEO and its stakeholders in the region to work with the winner schools to scale up their ESD programs from the district level to the national and regional levels for the sake of a sustainable Southeast Asia.

SEAMEO and its partners, such as UNESCO Bangkok, shall continue to strengthen their SEA-ESD network by introducing the members to the SEAMEO-Japan ESD Awards participating schools so that they can work collaboratively and systematically in promoting ESD in the region through their joint programs.

Finally, we thank the various Ministers of Education for their joint statement to ensure and sustain the development and advancement of knowledge resources through the initiative of the SEAMEO centers and secretariat, such as online and open educational resources being available to every learner wherever they may be and in whatever condition they may be in. This supports our efforts to sustain ESD.

References

Khor, S. S. (2019). Role of SEAMEO RECSAM in inclusive and STEM education development. *Journal of Southeast Asian Education, 1*, 97–104.

Leicht, A., Heiss, J., & Byun, W. J. (2018). *Issues and trends in education for sustainable development*. UNESCO.

SEAMEO CED. (2019). *Report on the 2nd international seminar on education for rural communities: Education and learning for building sustainable livelihoods*. SEAMEO CED.

SEAMEO SEARCA. (2019). School and home gardens can help achieve sustainable development goals. *SEAMEO Education Agenda Magazine*, SEAMAG.

SEAMEO Secretariat. (2018). *Final report: Sustainability begins with teachers in Southeast Asia*. SEAMEO Secretariat.

SEAMEO SPAFA. (2017). Update on mangrove conservation and education programme in schools and communities: Inter-centre collaboration. *Final Report of the 40th SEAMEO High Officials Meeting*, SEAMEO SPAFA.

UNESCO. (2015). *Education 2030: Incheon declaration and framework for action – Toward inclusive and equitable quality education and lifelong learning for all*. UNESCO.

UNESCO. (2016). *What is global citizenship education?* https://en.unesco.org/themes/gced/definition

UNESCO. (2018). *Southeast Asia education for sustainable development teacher educators' network launched.* https://bangkok.unesco.org/content/southeast-asia-education-sustainable-development-teacher-educators%E2%80%99-network-launched

UNESCO. (2019). *UNESCO and sustainable development goals.* https://en.unesco.org/sustainabledevelopmentgoals

UNESCO, & SEAMEO Secretariat. (2017). *Final report of sub-regional workshop on education for sustainable development (ESD) for teacher education institutions.* UNESCO & SEAMEO Secretariat.

Part II
Science Curriculum for Sustainable Development in Formal Schooling

Part II
Science Curriculum for Sustainable Development in Formal Schooling

Chapter 8
Environmental Education in Science Curricula in Korea: Analysis of the 2015 National Science Curricula and Suggestions for the Future

Sun-Kyung Lee, Namsoo Kim, Hyeongson Ju, and Youngsun Kwak

8.1 Introduction

At the center of networks among global risks estimated by scientists in 2019 are climate change and environmental issues (Future Earth, 2020). The Global Risks Report annually announced by the World Economic Forum contains information on global risks that global economic leaders think are highly likely to occur and that will cause the greatest damage if not properly handled. According to the 2020 report, the top five risks include extreme climatic phenomena, failure in climate action, natural disasters, biodiversity loss, and human-induced environmental disasters, all of which are deemed to belong to the category of "environment" (World Economic Forum, 2020).

Against this backdrop, the global community as a whole has suffered from the COVID-19 pandemic since 2020, heralding a new era of multiple disasters. Experts have cited environmental problems including the destruction of wildlife habitats, as well as climate change, as the main causes of the pandemic. As a result, One Health approach, in which the health of humanity, animals, and the earth is handled via an integrated system, has been highlighted as an approach to prevent and to effectively deal with such zoonoses (FAO, 2008). This approach is based on the premise that

S.-K. Lee (✉)
Cheongju National University of Education, Cheongju, Republic of Korea
e-mail: sklee@cje.ac.kr

N. Kim
Korea Research Institute for Environment and Development, Seoul, Republic of Korea

H. Ju
Korea National Open University, Seoul, Republic of Korea

Y. Kwak
Korea National University of Education, Cheongju, Republic of Korea

H. Fujii, S.-K. Lee (eds.), *Science Education for Sustainable Development in Asia*, Education Innovation Series, https://doi.org/10.1007/978-981-99-8711-5_8

individuals, society, animals, the global environment, and so forth are the components of the geo-ecological system and that the health and safety of each element directly relate to other components and the whole system. Based thereon, it can be understood that the "environment" is not considered as a sphere added separately any longer but as the basis of people's daily lives and is closely intertwined with life as a whole. Amid the spread of the COVID-19 pandemic, extreme climatic phenomena attributed to climate change have occurred in various fashions, worsening the situation and thereby making it necessary to work hard to relieve and effectively react to future complicated disasters not only at a national level but also in every realm.

In the ESD for 2030 Roadmap, Stefania Giannini, UNESCO Assistant Director-General, pointed out, "We have more frequently asked whether what people are learning is really connected with our lives and contributes to our survival on the earth" (UNESCO, 2020). In other words, we are asking the following questions: To what degree and how is the environment being dealt with in schools and in the sector of education? Or to what degree and how should it be handled? Answering these questions requires us to consider the various layers of school education because EE for specific individuals and groups can be offered in diverse ways depending not only on national curricula but also on educational orientation and support of metropolitan and provincial offices of education, on curricula and subject operation at a school level, on teachers' expertise at a classroom level, and on many other contexts (Kim & Lee, 2020). This chapter focuses on national curricula among various layers where EE is provided. It is because it is difficult for classes to offer feedback to national curricula but is certain that national curriculum documents affect actual classes via various channels. This is all the more so, especially in Korea where national curriculum documents do not have a recommendation/guidance-oriented nature but a law-oriented nature (Gwon et al., 2018).[1]

EE in schools is mainly handled via an integrated approach and a separate subject approach. EE in schools in Korea is decentralized in related subjects for kindergartens and elementary, middle, and high schools, is provided as a cross-curricular learning topic, or is offered as an elective subject in middle and high schools. In elementary school, EE is included in related subjects such as science, society, ethics, and high schools; since "Environment" and "Environmental Science," and others, were introduced as elective subjects in 1995 with the sixth revision of the National Curricula, a separate subject approach has been chosen where EE is comprehensively offered, focusing on separate subjects. However, in reality, so forth,

[1] The law-oriented nature means requiring the normative implementation of curricula with legal binding force while the recommendation–/guidance-oriented nature signifies that local governments or schools can determine specific curricula in case the central government presents curriculum guidelines (Gwon et al., 2018).

which is regarded as an integrated approach and decentralized system (Nam, 1995).[2] In middle school, EE is included in other subjects, as well as being offered as separate subjects, in middle and high schools, which can be considered as a combined approach.

The status of the existing curricula should be analyzed before coming up with strategies for enhancing EE in the next national curricula. Therefore, this chapter establishes EE analysis frameworks to analyze EE in curricula in terms of content and integration based on the exploration of key frameworks in local and global research and education connected to EE analysis. Then, the 2015 revised science curricula are analyzed by using such frameworks. This is designed to provide an opportunity to come up with strategies for improving EE in the next national curricula.

This chapter includes various systems emerging in the history of EE such as on-site surveys, education for preservation, and education for development (Palmer & Neal, 1994), reflecting key trends in EE in the post-1990s era and focusing on education that offers opportunities for action and contributions and includes the extended aspects of Education for Sustainable Development (ESD). In other words, this is based on EE in a wide sense or education for environment and sustainable development including or promoting sustainability, rather than on EE in a narrow sense.

8.2 Environmental Education Analysis Framework

Even in the areas of general education and education for environment and sustainable development, various frameworks are proposed. These not only provide information on the status of the existing education but also present the direction of future development.

8.2.1 OECD's Learning Framework 2030

The OECD has promoted the project titled "The Future of Education and Skills: Education 2030" to identify and develop core competencies for a future society (OECD, 2018). This project proposes the OECD's Learning Framework 2030

[2] The "integrative approach" currently means not only including EE in related subjects such as science, society, ethics, and so forth but also connecting and integrating the subjects (science, society, ethics, etc.) to and into EE, respectively. In other words, it signifies a convergence-based approach in many cases. Based thereon, some scholars argue that such an approach is not suitable for explaining a decentralized system. However, this paper respects the position that EE is added to existing subjects, using information from Nam (1995) as it is.

containing the future, vision, and principles of education. The framework presents well-being at an individual and social level as the objective of education. In order to achieve such an objective, it requires transformative competencies including "creating new values," "reconciling tensions and dilemmas," and "taking responsibility."

In addition, this initiative requires the development of knowledge, functions, attitudes, and values to cultivate transformative competencies, highlighting that these attributes are not simply acquired but should be used to resolve complicated problems. This framework emphasizes that learners should have learner agency and that colleagues, parents, teachers, and communities should cooperate with one another. The fact that learners should have learner agency means that they participate in social affairs, taking responsibility, and are able to find behaviors that are required to achieve their goals.

The fact that the OECD's framework more comprehensively defines well-being at an individual and collective level as the objective of education is based on the recognition that education should make contributions to preventing the rapid development of science and technology from causing inequalities, social segmentation, and resource depletion. In this context, well-being is related not only with material elements including occupations, health, income, and residence but also with the quality of life including democratic participation, social connection, education, safety, satisfaction with life, and the environment. As a result, it is also closely connected with the United Nation's Sustainable Development Goals (SDGs).

Transformative competences highlighted by the OECD's framework mean the OECD's core competences (OECD, 2005) plus the aforementioned three competencies. According to the framework, cultivating such competencies requires multidisciplinary knowledge for crossing the boundaries of studies; cognitive knowledge for thinking like mathematicians, historians, and scientists; and procedural knowledge developed via design and systems thinking. It also needs cognitive and metacognitive functions, social and emotional functions, and substantial and physical functions that are required to apply knowledge in unknown and changing circumstances. Moreover, the framework reflects the position that the use of comprehensive knowledge and functions can be mediated via values and attitudes including motives, confidence, and respect for diversity.

8.2.2 Sustainability Curriculum Framework

The Department of the Environment, Water, Heritage and the Arts of Australia developed the Sustainability Curriculum Framework (DEWHA, 2010) proposing what students should learn, as well as appropriate time and circumstances for such learning, in order to promote sustainability, one of the three priorities, in their own lives. The framework includes three different age groups (K-grade 2, grades 3–6, and grades 7–10) and is based on the premise that because education for sustainability has interdisciplinary and cross-curricular characteristics, a single learning realm cannot offer knowledge and opportunities that are required to enable students to contribute to sustainability. Therefore, the framework emphasizes that appropriate knowledge and functions are mutually connected beyond the boundaries of grades and subjects and that education for sustainability is a holistic approach leading to motives for and contributions to acting to improve performances for a more sustainable world, rather than simply acquiring knowledge and functions. Ultimately, this framework highlights learning that helps people live a more sustainable life, thereby stressing that knowledge and functions are meaningless if they do not lead to effective behaviors.

The sustainability curriculum framework consists of sustainability action processes, the knowledge of systems, and the repertoires of practice (Fig. 8.1). These are interdependent and overlap with one another, each of which is considered to be insufficient. The knowledge of systems includes knowledge of ecosystems and human social systems, while the repertoires of practice are composed of perspectives of the world, systematic thinking, and future- and design-related elements. Sustainability action processes consist of determining objects for change, determining the scope of actions, action planning, implementing action plans, and evaluation and self-reflection. Given that the framework aims to help students live a sustainable life, sustainability action processes are deemed to be the most crucial element. Moreover, rather than normatively highlighting sustainability actions, it has specific action plans by phase. The repertoires of practice include competencies, attitudes, and values for actions.

8.2.3 *Environmental Literacy Framework*

The North American Association for Environmental Education (NAAEE) developed the framework for assessing environmental literacy (Fig. 8.2) (Hollweg et al., 2011), in which experts in social sciences, science education, and EE participated.

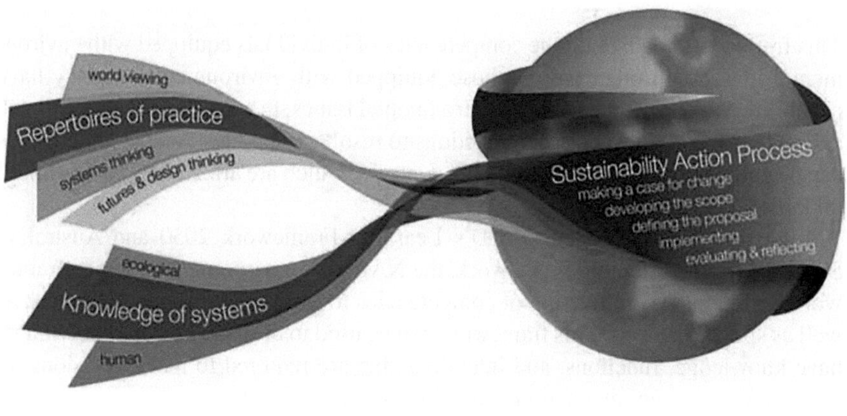

Fig. 8.1 Sustainability curriculum framework of Australia. (DEWHA, 2010)

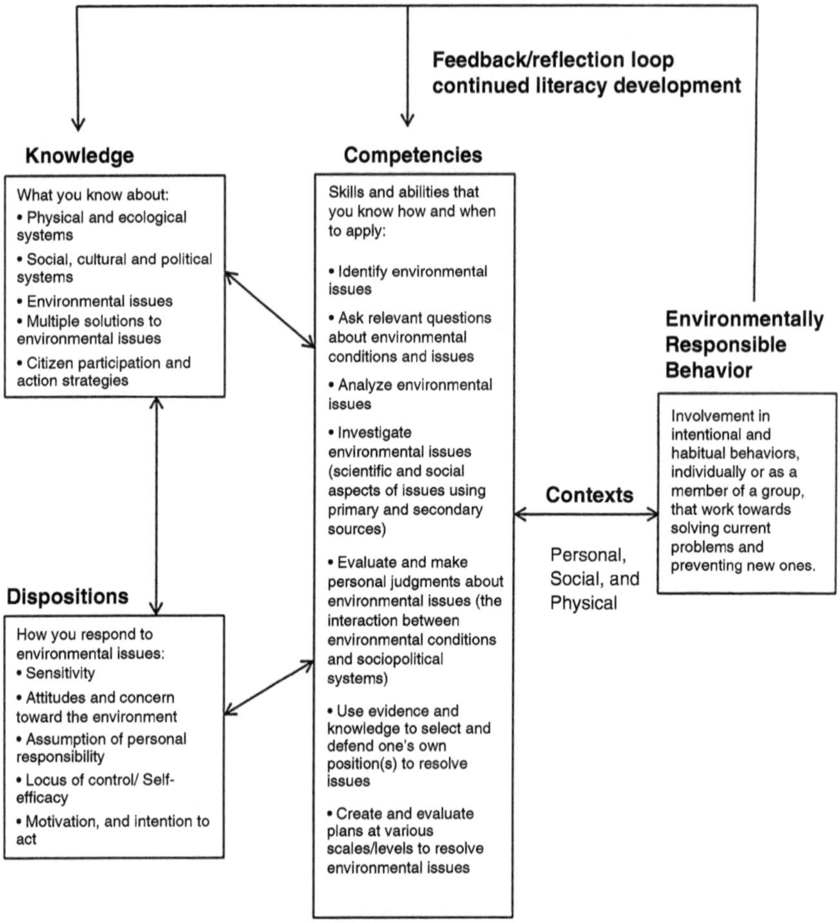

Fig. 8.2 NAAEE's environmental literacy framework. (Hollweg et al., 2011)

This framework focuses on the competencies of individuals equipped with environmental literacy. In other words, those equipped with environmental literacy have capabilities to check and analyze environmental issues, to assess potential solutions, and to propose and justify specific actions to resolve the issues. Such competencies can be cultivated in regional and global contexts, which are affected by environmental knowledge and orientation.

Like the aforementioned OECD's Learning Framework 2030 and Australian Sustainability Curriculum Framework, the NAAEE's environmental literacy framework highlights the cultivation of competencies to resolve environmental issues, as well as specific actions. This framework can be used to evaluate whether individuals have knowledge, functions, and intentions that are required to make decisions on environmental issues at a national, regional, or global level.

8.2.4 Lee et al.'s Analysis Framework for Environmental Education

Based on the OECD's Learning Framework 2030, Australian Sustainability Framework, and the NAAEE's Environmental Literacy Framework, Lee et al. (2020a) developed an environmental education analysis framework. This analysis framework consists of environmental knowledge, environmental competencies, environmental values and attitudes, and environmental action and participation. These four categories are deemed to mutually affect one another, rather than be separated and isolated (Fig. 8.3). Environmental competencies are required for environmental action and participation and can be cultivated in the process of promoting environmental action and participation.

Environmental knowledge and values and attitudes are also required for environmental action and participation and can be further enhanced in the aforementioned process. Environmental knowledge, attitudes, and values are also considered to mutually affect one another. In other words, it should be noted that this framework highlights environmental action and participation, presenting a series of processes for them as sub-elements. According to this framework, encountering the natural environment and social systems themselves along with environment- and sustainability-related issues in a real world (a kind of experience) can be regarded as a starting point for the establishment of relationships with the environment, as well as for environmental action and participation. Furthermore, this analysis framework is based on the premise that learners should be able to explore and resolve environment- and sustainability-related problems or issues by using interdisciplinary knowledge and action competencies via education for environment and

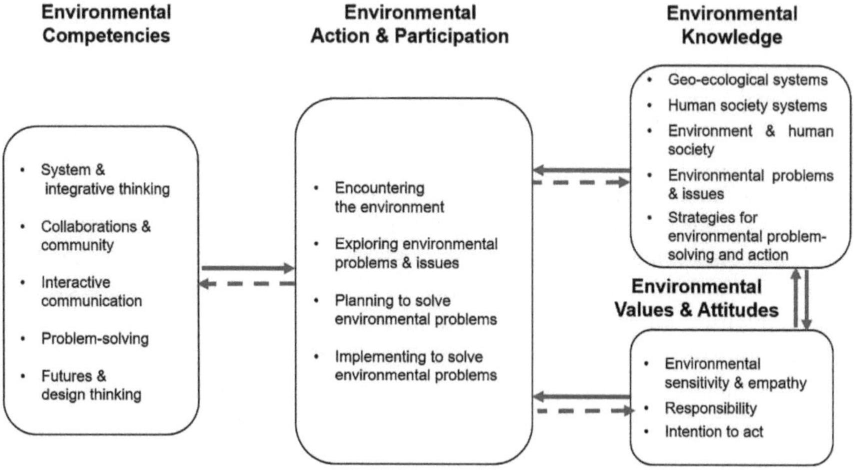

Fig. 8.3 Lee et al.'s analysis framework for environmental education. (Lee et al., 2020a)

sustainable development. To this end, learners should be able to experience the process of crafting, implementing, and monitoring specific action plans.

The aforementioned frameworks explain environmental action and participation as follows. The NAAEE's Environmental Literacy Framework describes as competencies a series of processes of checking and analyzing environmental issues, assessing potential solutions, and proposing/justifying specific actions. According to this framework, only the process of planning and assessing solutions to environmental issues is included in competencies. Contrarily, the Sustainability Curriculum Framework sets the implementation of sustainability as the goal of education for sustainability, presenting five specific steps. Lee et al.'s Analysis Framework has something in common with Australian Sustainability Curriculum Framework, in that it is not limited to exploring environmental problems or issues and planning solutions thereto but includes implementing and monitoring action plans.

Environment subjects for middle and high schools in the 2015 revised curricula of Korea set as one of educational goals the integrative exploration of environmental problems or environmental exploration. Environment subjects for high schools include "participation in activities for sustainable lives" in their educational goals. Considering how educational goals are described in curricula, it is deemed to be a little hard to specifically present a series of processes for participation.

Table 8.1 shows descriptions of and keywords for the four categories and subcategories in the framework. The keywords added or revised via the analysis of curricula following the development of the draft are expected to help understand categories and can be used to improve accuracy when analyzing achievement standards for curricula in the future.

8.3 Environmental Education in Science Curricula

In order to identify the status of EE in science curricula, subjects "Science" and "Integrated Science" for elementary/middle schools and high school, respectively, in the 2015 revised science curricula of Korea were analyzed by using Lee et al.'s framework (Lee et al., 2020b). Tables 8.2 and 8.3 show areas and core ideas included in "Science" and "Integrated Science" in the 2015 revised curricula.

Herein, achievement standards presented in the science curricula are mainly analyzed, and the numbers of achievement standards for elementary, middle, and high schools are 113, 92, and 32, respectively. In the process of analyzing the criteria, this study uses information presented in the sections of "learning elements," "descriptions of achievement standards," "teaching/learning methods and notes," and "assessment methods and notes." The science curricula present overview by unit and achievement standards, offering information in the following order: (A) learning elements, (B) descriptions of achievement standards, (C) teaching/learning methods and notes, and (D) assessment methods and notes.

How much EE is integrated in achievement standards is evaluated, allocating scores (0, 1, 2, and 3 points). Specific score allocation criteria are as follows: 3

Table 8.1 Category descriptions and keywords in Lee et al.'s analysis framework for environmental education

Categories		Descriptions	Keywords
Environmental knowledge	Geo-ecological systems	(In relation to the environment) knowledge of life and physical systems supporting the earth's biosphere in process and interaction	Earth systems (water, atmosphere, soil, etc.), resources/energy, principles of ecosystems (structures/functions, changes, patterns, diversity, and interdependency), biodiversity, cycle of matter, climate, and climate change (in relation to the environment)
	Human society systems	(In relation to the environment) social, cultural, political, economic, and technological systems set up by humanity	Politics, cities, architecture, transportation, agriculture, lifestyles, consumption, ethics, energy systems, science/technology, population, and climate change (in relation to the environment)
	Relations between the environment and human society	(In relation to the environment) interaction and interdependency between ecosystems and human society systems	Impact of the natural environment on human society, ecosystem services, bio-cultural diversity, climate change, social/ecological systems, and social/technological systems
	Environmental problems and issues	Environmental issues and problems caused by social, cultural, political, economic, and technological systems	Environmental problems or conflicts caused by cities, architecture, transportation, agriculture, lifestyles, consumption, industries, and science/technology
	Strategies for environmental problem-solving and action	A diversity of personal/social values and action strategies contributing to resolving environmental problems (issues); sustainable (ecological and ecofriendly) society, culture, politics, economy, and technology	Environmental philosophy/ethics (including perspectives of the environment), environmental rights, environmental policies, environmental laws, environmental technology, appropriate technology, sustainable production and consumption, social economy, energy transition, principles of sustainable development (equality and environmental justice), locality, sustainable residence, sustainable transportation, green consumption, energy-saving, environmental protection, environmental management, and the SDGs

(continued)

Table 8.1 (continued)

Categories		Descriptions	Keywords
Environmental values and attitudes	Environmental sensitivity and empathy	Recognizing the values of the natural environment; accepting environmental changes in a sensitive way; having interest in and having an understanding of environmental victims, as well as the seriousness of environmental problems	Environmental/ecological sensitivity, appreciation of nature, change recognition, empathy, and consideration
	Responsibility	Responsibility for contributing to resolving environmental issues (problems) individually or collectively	(In relation to the environment) personal and social responsibilities
	Intention to act	Willingness or intention to resolve current or future environmental issues (problems)	Willingness and intention to act
Environmental competencies	Systems and integrative thinking	Diagnosing and predicting environmental issues (problems) as a whole in an integrative way, paying attention to interaction and interdependency among individual elements and considering complexity, uncertainties, and risks	Understanding/consideration of the complexity (socioeconomic, cultural, and environmental elements), uncertainties (definition, causes, outlook, solutions, etc.), and risks, systems thinking, and integrative understanding
	Cooperation and community	Learning together and communicating with one another to diagnose environmental issues (problems), mediate conflicts, and come up with proper solutions	Learning together, mediating conflicts, cooperation, collaboration, and community spirit
	Interactive communication	Voice opinions on environmental issues (problems), listening to others' opinions and communicating with one another	Presentation, expressions, dialogues, and communication
	Problem-solving	Ability to present, select, and apply alternatives/solutions to environmental issues (problems) based on various knowledge and information	Proposal, selection, and application of alternatives/solutions/ solutions to environmental issues (problems)
	Futures/design thinking	Ability to specifically forecast and compare various future scenarios. Ability to implement ideas for a desirable future into concrete results through visualization and modeling	Future prediction and forecast (greenhouse gas emission scenario, future climate change, future fine dust concentration, future green area forecast, population change and environmental change, etc.), design and conception, prototype, and technology development

Environmental action and participation	Encountering the environment	Meeting and understanding the natural environment and social systems and the people within them. Facing real issues related to environment and sustainability. Considering other people's ideas and worldviews about the environment and sustainability from different perspectives. Finding potential for change	Experience the natural/artificial environment, visit places/ society/facilities, meet and understand diverse people, and face real-world issues related to environmental/sustainability. Consider other people's ideas about the environment and sustainability from different perspectives, examining other people's worldviews. Find the possibility of change
	Exploring environmental problems/issues	Critical reasoning, investigation, and exploration of environmental and sustainability problems or issues using interdisciplinary knowledge and competences	Investigate and explore environmental and sustainability problems or issues
	Planning to solve environmental issues (plan for change)	Targeting change for addressing environmental issues and sustainability, defining the scope of action, and developing an action plan	Targeting change for addressing environmental issues and sustainability, defining the scope of action, and developing an action plan
	Implementing to solve environmental problems (make change)	Putting action plans into action for environmental issues and sustainability. Recording, monitoring, evaluating, and reflecting on the implementation process and results. Moving on to another environmental practice	Putting (action plans) into action for environmental issues and sustainability. Recording, monitoring, evaluating, and reflecting on the implementation process and results. Moving on to another environmental practice

Table 8.2 Areas and core ideas of the 2015 revised science curricula for elementary and middle schools

Area	Core ideas	Area	Core ideas
Forces and motion	Space-time and motion	Life sciences and human life	Biotechnology
	Force	Structure and energy of living organisms	Units of life
	Mechanical energy		Structures and functions of animals
Electricity and magnetism	Electricity		Structures and functions of plants
	Magnetism		Photosynthesis and respiration
Heat and energy	Heat balance	Homeostasis and body control	Stimulation and reactions
	Laws of thermodynamics	Continuity of life	Reproduction
	Energy conversion		Heredity
Waves	Types of waves		Evolution and diversity
	Properties of waves	Environment and ecosystems	Ecosystems and interaction
Structure of matter	Particles of matter	Solid earth	Earth systems and field of force
Properties of matter	Physical and chemical properties		Plate tectonics
	State of matter		Matter constituting the earth
			History of the earth
Changes of matter	Changes in the state of matter	The atmosphere and ocean	Properties and cycle of seawater
	Chemical reaction	The universe	Motion and cycle of the atmosphere
			Composition and motion of the solar system
	Energy release and absorption		Characteristics and evolution of stars
			Structure and evolution of the universe

Table 8.3 Areas and core ideas of the 2015 revised science curriculum for high school

Area	Core ideas
Matter and regularity	Regularity and combination of matter
	Matter constituting nature
Systems and interaction	Mechanical system
	Earth systems
	Life systems
Changes and diversity	Chemical changes
	Biodiversity and maintenance
Environment and energy	Ecosystems and environment
	Development and new/renewable energy

Table 8.4 Examples of scoring for EE elements in science curricula: category of environmental knowledge

Score	Achievement standards	Elements of EE
3 points	[Section 4, 17–01] able to explain the process of water moving, changing, and circulating via various phenomena occurring among living organisms, the surface of the earth, and the air. *Knowledge on the elements of earth systems and their interaction	Knowledge on geo-ecological systems
	[Section 6, 05–01] able to explain that elements of ecosystems affect one another, understanding that ecosystems consist of biological and nonbiological elements	Knowledge on geo-ecological systems
	[Integrated section 10, 07–01] able to reason that the global environment has consistently changed going through the geographical age and that the current biodiversity has formed in the process of adapting to such environmental changes	Knowledge on geo-ecological systems
2 points	[Section 4, 17–02] able to study cases where the issue of water shortages is resolved via creative methods, clearly understanding the importance of water	Environmental problems and issues
	[Section 9, 16–01] able to study disasters–/calamities-related cases and to analyze the causes and damage thereof in a scientific way. *Achievement criterion: Disasters–/calamities-related cases are studied, selecting one of the following: Leakage of chemicals, spread of infectious diseases, meteorological disasters, quakes, volcanic activities, and transport accidents	Environmental problems and issues
1 point	[Section 9, 01–01] able to explain the layer structure and characteristics of the geosphere, understanding the elements of earth systems	Knowledge on geo-ecological systems
	[Integrated section 10, 07–02] able to reason the emergence of germs resistant to antibiotics and insecticides, understanding the principles of evolution via modifications and natural selection	Environment and human society

points for cases where EE elements are presented as key learning themes and content by unit in achievement standards by sphere (core content); 2 points for cases where EE elements are presented as part of key learning themes and content by unit in achievement standards or descriptions by sphere (main content); 1 point for cases where EE elements are mentioned in the descriptions of achievement standards but are not handled as important information (related content); and 0 for cases where EE elements are not clearly presented in achievement standards or description by sphere (N/A). Moreover, in cases where one achievement criterion relates to several EE elements, different scores are allocated by EE element. Specific examples are presented in Table 8.4.

In order to improve the reliability of assessment results, three experts in EE and two science subject professionals discussed the analysis framework together, analyzing contents, allocating scores, and implementing cross-checking. Then, the results were checked and revised again. Assessment results consist of the sum of scores and integration rates. The integration rates are calculated as below:

$$\text{Integration rate} = \left\{ \begin{array}{l} \text{Points for the integration of EE by achievement criterion /} \\ \text{number of achievement standards} \times 3 \\ \left(\text{points for cases where EE is well – reflected} \right) \end{array} \right\} \times 100$$

Integration rates for elements in each category (environmental knowledge, environmental values and attitudes, environmental competencies, and environmental action and participation) are summed up and divided by the number of elements to calculate mean values.

According to analysis results, EE is mainly offered in the form of environmental knowledge in the 2015 science curricula regardless of school level, providing less information on environmental values and attitudes, environmental competencies, and environmental action and participation (Table 8.5).

First, the average shares of categories in EE were examined. Environmental knowledge accounts for the highest percentage of EE (4.7% for elementary school; 4.5% for middle school; and 25.0% for high school), followed by environmental competencies (0.3–6.7%), environmental action and participation (0.9–4.5%), and environmental values and attitudes (0.4–3.1%). In the category of environmental knowledge, knowledge on geo-ecological systems is mainly offered (47.9% for high school), providing information on environment and human society (27.1% for high school). Much less information on strategies for environmental problem-solving and action is offered in elementary and middle schools (1.8% and 2.2%) than in high school.

Environmental values and attitudes are found to account for the lowest percentage of EE in elementary, middle, and high schools (1.1% for elementary school, 0.4% for middle school, and 3.1% for high school). In particular, no information on responsibility and intention to act is offered in elementary, middle, and high schools. This is deemed to be attributed to the fact that the science subjects focus on scientific knowledge and exploration, failing to be connected to responsibility or intention to act.

The share of environmental competencies in EE is much less in elementary and middle schools (0.3 and 0.6%) than in high school. Interactive communication competencies are found to be handled in the subject for elementary and middle schools (1.2 and 2.9%), and no information on cooperation and community-related competencies is offered regardless of school level. This shows that the science subjects for elementary and middle schools focus on research/presentation-related communicative competencies via group activities but fails to effectively deal with conflict mediation, community spirit, systems/integrative thinking, and problem-solving. The subject of integrated science for high school is found to contain information on systems/integrative thinking (15.6%) and problem-solving (6.3%) due to its characteristics.

In the category of environmental action and participation, the subject for elementary school deals with encountering the environment and exploring environmental problems/issues, while the subject for middle school contains information on planning solutions to environmental problems. The subject of integrated science for high

Table 8.5 Integration of environmental education in the 2015 science curricula in Korea[a]

Classification		Elementary school (113)[b]		Middle school (92)		High school (32)	
		Score[c]	Ratio (%)[d]	Score	Ratio (%)	Score	Ratio (%)
Environmental knowledge	Geo-ecological systems	52	15.3	34	12.3	46	47.9
	Human society as a system	1	0.3	7	2.5	15	15.6
	Environment and human society	15	4.4	10	3.6	26	27.1
	Environmental problems and issues	6	1.8	4	1.4	14	14.6
	Strategies for environmental problem-solving and action	6	1.8	6	2.2	19	19.8
	Subtotal	**80**	**23.6**	**61**	**22**	**120**	**125**
	Average	**16**	**4.7**	**12.2**	**4.4**	**24**	**25.0**
Environmental values and attitudes	Environmental sensitivity and empathy	10	2.9	3	1.1	9	9.4
	Responsibility	1	0.3	–	–	–	–
	Intention to act	–	–	–	–	–	–
	Subtotal	**11**	**3.2**	**3**	**1.1**	**9**	**9.4**
	Average	**3.7**	**1.1**	**1**	**0.4**	**3**	**3.1**
Environmental competencies	Systems/integrative thinking	–	–	–	–	15	15.6
	Cooperation and community	–	–	–	–	–	–
	Interactive communication	4	1.2	8	2.9	5	5.2
	Problem-solving	1	0.3	–	–	6	6.3
	Futures/design thinking	–	–	–	–	6	6.3
	Subtotal	**5**	**1.5**	**8**	**2.9**	**32**	**33.4**
	Average	**1**	**0.3**	**1.6**	**0.6**	**6.4**	**6.7**
Environmental action and participation	Encountering the environment	10	2.9	5	1.8	6	6.3
	Exploring environmental problems/issues	2	0.6	5	1.8	6	6.3
	Planning solutions to environmental problems	0	–	3	1.1	4	4.2
	Implementing plans to address environmental problems	0	–	–	–	1	1.0
	Subtotal	**12**	**3.5**	**13**	**4.7**	**17**	**17.8**
	Average	**3**	**0.9**	**3.3**	**1.2**	**4.3**	**4.5**

[a]Source: Lee et al. (2020b), Kwak et al. (2021)
[b]The figures are the sum of the numbers of achievement standards. In other words, the numbers of achievement standards for elementary, middle, and high schools are 113, 92, and 32, respectively
[c]Scores: Sum of evaluation scores for EE included in each achievement criterion. The highest score for each achievement criterion is 3 points. Therefore, the maximum scores for elementary, middle, and high schools are $113 \times 3 = 339$ points, $92 \times 3 = 276$ points, and $32 \times 3 = 96$ points, respectively
[d]Ratio(%): (sum of evaluation scores for EE included in each achievement criterion/number of achievement standards $\times$ 3) $\times$ 100

school is found to have information on encountering the environment, exploring environmental problems/issues, and environmental problem-solving, but whose share is deemed to be very low. Therefore, the science subjects for elementary, middle, and high schools are presumed to focus on encountering the environment, almost failing to deal with planning solutions to environmental problems and implementing plans to address environmental problems.

As a result, considering the educational goals of the science subjects highlighting scientific knowledge, attitudes, exploration, STS, and so forth, the subjects are found to focus on environmental knowledge among the four categories, partially dealing with competencies corresponding to exploration. In addition, although the subjects also focus on interest in and curiosity about natural phenomena and problem-solving, it is found to have little information relating to the category of environmental values and attitudes including environmental sensitivity and empathy, responsibility, and intention to act.

Looking into characteristics by school levels, the degree to which EE is integrated into achievement standards of science curricula for elementary and middle schools is found to be very low. EE is integrated into part of achievement standards, but achievement standards containing all of the four categories are found to number only two for elementary and middle schools, respectively, which is deemed to be very small. Given the characteristics of science curricula for elementary and middle schools, it is understandable that units relating to life and earth sciences deal with EE in connection with living things/the natural environment, energy/energy conservation, environmental issues, and so on. The science subjects for elementary and middle schools contain little information on environmental values and attitudes (1.1% for elementary school and 0.4% for middle school) among the four categories, having no information on planning solutions to environmental problems (0%) and implementing plans to address environmental problems (0%) belonging to the category of environmental action and participation.

The subject of "Integrated science" for high school contains more information on EE than the science subjects for elementary and middle schools. However, even in this case, the subject focuses on environmental knowledge (25.0%), containing much less information on environmental competencies (6.7%), environmental action and participation (4.5%), and environmental values and attitudes (3.1%). The subject of "integrated science" highlighting core ideas in the 2015 revised curricula is found to have more information on EE than science curricula for elementary and middle schools. In particular, focusing on core ideas (units) including "ecosystems and environment," "development and new/renewable energy," "bio-diversity and maintenance," and "earth systems," the subject of "integrated science" contains EE-related information more comprehensively. In other words, specific units therein present achievement standards, highlighting the aforementioned core ideas, acquiring high scores in environmental knowledge. However, even their achievement standards focus on environmental knowledge, containing insufficient information on the other categories (environmental values and attitudes, environmental action and participation, etc.).

These results show that the science subjects receive much lower scores in environmental values and attitudes, environmental competencies, and environmental action and participation than in environmental knowledge. This is deemed to be attributed to the fact that achievement standards are described focusing on knowledge and paying less attention to values and attitudes as well as action and participation. Moreover, because achievement standards from curriculum documents are analyzed, it may be hard to find information on values and attitudes or action and participation that should accompany when offering scientific knowledge via teaching and learning. Korea's science curricula are found to highlight STEAM and science/technology in convergence and to concentrate on positive aspects. This seems to be because the roles of science and technology, in connection with environmental problems, focus on addressing problems (e.g., appropriate technology, etc.) and improving situations (e.g., preventing global warming, etc.).

8.4 Recommendations for Environmental Education in the Next Science Curricula

Based on these research results, the following plans to improve EE via science curricula can be presented in connection with the revision of new national science curricula.

The science subject aims to cultivate learners' scientific literacy to resolve personal and social problems in a scientific and creative way by enabling them to understand the concepts of science and to foster scientific research competencies and attitudes. It is also designed to help students recognize mutual relations between science and society and grow to become desirable democratic citizens by leading them to learn scientific knowledge and research methods in a pleasant manner and to cultivate scientific literacy (Ministry of Education, 2015).

The 2015 revised science curricula focus on cultivating scientific literacy, highlighting the development of five competencies (scientific thinking, scientific research, scientific problem-solving, scientific communication, and scientific participation and lifelong learning). In particular, the 2015 revised science curricula are organized with a focus on core ideas including energy and systems. Selecting and presenting research activities by unit (elementary, middle, and high schools) and helping students carry out research activities to satisfy achievement standards, they aim to improve students' research competencies.

EE in the science subjects focuses on scientific knowledge and consists of four categories including environmental knowledge, environmental competencies, environmental values and attitudes, and environmental action and participation. However, in terms of documents, science curricula are deemed to significantly lack information on environmental competencies, environmental values and attitudes, and environmental action and participation. Therefore, highlighting core ideas such as sustainable development, systems, and energy in terms of achievement standards, research topics, and activities (sample) and clearly including related information in the criteria, EE should be strengthened in connection with other subjects. In addition, integrating into the science subjects the competencies for systems/integrative thinking, cooperation/communities, futures/design thinking, and so forth, it needs to be reorganized.

Among elements constituting the science subjects, life and earth sciences directly relate to EE. Therefore, related units (e.g., Journey of Water and Life & Environment for elementary school; Disasters & Safety and Energy Conversion & Conservation for middle school; and Earth Systems, Life Systems, Biodiversity & Maintenance, and Ecosystems & Environment for high school) should clearly contain information on environmental competencies and action and participation, providing experiences in terms of cognitive aspects, values and attitudes, and action via research activities through which learners can encounter the environment, addressing environmental problems.

In terms of strengthening EE by school levels, the quantity and level of integrated units for EE for elementary and middle schools can be increased. In the 2015 revised curricula, the number of integrated units for elementary school is two (Journey of Water and Energy & Life). The number of the integrated units should be increased in the future, reinforcing EE in other units. The number of integrated units for middle school is three (Science & My Future, Disasters & Safety, and Scientific Technology & Human Civilization). The number of the integrated units should be increased in the future, reinforcing EE in other units. The subject of "integrated science" for high school should be significantly revised, maintaining the section for environmental knowledge, in order to enable students to acquire information on environmental competences, values and attitudes, and action and participation.

Considering integrated science units based on core ideas, as well as integrated units for elementary and middle schools, a separate integrated unit for EE can be offered. For example, highlighting the big idea of sustainability, related information on science/environment/humanity can be consolidated to develop main units. It may be also necessary to seek ways to offer EE via capstone activities where environment-related information is ultimately provided, based on life and earth sciences closely linking to EE. Lastly, integrating education relating to STEAM and socio-scientific issues, environment-related information can be added to each unit for EE for elementary, middle, and high schools.

References

Department of the Environment, Water, Heritage and the Arts (DEWHA). (2010). *Sustainability curriculum framework: A guide for curriculum developers and policy makers*. Commonwealth of Australia.

Food and Agriculture Organization (FAO). (2008). *Contributing to one world, one health: A strategic framework for reducing risks of infectious diseases at the animal-human-ecosystems interface*. http://www.fao.org/3/aj137e/aj137e00.pdf

Future Earth. (2020). *Our future on earth.* https://drive.google.com/file/d/1chEx2Aewehp1_0nXYnERwUViJI6qR2hi/view

Gwon, J., Lee, K., Min, Y., Kim, H., Kim, K., & Kim, H. (2018). *International comparative study of subject curriculum on elementary and secondary school: Focused on Korean language, social studies, mathematics, physical education, and music subject.* Korea Institute for Curriculum and Evaluation. (in Korean).

Hollweg, K. S., Taylor, J. R., Bybee, R. W., Marcinkowski, T. J., McBeth, W. C., & Zoido, P. (2011). *Developing a framework for assessing environmental literacy.* North American Association for Environmental Education.

Kim, N., & Lee, S. (2020). A review on the general guidelines of the 2015 revised curriculum from the perspective of education on environment and sustainable development and proposals for a plan to improve the next curriculum. *Environmental Education, 33*(4), 425–442. (in Korean).

Kwak, Y., Lee, S., Kim, N., & Ju, H. (2021). Current status and revision plan of environmental education in the 2015 revised 'integrated science' curriculum. *Biology Education, 49*(4), 521–532. (in Korean).

Lee, S., Kim, N., Joo, H., Gwak, Y., Park, Y., Park, H., & Jeon, P. (2020a). Strategies for strengthening environmental education in the next national curriculum: A preliminary study on the development of a framework for content analysis of environmental education. *Environmental Education, 33*(3), 247–261. (in Korean).

Lee, S., Kim, N., Joo, H., Gwak, Y., Park, Y., Park, H., Seo, W., Yeom, E., & Jeon, P. (2020b). Content analysis of environmental education in the 2015 revised national curriculum and implications for the next curriculum: Focus on social studies, science, ethics, practical arts, and Korean language subjects. *Environmental Education, 33*(4), 443–462. (in Korean).

Ministry of Education. (2015). *Science curricula. Ministry of Education announcement* 2015–74 [Separate Volume 9].

Nam, S.-J. (1995). *Environmental education.* Daehaksa Publishing. (in Korean).

OECD. (2005). *The definition and selection of key competencies: Executive summary.* https://www.oecd.org/pisa/definition-selection-key-competencies-summary.pdf

OECD. (2018). *The future of education and skills: Education 2030.* Position paper. http://www.oecd.org/education/2030/E2030%20Position%20Paper%20(05.04.2018).pdf

Palmer, J., & Neal, P. (1994). *The handbook of environmental education.* New Fatter Lane.

UNESCO. (2020). *Education for sustainable development: A roadmap.* UNESCO.

World Economic Forum. (2020). *The global risks reports 2020.* World Economic Forum.

Chapter 9
Infusing Education for Sustainable Development into the Curriculum: Best Practice from Indonesia's Science Curriculum Implementation

Eko Hariyono, Agus Fani Candra Wijaya, and Dadi Rusdiana

9.1 Indonesia Education Profile

9.1.1 Human Development and Students' Academic Achievement

Even though it is a developing country, Indonesia has shown a relatively fast economic growth, with a high rate of expansion (Elias & Noone, 2011). Based on data from the National Development Agency (Bappenas), Indonesia's average economic growth of the last 3 years was 5.033 (Deputi Bidang Ekonomi, 2020). This relatively good economic growth has contributed to increasing the welfare of the community. However, Indonesia's economic growth slowed down slightly in the third quarter of 2020 due to the COVID-19 pandemic, which weakened people's purchasing power, negatively affecting community welfare and requiring immediate and focused action.

Improved economic growth is accompanied by an increase in Indonesia's Human Development Index (HDI). As shown in Fig. 9.1, Indonesia's HDI score showed an increasing trend from 1990 to 2019; in 2019, Indonesia was included in the category of high human development, with a growth index of 0.718. Based on the 2019 HDI data, Indonesia ranks 107th out of 189 countries – an improvement from 2015, when Indonesia's growth index was around 0.689, ranking it 113th out of 188 countries. This reinforces the need for improving the quality of Indonesian human

E. Hariyono (✉)
Universitas Negeri Surabaya, Surabaya, Indonesia
e-mail: ekohariyono@unesa.ac.id

A. F. C. Wijaya · D. Rusdiana
Universitas Pendidikan Indonesia, Bandung, Indonesia

H. Fujii, S.-K. Lee (eds.), *Science Education for Sustainable Development in Asia*, Education Innovation Series, https://doi.org/10.1007/978-981-99-8711-5_9

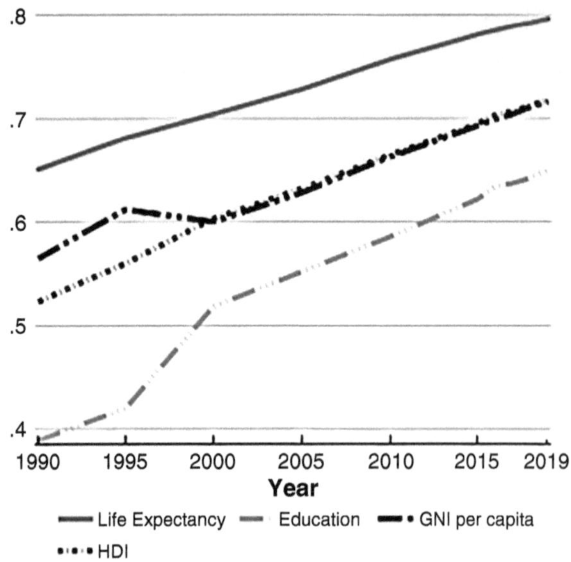

Fig. 9.1 Tendency for Indonesia's Human Development Index components. (Human Development Report, 2020)

resources so that Indonesia can be comparable to other developed countries. One approach to achieve this is to optimally manage education to support the nation's sustainable development (Sukasni & Efendy, 2017).

Human resources play a key role in accelerating social-economic development and providing efficient services (Omebe, 2014). The quality of human resources is assessed from the individual's knowledge, skills, and behavior and can be determined from the perspective of health, intellectual, emotional, social, and spiritual intelligence and morality, as a form of interaction between human beings, their environment, and their responsibilities (Ali, 2017). The quality of human resources is improved through the development of people's knowledge and skills (Kumar, 2018). Education is the single most effective strategy for improving the quality of human resources.

Education and human resource development can be likened to two sides of a coin: they are closely related and are difficult to conceptualize separately. It is impossible to obtain good human resources if education is not prioritized; a good education is required to realize good human resources. Education is an important driver of change and has become an effective instrument for large-scale, radical achievement in all fields (Rena, 2006). It is the only factor that plays an important role in addressing all challenges to improving human resource quality and must be used optimally. The success and limitations of a state's education system in each state are determined by the ability of human resources to implement it (Omebe, 2014). Education is a basic priority for realizing the progress of a nation, which is why the government continuously strives to improve the quality of education (Zulfikar, 2009). Through superior education, Indonesia is expected to be able to develop qualified individuals who are able to address both current and future developmental challenges (Hariyono et al., 2019a, b).

The quality of education in Indonesia is currently improving; however, faced with the realities of a rapidly and continuously changing world, education must also be continuously improved. Compared to other developed countries, Indonesia is still relatively lagging in terms of education. Based on the results of the Programme for International Student Assessment, conducted by countries incorporated in the Organisation for Economic Co-operation and Development (OECD), it can be seen that the average score of Indonesian students' science ability reached 389, while the OECD's average score was 489 (OECD, 2019). This shows that the students' science ability in Indonesia is relatively lower than that in other OECD member countries.

Based on the data from the Global Human Capital Forum, Indonesia ranks 65th out of 130 countries in education, still lagging behind Singapore, Malaysia, Thailand, and the Philippines (Arifa & Prayitno, 2019). In terms of the annual rankings, Indonesia scored relatively lower in 2018 compared to 2015 based on trends in reading and mathematics performance (OECD, 2019), as more than 50% of Indonesian students under the age of 15 had not mastered basic reading or math skills (OECD and Asian Development Bank, 2015). Guided by these challenges, it is time to revolutionize Indonesian education, so that its educational products will be on par with that of other developed countries. Such revolution should emphasize the concept of future education, as enshrined in the 1945 Constitution Republic of Indonesia, which proposes to develop the entire Indonesian population and strive for excellence in science, spirituality, skills, productivity, and competitiveness.

9.1.2 Competencies Required of Teachers in Modern Society

In Indonesia, the education challenge is growing in line with the rapid development of the era and increasing global demands but has not yet been addressed by teachers as agents of reform in education and learning. Teachers tend to prefer conventional learning methods because they are considered relatively easy to implement and they ensure that the curriculum is delivered as planned, thus facilitating the transfer of information between teachers and students (Zulfikar, 2009). Teachers prioritize the development of students' cognitive abilities over other competencies needed in the twenty-first century, such as problem-solving, decision-making, critical and high-level thinking, and collaboration skills. This contributes to students' low learning competence but is very difficult to change, as the most Indonesian teachers' achievement is still far below the target set by the mandate of the Laws of Teachers and Lecturers (Undang-undang Guru dan Dosen), both in terms of educational qualifications and competencies (Arifa & Prayitno, 2019).

One way of sparking education reforms is by changing the learning strategy from a simplistic to a complex approach. This can be achieved through a shift in orientation from the teacher as a source of knowledge to the student as the focus of learning (Pana & Escarlos, 2017). Traditional learning patterns that promote memorization of theory should be abandoned, and teachers must be able to equip students with more meaningful learning strategies in the hope that students can use the knowledge gained to solve problems in their everyday lives. From this perspective, students are equipped with research experience that will help them develop critical thinking skills and construct their own knowledge (Corlu & Corlu, 2012). Teachers should also be brave enough to assure that the learning experience can serve as preparation for the future. To achieve this, teachers need to develop greater professional capacity and responsibility (OECD and Asian Development Bank, 2015) and be equipped with strong and efficient professional competencies (Mandal, 2018).

Education is strongly related to a community's assumption of responsibility for sustainable future development (Rauch & Steiner, 2013) and, thus, needs to be revised to ensure that students succeed in becoming innovators (Alismail & McGuire, 2015). To realize a more systematic learning process that follows the principles of education and learning, changes to the curriculum need to be made. These are ways to empower education (Zulfikar, 2009). Such a shift would acknowledge that not all students can understand the content by teachers using current strategies and most teachers are cautious about implementing creative and innovative learning strategies (Sukasni & Efendy, 2017).

Learning that leads to knowledge mastery must be centered on the achievement of competencies, especially in terms of thinking and problem-solving skills. Teachers play an important role in developing the students' attitudes, academic performance, and conformity to twenty-first-century learning skills (Pana & Escarlos, 2017). Moreover, teachers have a responsibility not only to deliver teaching content but also to contribute to the identity development process of their learners.

American Association of Colleges for Teacher Education and the Partnership for twenty-first Century Skills (2010) informs that all students must have four basic skills: communication, collaboration, critical thinking and problem-solving, and creativity and innovation. Communication is the ability to articulate thoughts and ideas effectively using oral and written communication skills in various formats and materials. Collaboration is the ability to work effectively and respectfully with a diverse group of people. Critical thinking and problem-solving are used to analyze and evaluate evidence, arguments, claims, and beliefs effectively and to solve unorthodox problems, both in conventional and innovative ways. Innovation and creativity involve the use of a wide range of techniques to develop new and valuable ideas. These skills are indispensable and must be mastered by students to excel in the twenty-first century. Several strategies are needed to realize the broad-based development of these skills, including (i) providing alternative routes for high-quality learning, (ii) building partnerships to advance education reform, and (iii) preparing prospective teachers with better teaching experiences through the partnership with urban schools. Therefore, twenty-first-century skills requirements should be integrated into the national standard curriculum not only for the benefit of students and teachers but also to facilitate the career preparation of future generations (Alismail & McGuire, 2015).

Advanced education strategies are always successful in helping students respond to present and future challenges, and education for sustainable development (ESD) has proven to be a successful approach in this regard. The ESD approach endeavors to empower students by developing their competence in dealing with change. According to Boeve-de Pauw et al. (2015), ESD aims to provide learning services to students through experiences with the world around them based on their observations and encourages them to take action to ensure a sustainable future. This requires teachers with a high level of education competency. Teachers are instrumental in realizing sustainable development, as they are strong agents of change, equipped to deliver the required educational response (Rieckmann, 2019).

9.2 Why ESD Is Important

Sustainable development has become a major political concern worldwide (Burmeister & Eilks, 2013). The concept can be interpreted as a paradigm for thinking about the future by considering environmental, social, and economic aspects in a balanced manner, thereby improving the population's quality of life (UNESCO, 2012). ESD is an important consideration in preparing for global environmental changes in the present and future (Hariyono et al., 2018). It has the potential to accelerate sustainable development because it can engage all global stakeholders to work toward change (Sutanto, 2017).

Conceptually, ESD rests on three mutually reinforcing pillars: environmental, social, and economic. ESD provides knowledge and skills that encourage lifelong learning, thereby helping people find solutions to environmental, economic, and social problems (UNESCO, 2006). These principles are in line with the Rio Declaration, which includes the right to a healthy and productive life in harmony with nature; development that meets current and future environmental needs fairly, eradicates poverty, and reduces disparity; and environmental protection (UNESCO, 2012).

The relationship between education and ESD is complex. Based on the research, results have shown that basic education is key to citizens' ability to develop and achieve their sustainability targets (UNESCO, 2006). Most people with a good education will find exactly a solution in their life. Various problems related to the inefficient use of energy, low water conservation, increased pollution, and lifestyle-related challenges must be solved. ESD presents a solution. We can prove that ESD constitutes a program that is appropriate for addressing problems in our surroundings. Dannenberg and Grapentin (2016) declared ESD to be a very effective learning strategy for addressing these problems, as it can provide students with the opportunities for acting strategically in situations of uncertainty. This raises an important point – ESD encourages action based on decisions rather than to think merely about the subject matter.

ESD is not a new concept. It was first presented at the Conference on Environment and Development in 1970 and again at the Earth Summit in Rio de Janeiro in 1992, known as Agenda 21. Agenda 21 contains strategic plans related to local, national, and global actions for reviewing human and environmental impacts. These actions consist of (i) encouraging and improving the quality of education, (ii) curriculum reorientation, (iii) fostering public awareness of the concept of sustainable development, and (iv) training the workforce (UNESCO, 2009). The program provides important support for upgrading the quality of life on a wide scale. At the international level, ESD is increasingly recognized as a major element of high-quality education (Dannenberg & Grapentin, 2016).

According to the UNESCO declaration in Ahmedabad in 2007, education in the context of sustainable development should encourage a lifestyle where each component is interconnected in a mutually beneficial and strengthening way (Gokool-Ramdoo & Rumjaun, 2017). Therefore, education in Indonesia should be oriented toward ESD to build a resilient society in the face of changing natural phenomena (Hariyono et al., 2018). In the long term, an ESD program will raise awareness of natural disasters among Indonesian who can use their acquired knowledge to optimize their lives.

A multidisciplinary ESD emphasizes the value of a sustainable, fair, and contextually relevant society, instilled and realized through various disciplines (Perello-Marin et al., 2018). This transformational and holistic model is directed at content and learning beyond the existing pedagogical approaches and learning environments by not only integrating content on climate change, poverty, and sustainable consumption into the curriculum but also developing interactive student-centered learning settings (UNESCO, 2017). ESD is seen as a type of education in which environmental values and attitudes are directly explored, thus encouraging active learning and ensuring its application through participatory approaches (Anyolo, 2018). ESD employs a teaching and learning process that facilitates the development of problem-solving skills related to environmental, economic, and social issues (Santone et al., 2014). ESD-based learning can potentially be applied to science education by combining the fields of chemistry, biology, and physics and framing it in the context of social fields such as economics, social sciences, and the humanities (Burmeister & Eilks, 2013). This way, a new and comprehensive science education can be provided in accordance with student's anticipated need to solve problems in the interest of societal development.

UNESCO's ESD approach produces major changes in social engineering. Through ESD, students will receive an education that is relevant to today's challenges (UNESCO, 2017). However, thus far, the concept of sustainability has not been clearly conveyed to most of the community; moreover, the learning outcomes related to sustainability are also considered unclear (Kioupi & Voulvoulis, 2019).

To understand the key concepts of ESD better, its characteristics need to be understood. Singhal and Shrivasthava (2004) described several ESD characteristics: (i) understanding the interdependence of creatures with one another, (ii) increasing awareness of environmental problems and complex systems related to sustainable development, and (iii) developing abilities, attitudes, and values associated with sustainable development. Learning in ESD occurs through facilitating changes in students' behavior and ways of thinking (Adomssent & Hoffmann, 2013).

9.3 Applying ESD to Curricula

9.3.1 Teachers' Role in ESD

ESD learning is crucial for addressing current and future life challenges. In practice, teachers are required not only to introduce information about ESD but also to understand how to entrench ESD into their students' lives. To apply ESD to science learning successfully, teachers must have a good understanding of ESD and its implications (Burmeister & Eilks, 2013).

ESD is often confused with environmental education (EE). In fact, both aim to create a better world with a balance between economy, ecology, and society. While EE has been used for a longer time, ESD has a broader scope (Kankovskaya, 2016). ESD emphasizes a multidisciplinary approach aimed at instilling the overarching values and principles of sustainability to achieve a more sustainable, fair, and responsive society (Perello-Marin et al., 2018). ESD does not only comprise knowledge sharing but also the development of a positive attitude toward its components, and it is necessary for developing good life skills (Bachati, 2015).

Over the past two decades, many ESD initiatives have highlighted the importance of integrating ESD in the curriculum (Balsiger et al., 2017). Transforming learning and education into ESD requires commitment and effort from faculties and individual academics (Filho et al., 2018).

Success in implementing ESD is largely determined by the teacher's knowledge of the concept of ESD and their role in it. In this approach, teachers play an important role as they can be effective agents in enacting change and addressing important issues in the learning community (Bachati, 2015). In addition, teachers are considered able to encourage and empower the next generation to play an active role in developing communities that can address present and future global challenges. Furthermore, ESD teachers are expected to organize learning opportunities for students both inside and outside traditional settings (Rauch & Steiner, 2013).

Learning objectives in ESD activities can be classified as follows: (i) to study the key issues of sustainable development from a multidisciplinary perspective and multiple viewpoints, (ii) to apply the essential elements of ESD to educational training and to assess and evaluate the development of learning sustainability competencies during training, and (iii) to collaborate and disseminate ESD practices among other schools and communities that are increasing their involvement in ESD (Okayama University ESD Promotion Centre, 2020). To meet these expectations, new patterns and strategies are needed to equip prospective teachers for sustainable education. This strategy is effective when supported by a series of facts and solutions available to solve the given problem (Dannenberg & Grapentin, 2016) and can be applied by encouraging communication and dialogue on sustainable topics and directing students to new strategies (Perello-Marin et al., 2018).

9.3.2 *Infusing ESD into the Curricula*

Educational institutions can contribute to sustainable development in several ways: (i) by including sustainable development in all curricular programs and educational endeavors and research, (ii) acting as a local knowledge center for sustainable development to assist communities in addressing these challenges at a local level, and (iii) making sustainable development a key principle in logistics and managerial processes (Peter et al., 2016). One of the simplest ways to develop the concept of ESD is to integrate it into the learning curriculum, especially science in higher education. This is a first step in introducing ESD as an innovative educational program relevant to the field of science education.

Integrating ESD into the higher education curriculum is crucial, as higher education plays a critical role in realizing ESD. Furthermore, higher education is where future leaders are cultivated and best environmental practices can be demonstrated (Freidenfelds et al., 2018). Despite increasing public awareness of the challenges of sustainable development and the importance of overcoming them, however, little headway has been made in the integration of sustainable development in the university curriculum (Filho et al., 2018).

The integration of ESD into the curriculum will ensure the future development of sustainable communities (Álvarez-García et al., 2015). ESD learning not only emphasizes the acquisition of knowledge, values, and theories related to sustainable development but also various criteria such as learning to ask critical questions, explaining one's values, imagining a more positive and sustainable future, thinking systematically, responding to learning application, and exploring the dialectic between tradition and innovation (UNESCO, 2011). For optimum results, Dambudzo (2015) suggested that the ESD-based curriculum should be divided into the application of project-based activities, problem-solving, and investigation.

In integrating ESD into the science education curriculum, three key aspects of ESD must be noted – key knowledge, key skills, and key attitudes. All this is required to meet future challenges and ensure a sustainable future (Maidou et al., 2019). ESD Consulting (EAUC, 2021) describes all keys to develop knowledge, skills, and attitudes through ESD. Key knowledge refers to (i) the interdependence between people and life on earth; (ii) the limited carrying capacity of the planet; (iii) the values of biodiversity, along with social and cultural diversity in maintaining the well-being of our planet and society; (iv) the important role of rights and responsibilities in a sustainable society; (v) the importance of equality and justice in a sustainable society; and (vi) the presence of uncertainty and necessity to always be careful in making decisions about our planet. These concepts must be understood by a trained designer who can integrate ESD into the curriculum.

In addition to knowledge-related aspects, key skills must also be considered, including skills for understanding the relationship between problems and the ability to make decisions toward problem-solving collective and collaborative decision-making and critical thinking skills. These skills are especially needed in response to twenty-first-century challenges.

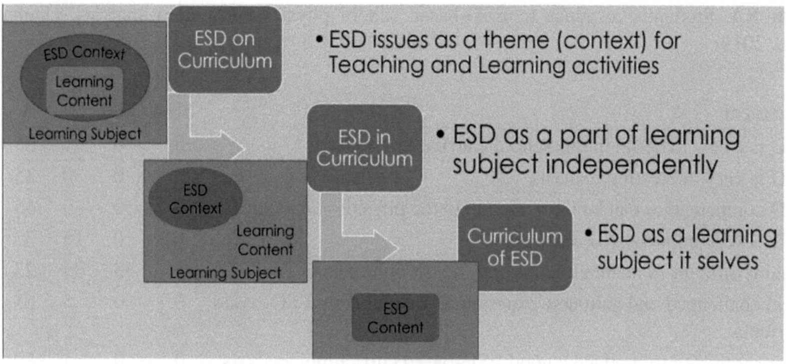

Fig. 9.2 Approaches to infusing ESD into the curriculum

Key attitudes include, among others, belief in taking action; respecting differences in individual behavior and responsibilities; respecting nature with all its limitations and harmonization; respecting differences and biological, social, and cultural diversity; and an attitude of caring for ourselves, the environment, and the earth.

Models for integrating the curriculum are presented in Fig. 9.2. There are three approaches for integrating ESD into the curriculum: (1) ESD as a contextual framework for predetermined lesson content, (2) ESD as a contextual framework with the inclusion of limited ESD content, and (3) as mainly ESD-focused content, within the contextual framework of ESD. In the first approach, ESD is a central theme in the learning process. Fogarty and Stoehr (2008) provides ten integrated curriculum models to facilitate this approach: fragmented, connected, nested, sequenced, shared, webbed, threaded, integrated, immersed, and networked (Fogarty & Stoehr, 2008). The purpose of this approach is to identify current competencies and content in the science curriculum that are aligned with ESD.

For example, in the Earth and Space Sciences course we developed, students are required to learn about the characteristics of volcanoes and efforts to minimize the risk of volcanic disasters based on the conditions of the volcanoes around them. This approach promotes knowledge about volcanoes and provides opportunities for students to solve problems related to disaster risk reduction based on ESD principles. Specifically, students conduct field observation activities to observe the activities of the community around Mount Kelud in building relatively small but sturdy housing with a roof angle of no more than 30°. This design is informed by local wisdom and is very safe because volcanic ash fallings on the roof will immediately slide off, thus preventing the roof from collapsing with potentially fatal consequences. This approach is very effective in training students to solve authentic problems encountered in daily life based on ESD. In addition, this approach is also suitable for the subject matter. ESD is not presented as content, but the characteristics of ESD form the basis for solving problems.

The second approach is characterized by incorporating ESD content into the curriculum or establishing ESD as part of the subject content. The purpose of this

Table 9.1 Student's response to ESD-based school physics curriculum analysis. (Jauhariyah et al., 2019)

Statement	Attitude scale (%)				
	SDA	DA	N	A	SA
ESD is a new material in the School Curriculum Study	5	0	5	65	25
ESD is very interesting material	0	0	0	35	65
ESD competencies can be integrated in to the physics curriculum	0	0	0	65	35
ESD is very important for students	0	0	15	35	50
I find it difficult to relate global issues to curriculum analysis	5	35	35	25	0
I feel challenged and gain new experiences in analyzing ESD-based curricula	0	0	5	55	40
I was motivated to produce a good ESD-based curriculum analysis	0	0	5	80	15
I was frustrated when looking for solutions to global problems as a characteristics of an ESD-based curriculum	15	10	65	5	5

SDA strongly disagree, *DA* disagree, *N* neutral, *A* agree, *SA* strongly agree

approach is to identify ESD content that aligns the competence of science teachers. One example of this approach is the Studying School Physics Learning Curriculum course, where students specifically learn about the concept of ESD and then perform ESD-related tasks using different methods and materials consistent with the ESD approach, including journals and books. In this course, students perform project-based tasks to develop physics learning devices in accordance with the principles of ESD. This approach is suitable for establishing ESD information as learning content.

Based on the results of our research (Table 9.1), all students recognize the novelty and potential use of integrating ESD into physics, which shows that ESD can potentially help students identify global issues in curriculum analysis.

The third approach incorporates ESD into all aspects of the curriculum, firstly by using ESD as a subject, with courses or training programs delivered to the teacher community. When the opportunity arises for the design of new curricular materials, ESD would be a suitable proposition – either as a new subject or an independent part of the curriculum. For example, to improve the teachers' knowledge and understanding of ESD, the Southeast Asian Ministers of Education Organization on the Teacher Education Quality Improvement Program in Science organized the Environmental Education Course for Sustainable Development. In this course, tenured teachers have the opportunity to develop their understanding and knowledge of ESD.

9.4 Integrating ESD Content and Context: The Volcano Learning Project as an Example

The Volcano Learning Project (VLP) is an innovative lecture program in the field of space and earth science with a focus on volcanic materials. It was developed against the background of low knowledge, skills, and attitudes of communities on volcanoes, which is part of Indonesia's geological landscape. This program is potentially useful for developing students' competence as candidate physics teachers, who learn about the typical geological features of Indonesia through the integration of ESD. Indonesia is an island nation with hundreds of active volcanoes, and eruption activities cause distress among surrounding communities. However, not many people are aware of this issue.

The VLP program is presented in Fig. 9.3. Its unique characteristics included its focus on both knowledge and geoscience content and geoscience skills and scientific rigor. Through this program, prospective teachers not only receive instruction on the subject matter in specifically geoscience knowledge and skills but also get the opportunity to engage in activities outside the classroom. This way they can apply the knowledge and skills gained in the classroom to solve real problems in the field. At the same time, students get the opportunity to interact with communities in the mountainous regions to collaboratively solve their problems, such as utilizing the potential of the environment around the mountainous regions to support the local economy.

Furthermore, the VLP program also integrates information, communication, and technology (ICT) into the curriculum, using simulation software developed according to the requirements of volcano learning developed by Makrakis (2011). Learning that materials and ICT can assist students in exploring relevant concepts, engaging in authentic learning and problem-solving, improving their metacognitive skills, and engaging with diverse media when presenting information. The combination of ESD and ICT in the VLP is highly suitable for teachers, especially when engaging

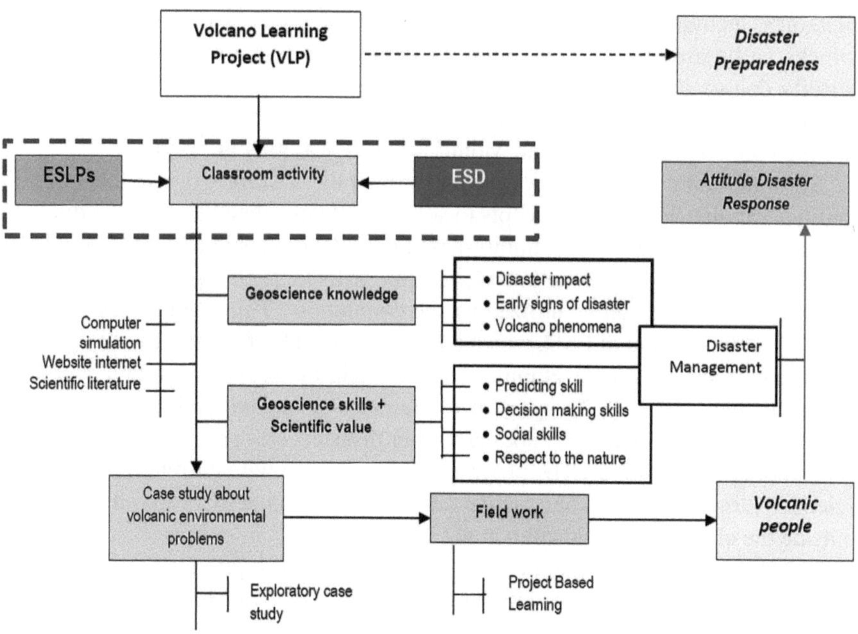

Fig. 9.3 Design of the VLP program

with inaccessible or abstract concepts. Volcanology is a domain of geoscience learning that requires a model or simulation media; the integration of ICT into the VLP will facilitate students' important learning experiences.

Makrakis and Kastoulas-Makrakis (2012) examined the relationship between ESD and ICT. First, ESD themes integrated into the school curriculum can benefit the application of ICT in education and learning. Second, ESD methods are conducive to constructivist and transformative learning, which can be amplified by using ICT-based learning tools such as concept mapping, modeling, and social networking. Third, the impact of ICT in areas such as cultural diversity and intercultural understanding, health, HIV/AIDS, governance, natural resources, climate change, rural development, sustainable urbanization, poverty alleviation, corporate responsibility, and accountability can be assessed. These are key areas of sustainable development. However, digital inequality, especially for women and other disadvantaged groups, remains a challenge for educators and policymakers.

Including ICT in the VLP program is useful for modeling and simulating abstract concepts and providing students with the opportunity to develop their thinking skills. This is in accordance with King's (2008) opinion that students have difficulty in learning geoscience concepts because they are abstract, so the inclusion of various media facilitates the process. Another advantage of ICT is that it provides an easily accessible source of information (Palma et al., 2014).

The VLP program is focused not only on disaster mitigation efforts but also on turning vulnerable communities into disaster-resistant and disaster-aware communities. One of the interesting applications of the project is shifting the way people view volcanic disasters. For example, people around Mount Kelud are grateful when an eruption occurs as they now understand that disaster brings with it many blessings, including the abundance of volcanic materials (sand, stone, and sulfur), which can be used by the community to support the economy, and improved soil fertility due to ash and volcanic gas, which increases agricultural productivity. In addition, for the people of Kelud, disasters can serve to strengthen social ties between the communities around Mount Kelud (Larasati et al., 2020).

An important competency developed through the VLP program is the student's ability to apply geoscience concepts to solve problems related to volcanic disasters in a way that is helpful to community-based geoscience problems. They should also be able to develop geoscience knowledge and skills through learning and research, the results of which can be utilized by the community. This competency is highly relevant to the concept of ESD, which emphasizes the ability to solve problems such that it contributes to the development of society in the social, economic, and environmental fields. The VLP program is future-oriented and tailored to the Indonesian context, considering that volcanoes and other disasters still pose a big challenge for surrounding communities. The program is also aligned with current trends in geoscience learning as it uses interactive and inquiry-based activities both inside and outside the classroom (Goldsmith et al., 2013).

Fieldwork is another strength of the VLP program, as it provides students with a first learning experience. They interact with the environment around the mountains, as well as the surrounding communities, and attempt to help solve them (Hariyono et al., 2019a, b).

Fig. 9.4 Interview activity with residents around mountain

The problem identified by the residents is used as the basis for developing student projects through research activities to produce a learning product that can be employed by the community. The field work process, depicted in Fig. 9.4, provides many advantages for students, including improved access to reliable scientific information; in addition to fostering a sense of responsibility for the natural environment among students, they also develop their prediction skills and capacity for enacting disaster mitigation, the ability to apply knowledge in informed decision-making, and the ability to explain basic data-based geoscience phenomena (Fermeli et al., 2014).

One of the interesting findings from the students' fieldwork activities was the use of the basalt and andesite rocks of Mount Kelud as water filter materials. According to the student's research results, the filter media purify and reduce odors in water samples by up to 80%. These results are likely to be further developed to support improved living conditions for the community.

In summary, VLP is a lecture program aligned with the concept of ESD and is a good example of the application of both ESD content and context in curriculum design. Through this program, students obtain real learning experiences. In addition to learning about volcanoes, students get the opportunity to solve direct economic, social, and environmental problems faced by the communities around volcanoes.

9.5 Conclusion

The key challenge for Indonesian science education is to build the meaning of student learning by using the knowledge, skills, and attitudes of science that is being learned. An educational program is needed to facilitate this process, and we contend that ESD provides an ideal solution for this problem. ESD has great potential for training and other opportunities for students to develop their competencies, especially in terms of solving problems. The three pillars of ESD – economy, environment, and society – are important considerations for measuring the ability to solve these problems.

To facilitate the understanding and application of ESD, it can be integrated into the science learning curriculum in three ways: by developing an ESD framework for delivering traditional science content, delivering limited ESD content as part of an ESD-informed curriculum, and completely integrating both ESD content and context into the curriculum with ESD as an independent, complex subject.

The VLP as a learning program is one example of integrating ESD into the curriculum. It provides students with the opportunity to interact with communities around volcanoes to learn about their everyday problems and think about how to solve them. Learning products resulting from fieldwork activities can be made available to the community and used to support the improvement of the local economy, environment, and society. Based on the outcomes of VLP activities, it seems that ESD programs have great potential for developing student problem-solving competence, so integrating ESD into other relevant materials is highly recommended.

References

Adomssent, M., & Hoffmann, T. (2013). *Concept paper. The concept of competencies in the context of education for sustainable development (ESD).* http://se-ed.co.uk/edu/wp-content/uploads/2010/09/130314-Concept-Paper-ESD-Competencies.pdf

Ali, M. (2017). *Curriculum development for sustainability education.* UPI Press.

Alismail, H. A., & McGuire, P. (2015). 21st century standards and curriculum: Current research and practice. *Journal of Education and Practice, 6*(6), 150–154.

Álvarez-García, O., Sureda-Negre, J., & Comas-Forgas, R. (2015). Environmental education in pre-service teacher training: A literature review of existing evidence. *Journal of Teacher Education for Sustainability, 17*(1), 72–85.

American Association of Colleges for Teacher Education and the Partnership for 21st Century Skills. (2010). *21st century knowledge and skills in educator preparation.* https://files.eric.ed.gov/fulltext/ED519336.pdf

Anyolo, E. O. (2018). Implementing education for sustainable development in Namibia: School teachers' perceptions and teaching practices. *Journal of Teacher Education for Sustainability, 20*(1), 64–81.

Arifa, F. N., & Prayitno, U. S. (2019). Peningkatan kualitas pendidikan: Program pendidikan profesi guru prajabatan dalam pemenuhan kebutuhan guru profesioal di Indonesia. *Aspirasi: Jurnal Masalah-masalah Sosial, 10*(1), 1–17. (in Indonesian).

Bachati, P. (2015). Promoting education for sustainable development: Role of teacher. *Journal of Training and Development, 1*(1), 21–26.

Balsiger, J., Förster, R., Mader, C., Nagel, U., Sironi, H., Wilhelm, S., & Zimmermann, A. B. (2017). Transformative learning and education for sustainable development. *GAIA-Ecological Perspectives on Science and Society, 26*(4), 357–359.

Boeve-de Pauw, J., Gericke, N., Olsson, D., & Berglund, T. (2015). The effectiveness of education for sustainable development. *Sustainability, 7*(11), 15693–15717.

Burmeister, M., & Eilks, I. (2013). An understanding of sustainability and education for sustainable development among German student teachers and trainee teachers of chemistry. *Science Education International, 24*(2), 167–194.

Corlu, M. A., & Corlu, M. S. (2012). Scientific inquiry based professional development models in teacher education. *Educational Science: Theory & Practice, 12*(1), 514–521.

Dambudzo, I. I. (2015). Curriculum issues: Teaching and learning for sustainable development in developing countries: Zimbabwe case study. *Journal of Education and Learning, 4*(1), 11–24.

Dannenberg, S., & Grapentin, T. (2016). Education for sustainable development – Learning for transformation. The example of Germany. *Journal of Futures Studies, 20*(3), 7–20.

Deputi Bidang Ekonomi Bappenas. (2020). Laporan Perkembangan Ekonomi Indonesia dan Dunia. Triwulan III Tahun 2020. *Edisi, 4*(3) (in Indonesian).

EAUC. (2021). *Embedding sustainable development in the curriculum*. https://www.sustainabilityexchange.ac.uk/files/embedding_sustainabilyt_in_the_curriculum_guide.pdf

Elias, S., & Noone, C. (2011). The growth and development of the Indonesian economy. *Bulletin*, December Quarter, 33–44. https://www.rba.gov.au/publications/bulletin/2011/dec/pdf/bu-1211-4.pdf

Fermeli, G., Steininger, F., Dermitzakis, M., Melendez, G., & Page, K. (2014). Literacy and students' interest on geoscience – Findings and result of GEOschools project. *Geophysical Research Abstracts, 16*. https://meetingorganizer.copernicus.org/EGU2014/EGU2014-8678.pdf

Filho, W. L., Raath, S., Lazzarini, B., Vargas, V. R., Souza, L., Anholon, R., Quelhas, O. L. G., Haddad, R., Klavins, M., & Orlovic, V. L. (2018). The role of transformation in learning and education for sustainability. *Journal of Cleaner Production, 199*, 286–295.

Fogarty, R., & Stoehr, J. (2008). *Integrating curricula with multiple intelligences* (2nd ed.). Corwin Press.

Freidenfelds, D., Kalnins, S. N., & Gusca, J. (2018). What does environmentally sustainable higher education institution means? *Energy Procedia, 147*, 42–47.

Gokool-Ramdoo, S., & Rumjaun, A. B. (2017). Education for sustainable development: Connecting the dots for sustainability. *Journal of Learning for Development, 4*(1), 72–89.

Goldsmith, S. T., Trierweiler, A. M., Welch, S. A., Bancroft, A. M., Von Bargen, J. M., & Carey, A. (2013). Transforming a university tradition into geoscience teaching and learning opportunity for the university community. *Journal of Geoscience Education, 61*(3), 280–290.

Hariyono, E., Abadi, L., Wijaya, A. F. C., & Fujii, H. (2018). Designing geoscience learning for sustainable development: A professional competency assessment for postgraduate students in science education program. *Jurnal Penelitian Fisika dan Aplikasinya, 8*(2), 61–70.

Hariyono, E., Madlazim, & Anggrayni, S. (2019a). The effectiveness of volcanology learning through inquiry based on education for sustainable development. *Journal of Science Education, 20*(2), 20–23.

Hariyono, E., Prastowo, T., & Hartati, M. (2019b). Physics student's research skills performance: A field-based approach in geoscience learning. *Journal of Physics: Conference Series, 1491*, 012003.

Human Development Report. (2020). *The next frontier: Human development and the anthropocene*. https://hdr.undp.org/sites/default/files/Country-Profiles/IDN.pdf

Jauhariyah, M. N. R., Hariyono, E., Abidin, E. N., & Prahani, B. K. (2019). Fostering prospective physics teachers' creativity in analysing education for sustainable development based curricula. *Journal of Physics: Conference Series, 1417*, 012086.

Kankovskaya, A. R. (2016). Higher education for sustainable development: Challenges in Russia. *Procedia CIRP, 48*, 449–453.

King, C. (2008). Geoscience education: an overview. *Studies in Science Education, 44*(2), 187–222.

Kioupi, V., & Voulvoulis, N. (2019). Education for sustainable development: A systemic framework for connecting the SDGs to educational outcomes. *Sustainability, 11*(21), 6104.

Kumar, B. P. (2018). Human resources development through quality education. *International Journal of Research in Economics and Social Sciences, 8*(2), 259–266.

Larasati, D. A., Prasetya, S. P., Segara, N. B., Imron, A., & Nurahma, F. (2020). Livelihood strategy after the eruption of Kelud Mountain in Sugihwaras Village Ngancar Subdistrict Kediri Regency (Case study: The society in Sugihwaras Village). *Advances in Social Science, Education and Humanities Research, 473*, 297–299.

Maidou, A., Plakitsi, K., Polatoglou, H., & M. (2019). Knowledge, perceptions and attitudes on education for sustainable development of pre-service early childhood teachers in Greece. *World Journal of Education, 9*(5), 1–15.

Makrakis, V. (2011). ICT-enabled education for sustainable development: merging theory with praxis. In M. Youssef & S. A. Anwar (Eds.), *Proceedings of the 4th Annual Conference on e-Learning Excellence in the Middle East 2011 – in Search of New Paradigms for re-Engineering Education* (pp. 410–419). Hamdan Bin Mohammed e-University.

Makrakis, V., & Kastoulas-Makrakis, N. (2012). Course curricular design and development of the M. Sc. programme in the field of ICT in education for sustainable development. *Journal of Teacher Education for Sustainability, 14*(2), 5–40.

Mandal, S. (2018). The competencies of the modern teacher. *International Journal of Research in Engineering, Science and Management, 1*(10), 351–360.

OECD. (2019). *Programme for International Student Assessment (PISA) Result from PISA 2018.* https://www.oecd.org/pisa/publications/PISA2018_CNIDN.pdf

OECD and Asian Development Bank. (2015). *Education in Indonesia: Rising to the challenge.* OECD Publishing.

Okayama University ESD Promotion Center. (2020). *Guide for the effective dissemination of the Asia-Pacific ESD teacher competency framework.* Okayama University ESD Promotion Center. http://ceteesd.ed.okayama-u.ac.jp/pdf/200511.pdf

Omebe, C. A. (2014). Human resource management education: Issues and challenges. *British Journal of Education, 2*(7), 26–31.

Palma, J. L., Courtland, L., Charbonnier, S., Tortini, R., & Valentine, G. A. (2014). Vhub: A knowledge management system to facilitate online collaborative volcano modeling and research. *Journal of Applied Volcanology, 3*(2). https://doi.org/10.1186/2191-5040-3-2

Pana, G. U., & Escarlos, G. S. (2017). Contemporary teaching strategies on students' attitude, academic performance and acquisition of the 21st century skills. *International Journal of Scientific & Technology Research, 6*(8), 332–337.

Perello-Marin, M. R., Ribes-Giner, G., & Diaz, O. P. (2018). Enhancing education for sustainable development in environmental university programmes: A co-creation approach. *Sustainability, 10*(1), 158.

Peter, C. J., Libunao, W. H., & Latif, A. A. (2016). Extent of education for sustainable development (ESD) integration in Malaysian community colleges. *Journal of Technical Education and Training, 8*(1), 1–13.

Rauch, F., & Steiner, R. (2013). Competences for education for sustainable development in teacher education. *CEPS Journal, 3*(1), 9–24.

Rena, R. (2006). Education and human resource development in Eritrea–an explanatory note. *Barbados (West Indies): International Journal of Education and Development using Information and Commutation Technology (IJEDICT), 2*(4), 67–81.

Rieckmann, M. (2019). Education for sustainable development in teacher education. An international perspective. In S. Lahiri (Eds.), *Environmental Education* (pp. 33–48). Studera Press.

Santone, S., Saunders, S., & Seguin, C. (2014). Essential element of sustainability in teacher education. *Journal of Sustainability Education, 6*. https://citeseerx.ist.psu.edu/viewdoc/download?doi=10.1.1.1081.8816&rep=rep1&type=pdf

Singhal, P. K., & Shrivasthava, P. (Eds.). (2004). *Challenges in sustainable development.* Anmol Publications Pvt. Ltd.

Sukasni, A., & Efendy, H. (2017). The problematic of education system in Indonesia and reform agenda. *International Journal of Education, 9*(3), 183–199.

Sutanto, H. P. (2017). Education for sustainable development in West Nusa Tenggara. *Cakrawala Pendidikan, 36*(3), 320–341. (in Indonesian).
UNESCO. (2006). *Education for sustainable development toolkit: Learning and training tool*. UNESCO.
UNESCO. (2009). *Review of contexts and structures for education for sustainable development 2009. Learning for a sustainable world.* UNESCO.
UNESCO. (2011). *Education for sustainable development An expert review of processes and learning*. UNESCO.
UNESCO. (2012). *Education for sustainable development. Source book.* UNESCO.
UNESCO. (2017). *Education for sustainability development goals. Learning objectives*. UNESCO.
Zulfikar, T. (2009). The making of Indonesian education: An overview on empowering Indonesian teachers. *Journal of Indonesian Social Sciences and Humanities, 2*, 13–39.

Chapter 10
Forest Fires, Animal Sacrifices, and School Garden: Local Contexts at the Intersection of Sustainability and Implemented Science Curriculum in Indonesia

K. F. Ardh

10.1 Introduction

10.1.1 Background

While various stakeholders have long recognized the importance of scientific literacy (SL), their approaches to operationalizing its definition vary. For instance, the 2018 Framework for the Programme for International Student Assessment (PISA) defined SL as possessing three competencies: "(1) scientifically explaining phenomena; (2) evaluating and designing scientific investigations; and (3) scientifically interpreting data and evidence" (OECD, 2019). PISA and other international assessment tests positioned SL as a litmus test for educational policy and practice.

The literature on the definition of SL (Aikenhead, 2006; Roberts, 2007) has developed two distinct but complementary visions of SL that correspond well to the PISA's definition. The Vision I, "Pipeline Science – preparing future scientists," involves using an epistemological method or acquiring scientific knowledge and procedures for later application. While Vision II, "Science for all," emphasizes the importance of scientific knowledge in everyday life and society through relevant contexts that students can identify and how it relates to economic growth and development.

Recently, the third vision of the SL was proposed following the adoption of the Sustainable Development Goals (SDGs) and the increased efforts to promote Education for Sustainable Development (ESD) (Sjöström & Eilks, 2018). The Vision III, "Science for Transformation," incorporates a critical discourse to advocate for sociopolitical causes, including democracy and social justice. Accepting

K. F. Ardh (✉)
Okayama University, Okayama, Japan
e-mail: Aldi@okayama-u.ac.jp

H. Fujii, S.-K. Lee (eds.), *Science Education for Sustainable Development in Asia*, Education Innovation Series, https://doi.org/10.1007/978-981-99-8711-5_10

Vision III entails acknowledging that the "local context," or culture, norms, values, worldviews, and the society in which educators live, influences content selection and teaching.

The ESD for 2030 roadmap (UNESCO, 2014) is arguably one of the most valuable resources for implementing Vision III of SL. The roadmap highlights the importance of increasing teacher capacity to expedite action at the local level. Because while agenda-setting for sustainable development continues at the global level, the local community remains the overarching foundation of all implementation (Vaughter & Tabucanon, 2020). It also implied necessitating transcending learning beyond the classroom by involving the local community.

Alternatively, the increased emphasis on including sustainability content in science education may have muddled the identity and placement of an already overburdened curriculum. Moreover, the current literature on sustainability, characterized by a lack of focus on curriculum and pedagogy, has further exacerbated the situation. Hence, teachers may be discouraged from incorporating sustainability into the science curriculum. Against the above background, this chapter features several case studies about how a local context can help bridge the gap between sustainability and implemented science curriculum.

10.1.2 Research Questions

This chapter has been organized using the following research questions:

1. *What is the present status of literature regarding the local context for sustainability and implemented science curriculum in Indonesia?*
2. *How do the participants perceive the position of sustainability content in the science curriculum?*
3. *How do the participants implement sustainability within and beyond the science curriculum?*

10.2 Literature Review

10.2.1 Tripartite Model of Curriculum

This chapter follows the operational definition of curriculum as per the tripartite model used by the IEA (International Association for the Evaluation of Educational Achievement) in TIMSS (Trends in International Mathematics and Science Study). The model divides the curriculum into three dimensions (Table 10.1), as summarized from multiple sources (Valverde et al., 2002; Kridel, 2010; Mullis & Martin, 2017). The following section provides explanation of the intended science curriculum.

Table 10.1 Three dimensions of curriculum

Dimension	Definition
Intended curriculum	Include the national education's intention, goals, and aims
Implemented curriculum	Include potentially implemented curriculum, which includes the content of textbooks and other organized resource materials, while the implemented curriculum is principles that guide teachers' pedagogy (e.g., as reflected in the lesson plan)
Attained curriculum	Relates to the measurement of educational output in accordance with the curriculum (knowledge, skill, and attitudes)

Table 10.2 Domains of the main competencies in Curriculum 2013

Spiritual	Social	Cognitive	Psychomotor
Understanding and practicing religious beliefs and values	*Shared cultural values, such as honesty and tolerance*	*Analyze, evaluate, and create representation of understanding*	*Process, communication, problem-solving skills*

Modified from Faisal and Martin (2019)

10.2.2 Sustainability Content in the Indonesian Science Curriculum

The current curriculum of 2013 ("K-13") succeeded the school-based curriculum (SBC) of 2006, revised from the competence-based curriculum (CBC) of 2004. This revision aimed to tackle Indonesian students' poor performances on international comparative assessments and other issues (Hadi & Novaliyosi, 2019). Simultaneously, the revision allowed for a more decentralized curriculum, giving schools and teachers more autonomy to design learning relevant to their immediate communities. At the time of this chapter's writing, the new curriculum for 2022 was in the prototype stage.

The K-13 system requires science to be taught as an integrated subject from elementary to junior high school without distinguishing between its different subject domains. As a competency-based curriculum, the K-13 is organized around four domains of the main competencies (MC) (Table 10.2). The spiritual and social domains are addressed indirectly in the learning, while the cognitive and psychomotor domains are broken into more detailed, basic competencies (BC). It also acts as a basis for organizing chapters and units in the science textbook.

The BCs cover physics, chemistry, biology, earth science, and integrated science topics (Table 10.3). Keywords in bold represent topics related to sustainability content. The spiritual and social domains of the MCs in the K-13 correspond to normative competencies, while the cognitive and psychomotor domains relate to collaboration and problem-solving competencies, which are all among key competencies for sustainability as introduced by UNESCO (2017). Table 10.4 shows traces of sustainability themes (Gericke et al., 2018) in the BC of junior high school science curriculum.

Table 10.3 Basic competencies in junior high school science curriculum categorized by domain

Domain	Grade VII	Grade VIII	Grade IX
Physics	*Measurement, heat, and temperature, energy*	*Newton's law of motion, work, and simple plan*	*Static electricity,* ***electricity utilization****, and electromagnetism*
Chemistry	*Mixture*	*Additive substance, osmosis, and capillarity*	*Chemical particles*
Biology	*Classification and* ***biodiversity****, structure of a cell, ecosystem,* ***pollution***	*Musculoskeletal system, plant tissue, cardiovascular system, respiratory system, excretory system*	*Reproductive system, development of plants and animals, heredity, biotechnology*
Earth science	***Climate change****, earth's structure, and* ***disaster risk mitigation****, solar system*	*Vibration and sound waves, light and optics*	***Properties of soil and sustainability of life***
Integrated			***Eco-friendly technology***

Table 10.4 Detailed BCs related to sustainability in lower junior high science curriculum

Grade	Cognitive	Psychomotor
VII	*8. Analyze environmental pollution and its impact on the ecosystem*	*8. Write an essay on the solutions for beating pollution based on your observations*
	9. Analyze climate change and its impact on the ecosystem	*9. Write an essay about the ideas to adapt/solve issues caused by climate change*
VII	*Explain the earth's layers, volcano, earthquake, and local disaster risk mitigation*	*Communicate efforts about local disaster risk mitigation*
VIII	*Health-related topics*	*Health-related topics*
IX	*9. Relate soil's physical, chemical, and biological characteristics with its importance for sustainable living*	*9. Present the results of the investigation about soil's characteristics and their importance in life*
	10. Analyze the process and product of environmentally friendly technologies for sustainable living	*10. Present a simple project about the process and product of environmentally friendly technologies*

10.2.3 Local Sustainability Context and Science Education Research in Indonesia

Recent years have seen increased research on integrating local sustainability issues into science education. For instance, Surata and Vipriyanti (2018) investigated how Balinese students and teachers perceived an indigenous cultural landscape consisting of five rice terraces linked by a water temple and encompassing an area of 195,000 ha. The temple was the focal point of a sustainable water management system called Subak, constructed of canals and weirs. The participants had an overall positive attitude toward it despite their limited information and experiences.

Zidny and Eilks (2018) conducted an ethnographic study of indigenous knowledge in the Baduy community of Banten province, Indonesia. They discovered that the community employed scientific ideas and principles in their agricultural system in environmentally friendly ways. The authors believe that indigenous knowledge results in holistic learning, helping learners gain a global perspective and understanding of scientific information. This can help them understand the role of social or cultural backgrounds in developing scientific knowledge.

Hidayati et al. (2021) conducted a SWOT (strength, weakness, opportunity, threat) analysis in collaboration with multiple stakeholders (schools, teachers, and non-governmental organizations) to recommend strategies for conservation education in the primary school curriculum. Their examinations of the Indonesian national curriculum revealed that it is potential to enable students and teachers to engage in more locally contextualized learning. They also suggested that conservation educators should seize the opportunity to address the challenge of ineffective environmental education in Indonesia's formal education system.

Sumarni et al. (2022) investigated the use of Indonesian traditional medicines in an ethno-STEM (science, technology, engineering, mathematics) curriculum. They conducted field observations and interviews to understand the medicinal plants used as ingredients and the preparation method. By indigenizing the knowledge, their research implied a good science teaching and learning process, and new and meaningful learning experiences in the same. Additionally, they discovered that diverse indigenous cultures were remarkably effective at promoting and sustaining science education.

Subiantoro et al. (2021) described a case study of four biology teachers participating in a local context-based learning professional development program. Their findings indicated positive developments along four dimensions: knowledge about socio-scientific issues (SSIs) and SL, the necessity to include SSIs in science instructions, situational factors to address SSIs in class, and teachers' attitudes toward teaching SSIs. Further research is needed in Indonesian contexts to extend the SSI teacher professional development program to different regions.

Lastly, Nida et al. (2020) conducted a pertinent study on teachers. They elicited information about the experiences and attitudes of Indonesian science teachers toward local context-based learning utilizing SSIs. Their survey revealed that teachers were underconfident about their ability to teach content outside the official curriculum. Additionally, the result indicated the teachers' beliefs about the potential of sustainability-related topics (such as environmentally friendly technology) in the science curriculum.

While previous research has shown an increase in research on incorporating local sustainability issues into science education in Indonesia. They did not, however, provide any information on the specific pedagogical approaches or themes used by teachers when incorporating local sustainability contexts into everyday science lessons. The following section will introduce multiple approaches gathered through semi-structured interview of in-service science teachers in Indonesia.

10.3 Methodology

10.3.1 Participants' Recruitment Method

The following inclusion criteria were used to select potential participants: (1) taught science in public or private middle school in Indonesia between academic year of 2021–2022; (2) consented to participate in a recorded interview; and (3) had graduated from or be in graduate school in education. The participants were chosen through purposive sampling to ensure balance in representation across gender, age, and place of residence (rural, suburban, and urban within and outside Java Island). Data saturation (as explained in Bryman, 2016) was reached with the respondent size of 21 in-service science teachers. Table 10.5 displays the general demographics of teachers participating in the interview.

10.3.2 Data Collection, Processing, and Analysis Method

Table 10.6 shows the original interview guide used during data collection. The interview consisted of three sections examining the participants' awareness and views about the local sustainability context and how it may be extended beyond the

Table 10.5 Demographics of the participants ($N = 21$)

No.	Domicile	Alias	Gender	Years of service	School level	Educational level
1	Rural ($N = 8$)	R1	Male	10	Elementary	M.Ed ongoing
2		R2	Female	7	Elementary	M.Ed ongoing
3		R3	Female	3	Junior high	B.Ed
4		R4	Female	4	High	B.Ed
5		R5	Female	5	Junior high	M.Ed
6		R6	Male	14	Elementary	M.Ed ongoing
7		R7	Female	7	Junior high	B.Ed
8		R8	Male	8	Junior high	B.Ed
9	Suburban ($N = 7$)	S1	Male	10	Junior high	B.Ed
10		S2	Female	7	Junior high	B.Ed
11		S3	Female	10	Junior high	M.Ed
12		S4	Female	8	Junior high	B.Ed
13		S5	Male	26	Junior high	M.Ed ongoing
14		S6	Male	1	Junior high	B.Ed
15		S7	Female	4	Junior high	M.Ed ongoing
16	Urban ($N = 6$)	U1	Female	2	Junior high	B.Ed
17		U2	Male	7	Junior high	B.Ed
18		U3	Male	4	Junior high	B.Ed
19		U4	Female	7	Junior high	M.Ed
20		U5	Male	9	Junior high	B.Ed
21		U6	Male	10	Junior high	M.Ed

Table 10.6 Semi-structured interview guide

Section	No	Question topics	Question type
Awareness	1a	Understanding of the concept of sustainability theme	Yes/no
	1b	Application of the sustainability content	Open-ended
	1c	The sustainability content in science textbooks	Yes/no
	1d	The sustainability content in the science curriculum	Yes/no
View	2a	Introduction of sustainability content in the science curricula	Yes/no
	2b	Prioritizing between sustainability and other content	Yes/no
	2c	The current ratio of sustainability and other content	Yes/no
	2d	The ideal ratio of sustainability and other content	Yes/no
Beyond classroom	3a	Perception of fellow teachers of the sustainability content	Yes/no
	3b	The sustainability practices in schools	Open-ended
	3c	Training in education for sustainability	Yes/no

Table 10.7 Awareness of sustainability content in science curriculum

No	Questions	Yes	No
1a	*Can you define the term sustainability/sustainable development?*	8	13
1b	*Have you ever applied sustainability themes in your class?*	21	0
1c	*Have you ever read about sustainability-related content in science textbooks?*	3	18
1d	*Have you ever read about sustainability-related content in the science curriculum?*	2	19

science curricula. The 30-min interviews were conducted using a web-based conference application. Respondents answered all question in Indonesian, and the transcripts were translated into English for analysis. Responses to the yes/no questions were tabulated and processed descriptively, while the open-ended ones were analyzed for recurring themes.

10.4 Results

10.4.1 Awareness and Views on the Sustainability Content in the Science Curriculum

As shown in Tables 10.7 and 10.8, only half the participants knew the meaning of sustainability and that it was covered in the science curriculum. After the interviewer explained the concept of sustainability to the participants who were unfamiliar with it, all of the participants responded that they had implemented it inconsistently. Table 10.9 displays how participants expressed unanimous support for the inclusion of sustainability in the science education and were willing to

Table 10.8 Excerpt related with awareness of sustainability content

Summary	Detailed excerpts
Unaware of sustainability content in the curriculum	*"Is sustainable development, as the name implies, a long-term development? I'd never heard or read that term before, and I don't recall seeing it in a textbook."(S4)*
	"ESD, SDGs? Those are new terms to me, and I'm interested to learn what it means. Could you please explain what it is and how it relates to science education?" (U4)
Aware of sustainability content in the curriculum	*"I'm familiar with STEM because I've taken part in training sessions in the past, but I'd never heard of ESD until now. What are their differences?"* (U2)
	"I remember coming across one of the goals as 'eliminating world hunger' while scrolling the social media. It motivated me to include SDG in one of my class." (U1)
	"During my undergraduate years, The environment ministry at that time pioneered the initiative for sustainable development in Indonesia and lead its establishment in the legal norms and rules, which in turn incentivize environment education in Indonesia." (S5)
	"Some themes in the junior high school science curriculum directly mentions sustainability like 'green technology' in third grade."(S3)

Table 10.9 View about the sustainability content in the science curriculum

No	Questions	Yes	No
2a	*Do you agree with introducing sustainability content in science classes?*	21	0
2b	*Do you think that sustainability content is more important than other content?*	21	0

Table 10.10 Comparison of sustainability-related contents in the implemented science curriculum ($N = 21$)

	<50%	50%	>50%	Avg	Min	Max
2c current	9	4	8	49%	20%	80%
2d ideal	1	6	14	63%	40%	80%

prioritize them over other content. Finally, Tables 10.10 and 10.11 show how participants perceived the ideal ratio of sustainability and other scientific contents in the curricula and how it compared with their current practice in science education.

10.4.2 Approaches to Implement the Local Sustainability Context

In a question, the participants were asked to illustrate approaches to implementing sustainability content. Their answers were classified into three thematic approaches. The first approach was waste management activities (Table 10.12), such as making coconuts into eco-planters and transforming PET bottles into eco-bricks. The

Table 10.11 Excerpts of the views on the sustainability content

Summary	Detailed excerpts
Optimistic view on sustainability content in curriculum	*"Education for Sustainable Development (ESD) and similar ideas that seek science education to be more relevant with daily life are extremely important in this day and age, especially for 'generation-z' and generations after us millennials."* (S2)
	"The challenge at the implementation level is that not all students have the ability to follow such instructional model with a high context, which is a significant limitation." (R3)
	"Students show much more interest in the class when I include discussion about real-life topics rather than explaining the concept as it is presented." (R5)
	"Beyond understanding scientific concepts that may lead to a good test result, I believe that content that can foster character is more important, and I believe that implementing sustainability content has the potential to enable it." (R8)
Realistic view on sustainability content in curriculum	*"Organizing ESD to reach or exceed 50% of the science class is a difficult task for teachers; it necessitates curriculum reorganization at the national level."* (U6)

Table 10.12 Excerpt related to waste management activities

Summary	Detailed excerpts
DIY filtration kit	*"In a class about chemical separation, I gave student project to make a simple water filtration kit, I asked them to utilize readily available materials like PET bottle and coconut fiber. This activity increases my students' awareness on how polluted the nearby river was. I want to do more of such activity in the future."*(U4)
Turning plastic waste into eco-brick	*"I reminisced the moment when one of my students visited me and said, 'Thanks to you sir now I am continuing my study in environmental engineering. It is because you have introduced Ecobrick in one of our science class.' Ecobrick is a PET bottle packaged with clean and dry used plastic and made to a set density to crease reusable building blocks that sequester plastic."*(S5)
Turning organic waste into eco-planter	*"I did an activity with my student to reuse eggshell as an eco-planter. I want them to be familiarized with the practice of recycling. I wish they will have insight that help them to see waste as a resource that can be utilized."*(S7)
Upcycling activity	*"Me and my student did a project on creating lampshade from a recycled material. I believe that they were having fun, while learning about the concept of upcycling."*(R2)
Laundry waste	*"My students frequently told me that when they were helping their parents do laundry in the nearby river, they discovered that the water was polluted. It can be green, yellow, or even red at times. Talking about this topic in class allowed them to connect more easily with the lesson's content, resulting in higher grades."* (R2)

second approach (Table 10.13) was a nature study that included cases about field trips to a nearby lake and using social media for biodiversity surveys. The third approach (Table 10.14) was a cultural study about forest fires and animal sacrifices.

10.4.3 Extending Sustainability Beyond Classroom

Table 10.15 contains three questions about extending sustainability beyond classroom. Most participants believed that their peers would have a similar answer. Only two-thirds reported having an environmental program in their schools, while a third stated that they had received training or education in sustainability. When asked if they would be interested in getting trained and educated in sustainability, a majority responded affirmatively. Table 10.16 presents the excerpt of the teachers' responses explaining the whole-school activities that implement local sustainability such as gardening.

Table 10.13 Excerpts of the views on the nature study approach

Summary	Detailed excerpts
Field trip to lake	*"It takes 30 minutes by car to get into the lake, I usually allocate three hours for the field trip, I let them observe color, smell and other characteristics of water pollution. My students express their enthusiasm during this activity."*(U4)
Coastal observation	*"after taking several pictures of neighboring coast in a span of five years (2016 and 2021), I have found out that the sea level actually rises in a way that is different with regular tide patterns. I then showed those pictures to my students to increase their awareness about the global warming phenomenon."* (S5)
Biodiversity survey	*"Me and several other teachers from my school photograph and document flora inside our school. We then upload it to an online platform that aggregates data from other school around the world that were doing the same activity. I think it helped them to be more aware flora around them and can cultivate a spirit for conservation."* (S7)

Table 10.14 Excerpts of the views on the cultural study approach

Participants	Excerpts
Forest fire	*"I remember my student noticed me how they found forest around them went ablaze during their bus ride to school. In addition to affecting their physical health in form of acute respiratory infection it must also affect their mental health. It was a public secret that foreign company started the forest fire as their plantation never caught on fire. However, the event motivated local government to practice countermeasures against forest fire. They, even ban intentional burn that locals used to practice."*
Animal sacrifice	*"I discovered that one of my community's funeral traditions, which requires the family of a deceased noble tribesman to sacrifice an excessive amount of livestock (pigs, bulls, and chickens), is in direct conflict with the Sustainable Development Goals' goal of responsible consumption. As a result, I hope to witness the transformation of this unsustainable practice through my science class."*

Table 10.15 Responses to questions about the whole-school approach

No	Questions	Yes	No
3a	Do you think that your colleague will have a different opinion?	3	18
3b	Does your school have any environmental programs?	14	7
3c	Have you ever participated in a training/education program about ESD?	7	14

Table 10.16 Excerpt related to extending sustainability beyond classroom

Participants	Excerpts
R8	*"In our school we have a permaculture program called school kitchen garden* (Fig. 10.1). *Students not only helped planting the animal, they are also involved in selling the produce to local communities. Our school also utilized bioethanol as a source for electricity and process wastewater so it is usable again."*
U6	*"I invited architect of the new city to give insight about the new city development that supposedly implement sustainability practices. I found them to be enthusiastic with the activity. It motivated me to do a similar activity in the future."*
S4	*"My school participates in the Adiwiyata program* (Fig. 10.2) *by the environment. To earn the highest accolade of 'Adiwiyata-independent' a school need to fulfill requirement in the city/regency, provincial, and national level. Those requirements include among others: waste separation, resource management, composting and recycling activity. Adiwiyata-independent school will guide other school that aspires to be awarded with the Adiwiyata."*
S5	*"I once participated in a program called 'eco-camp', there we did a role playing in pairs: one act as a father and one act as the son. The father needs to answer his son's question on 'why you inherited me no nature to enjoy dad?' That question struck a chord with me. It made me reflect that it is our duty to protect this one earth for our generation after."*

Fig. 10.1 School's kitchen garden. (Courtesy of Bina Cita Utama School)

Fig. 10.2 Adiwiyata School. (Courtesy of SMP Lab school UPI Cibiru)

Table 10.17 Transforming local context into the sustainability discourse

Local context	Environment	Economy	Social
Forest fire (SDG 13, SDG 15)	*Climate change intensify drought that increase risk of forest fire, which itself releases CO_2, a major greenhouse gas*	*Forest burning is an easy and relatively cheap way to clear land into a palm oil plantation that gives 5% of the country's GDP*	*Indigenous communities had been practicing small-scale, controlled forest burning as a part of their culture*
Animal sacrifice (SDG 12, SDG 13)	*Cows raised as a livestock released methane, which also contribute to the global warming*	*Farming is an important part of the local economy, as well providing input to the leather industry*	*Traditional ceremonies that involve animal sacrifice can maintain a sense of community*
School garden (SDG 2, SDG 17)	*Vegetables require less resources than raising livestock while giving a comparable yield*	*Managing a garden can provide students with experience like managing their own business*	*Students can either sell or donate the vegetable that they have raised to the local community*

10.4.4 Local Contexts at the Intersection of Sustainability and Implemented Science Curriculum

In Indonesia, the local socioeconomic and political circumstances have resulted in ESD being comparable to environmental education in terms of its thematic breadth, which includes a variety of social concerns (Nomura, 2009). Consequently upon utilizing sustainability discourse into the implemented science curriculum, there is a need to simultaneously addresses environmental, economic, and social aspect of a particular issue. Table 10.17 exemplified representative three local contexts gathered from the interview and how they can be transformed into a sustainability discourse.

In many ways, the implementation of local sustainability context into science curriculum echoes a steadily growing interest in seeking to innovate education by beginning from the ground up in the community, an approach known as place-based education (PBE) (Santone, 2019). PBE shares an unambiguous connection to place "any locality imbued with meaning" differentiating it with other context-rich teaching and learning modalities (Semken et al., 2017). Supporters of Place-based education (PBE) argue that it synergizes well with the SDGs (Ito & Igano, 2020), as the latter provide a succinct conceptual framework for learners to understand local issues and expanding them into the global context (Demarest, 2020). Beyond the SDGs and ESD that present perspective of "sustainability as policy," PBE may also foster the development of "sustainability as a frame of mind" (Bonnett, 2002) by encouraging learners to emotionally reconnect themselves with their surrounding nature and environment (Ontong & Le Grange, 2014).

Advocates of PBE in science education argue that the lack of relevant local context associates with a low academic performance in elementary and secondary school level, as evidenced by multiple international metrics (Buxton & Provenzo, 2012; Stuckey et al., 2013). Thus, since PBE provides learners with local context to otherwise abstract and fragmented global (or even universal) scientific concepts (Coker, 2017), it can improve academic performance while increasing scientific literacy (Eijck, 2010). Moreover, several studies have found that PBE can improve social and personal competencies (responsibility, critical thinking, self-efficacy, and planning skills) (Cincera et al., 2019). Other related research mentioned how PBE is also useful in communicating complex, controversial, and difficult-to-teach concepts because it focuses on local and observable effects (Khadka et al., 2021).

The current national course of study in Indonesia, Curriculum 2013, aims to develop students' cognitive skills and spiritual, attitude, and psychomotor competencies through activity-based learning processes divided into themes. In the 2022 revision, which is currently in the prototype stage, the opportunities for sustainability concepts are much higher. One of the hallmarks of Indonesia's current and upcoming national curriculum is that schools are essential to their communities. As a result, it allows students to learn from their communities and apply what they have learned to help the community. Overall, it opened the door for implementing context-rich and value-laden pedagogical approaches, including ESD and PBE in science and other subjects.

10.5 Conclusion

The purpose of this chapter is to present a literature review and case studies on how a local context can help bridge the gap between sustainability and science curriculum implementation. Semi-structured interviews with participating in-service Indonesian science teachers ($N = 21$) were used to collect the cases. The majority of participants had never heard of the term "sustainability." However, after receiving an explanation of the concepts, they expressed optimism for their incorporation into the implemented science curriculum. Three approaches were discovered when participants' strategies for integrating sustainability into science curricula were examined. First, waste management activities were visible in excerpts about coconut fiber for water filtration and eco-bricks made from plastic trash. A field trip to a nearby lake and a biodiversity survey of the schoolyard were used to demonstrate the second approach, nature study. During the discussion of indigenous practices, the third approach, cultural study, is illustrated in two instances: the tradition of forest burning and animal sacrifices. The interview also included examples of how the school garden could be used to extend sustainability outside of the science classroom.

Acknowledgments I would like to express my gratitude to my primary supervisor, Hiroki Fujii, who guided me throughout this process. I would also like to show my deep appreciation for teachers who had concerned to take part in in this study as a respondent.

References

Aikenhead, G. S. (2006). *Science education for everyday life: Evidence-based practice*. Teachers College Press.

Bonnett, M. (2002). Education for sustainability as a frame of mind. *Environmental Education Research, 8*(1), 9–20.

Bryman, A. (2016). *Social research methods* (5th ed.). Oxford University Press.

Buxton, C. A., & Provenzo, E. F. (2012). *Place-based science teaching and learning*. SAGE Publications.

Cincera, J., Valesova, B., Krepelkova, S., Simonova, P., & Kroufek, R. (2019). Place-based education from three perspectives. *Environmental Education Research, 25*(10), 1510–1523.

Coker, J. S. (2017). Pedagogy and place in science education. In D. Shannon & J. Galle (Eds.), *Interdisciplinary approaches to pedagogy and place-based education* (pp. 71–83). Palgrave Macmillan.

Demarest, A. B. (2020). Learning locally, understanding globally. *Middle Grades Review, 6*(3), n3.

Eijck, M. V. (2010). Place-based (science) education: Something is happening here. In D. J. Tippins, M. P. Mueller, M. Eijck, & J. D. Adams (Eds.), *Cultural studies and environmentalism: The confluence of ecojustice, place-based (science) education, and indigenous knowledge systems* (pp. 187–191). Springer.

Faisal, & Martin, S. N. (2019). Science education in Indonesia: Past, present, and future. *Asia-Pacific Science Education, 5*(1), 1–29.

Gericke, N., Boeve-de Pauw, J., Berglund, T., & Olsson, D. (2018). The sustainability consciousness questionnaire: The theoretical development and empirical validation of an evaluation instrument for stakeholders working with sustainable development. *Sustainable Development, 27*(1), 35–49.

Hadi, S., & Novaliyosi, N. (2019). TIMSS Indonesia (trends in international mathematics and science study). In *Prosiding seminar nasional*. http://jurnal.unsil.ac.id/index.php/sncp/article/view/1096

Hidayati, E., Agustin, I. Y., Karisma, Y. A., Djola, W. I., Latifah, S., & Valentino, N. (2021). Place-based pedagogy and Indonesian national curriculum: A strategy analysis for forest and Javan Lutung conservation in the Bromo-Tengger-Semeru national park. *Proceedings of the 2nd annual conference on education on social science (ACCESS 2020)*. https://atlantis-press.com/proceeding.access-20/125956968

Ito, H., & Igano, C. (2020). Place-based environmental education to promote eco-initiatives: The case of Yokohama, Japan. *Regional Studies, Regional Science, 7*(1), 292–308.

Khadka, A., Li, C. J., Stanis, S. W., & Morgan, M. (2021). Unpacking the power of place-based education in climate change communication. *Applied Environmental Education & Communication, 20*(1), 77–91.

Kridel, C. (Ed.). (2010). *Encyclopedia of curriculum studies* (Vol. 1). Sage Publication.

Mullis, I. V., & Martin, M. O. (2017). *TIMSS 2019 assessment frameworks*. International Association for the Evaluation of Educational Achievement.

Nida, S., Rahayu, S., & Eilks, I. (2020). A survey of Indonesian science teachers' experience and perceptions toward socio-scientific issues-based science education. *Education Sciences, 10*(2), 39.

Nomura, K. (2009). A perspective on education for sustainable development: Historical development of environmental education in Indonesia. *International Journal of Educational Development, 29*(6), 621–627.

OECD. (2019). PISA 2018 science framework. In OECD. (Ed.), *PISA 2018 assessment and analytical framework*. OECD Publishing.

Ontong, K., & Le Grange, L. (2014). The role of place-based education in developing sustainability as a frame of mind. *Southern African Journal of Environmental Education, 27*(1), 27–38.

Roberts, D. A. (2007). Scientific literacy/science literacy. In S. K. Abell & N. G. Lederman (Eds.), *Handbook of research on science education* (pp. 729–780). Lawrence Erlbaum.

Santone, S. (2019). Reclaiming education as a public good: Place-based teacher education. *Childhood Education, 95*(5), 66–69.

Semken, S., Ward, E. G., Moosavi, S., & Chinn, P. W. (2017). Place-based education in geoscience: Theory, research, practice, and assessment. *Journal of Geoscience Education, 65*(4), 542–562.

Sjöström, J., & Eilks, I. (2018). Reconsidering different visions of scientific literacy and science education based on the concept of Bildung. In Y. J. Dori, Z. R. Mevarech, & D. R. Baker (Eds.), *Cognition, metacognition, and culture in STEM education: Learning, teaching and assessment* (pp. 65–88). Springer.

Stuckey, M., Hofstein, A., Mamlok-Naaman, R., & Eilks, I. (2013). The meaning of 'relevance' in science education and its implications for the science curriculum. *Studies in Science Education, 49*(1), 1–34.

Subiantoro, A. W., Treagust, D., & Tang, K. S. (2021). Indonesian biology teachers' perceptions about socio-scientific issue-based biology instruction. *Asia-Pacific Science Education, 1*(aop), 1–25.

Sumarni, W., Sudarmin, S., Sumarti, S. S., & Kadarwati, S. (2022). Indigenous knowledge of Indonesian traditional medicines in science teaching and learning using a science–technology–engineering–mathematics (STEM) approach. *Cultural Studies of Science Education, 17*, 467–510.

Surata, S. P. K., & Vipriyanti, N. U. (2018). The subak cultural landscape as environmental education: Knowledge, attitudes, and experiences of Balinese teachers, student teachers, and students. *The Journal of Environmental Education, 49*(1), 59–70.

UNESCO. (2014). *UNESCO roadmap for implementing the Global Action Programme on Education for Sustainable Development*. https://unesdoc.unesco.org/ark:/48223/pf0000230514

UNESCO. (2017). *Education for sustainable development goals: Learning objectives*. UNESCO.

Valverde, G. A., Bianchi, L. J., Wolfe, R. G., Schmidt, W. H., & Houang, R. T. (2002). *According to the book: Using TIMSS to investigate the translation of policy into practice through the world of textbooks*. Kluwer.

Vaughter, P., & Tabucanon, M. (2020). Introduction. In UNESCO (Ed.), *Multi-stakeholder approaches to education for sustainable development in local communities: Towards achieving the SDGs in Asia* (pp. 1–8). UNESCO.

Zidny, R., & Eilks, I. (2018). Indigenous knowledge as a socio-cultural context of science to promote transformative education for sustainable development: A case study on the Baduy community (Indonesia). In I. Eilks, S. Markic, & B. Ralle (Eds.), *Building bridges across disciplines for transformative education and a sustainable future* (pp. 249–255). Shaker.

Chapter 11
Curriculum Development of Japanese Elementary School Science Regarding Biodiversity Conservation Focusing on Alien Species

Toru Doi and K. F. Ardh

11.1 Introduction

Humans benefit from biodiversity because it provides various ecosystem services, including oxygen production; food, fuel, and lumber supply; water retention and purification in forests; and the creation of visually appealing landscapes (Millennium Ecosystem Assessment, 2005). In other words, modern society must conserve biodiversity as it is critical to its existence. However, unsustainable practices, like indiscriminate hunting and overfishing, have put biodiversity in an ongoing crisis.

One of the most critical biodiversity crises is the invasion of alien species (Ministry of Environment, Japan, 2012). Individuals who were uneducated about their potential harm imported them from abroad. People often act after the invasive alien species wreak havoc on the native ecosystem, which in most cases is too late. Education has the potential to transform thinking and behavior and is part of the solution to this crisis (Secretariat of the Convention on Biological Diversity, 2010).

There are numerous studies teaching students about this topic, mostly focusing on individual causes of the biodiversity crises disregarding future consequences of the present action. However, there are a few practical studies about teaching invasive species (e.g., Umezu et al., 2017), especially in Japanese elementary school settings. This chapter discusses a study that developed elementary school curricula to fill a research gap on biodiversity conservation and invasive alien species. The following research questions determined the structure of this chapter.

T. Doi (✉)
Yasuda Women's University, Hiroshima, Japan
e-mail: doi-t@yasuda-u.ac.jp

K. F. Ardh
Okayama University, Okayama, Japan
e-mail: Aldi@okayama-u.ac.jp

H. Fujii, S.-K. Lee (eds.), *Science Education for Sustainable Development in Asia*, Education Innovation Series, https://doi.org/10.1007/978-981-99-8711-5_11

1. *What are the requirements to develop a curriculum about biodiversity conservation and alien species?*
2. *How do students perceive alien species before and after the intervention lesson?*
3. *How does the developed curriculum foster students' abilities to predict future consequences of present actions to deal with alien species?*

11.2 Literature Review

11.2.1 Biodiversity Conservation in Science Education

Lindemann-Matthies (2002) utilized an active environmental education program titled "Nature on the Way to School," developed by Pro Natura, an environmental organization in Switzerland. The study involved over 4000 Swiss students aged 8 to 16 from 248 classes. Students chose their favorite animals and plants from the species they saw to and from school and then explained their choices to their peers or parents. The program demonstrated that observing local wildlife had long-lasting effects on positive behavior toward biodiversity conservation, regardless of the participant's age and gender.

Framer et al. (2007) simulated a field trip to the Great Smoky Mountains National Park. The participants were fourth graders, and the main activities were hiking a short section of the Appalachian Trail and completing an interactive learning experience using straws, cups, and thumbtacks to understand the effect of the balsam woolly adelgid insect (*Adelges piceae*), an invasive alien species on the Frasier fir (*Abies fraseri*), a native plant species. Postactivity interviews revealed that the activities had increased students' knowledge about biodiversity conservation and environmentally friendly behaviors.

Alexandar and Gopalsamy (2011) developed modules about biodiversity conservation for high school students in India. Each module required students to triangulate data, conduct analyses, and write a comprehensive topical paper. Per the follow-up survey results, active learning participatory methods helped students learn about their local biodiversity in a way that was relevant to their lives. Simultaneously, it promoted students' cognitive, affective, and behavioral development.

Schönfelder and Bogner (2017) evaluated the effectiveness of two methods for teaching environmental knowledge about bee conservation to middle school students in Germany. The first method involved direct observation of the beehive, whereas the second involved indirectly observing through an e-learning tool. Their findings revealed that both methods produced comparable results even 6–9 weeks after the program ended. An e-learning tool could be an excellent substitute for potentially dangerous subjects, such as beehives.

In summary, the previous studies reveal three critical steps to develop the content and goals of a biodiversity conservation curriculum. To begin, students must be familiar with the biodiversity of their immediate environment. Following that, they must comprehend the importance of biodiversity and how human activities may impact it. Finally, learners should gain the knowledge and behaviors necessary for conservation. Teachers need to adopt appropriate learning methods based on the students' profiles and the nature of the subject.

11.2.2 Alien Species in Science Education

Schreck et al. (2013) studied a museum workshop about invasive alien species for teenage students (13–15 years) in Portugal. In the workshop, students learned to identify invasive plants, competition between native and invasive plants, and ways to control invasive plants. Results of the follow-up survey a year later demonstrated how learners could retain the knowledge.

Waliczek et al. (2018) investigated the impact of curriculum models on college students' knowledge and attitudes toward invasive species in Texas. The curriculum included lectures using electronic presentations and laboratory activities, such as case studies, visual aids, and scavenger hunts. Comparing pre- and posttest results revealed that students receiving both the lecture and the laboratory activity performed better than those receiving either or neither.

Skukan et al. (2020) developed a game-based curriculum about marine invasive alien species for students (11–12 years old) in Spain. The curriculum contained a lecture, card game, and three-month field activity to search for invasive seaweed species on a nearby beach. Through the activities, students identified four invasive alien seaweed species. The questionnaire survey results revealed that students' knowledge and their abilities to make appropriate decisions regarding alien species had improved.

In summary, the studies reveal four requirements for the alien species curriculum. Firstly, students should interact with familiar and unfamiliar organisms. Secondly, they should have a firm grasp on the origins of alien species and the impacts they bring. Thirdly, they must evaluate the best course of action and base their decisions on it. Fourthly, the developed curriculum must reflect the context of each country.

11.2.3 Japanese Context in Biodiversity Conservation and Alien Species

Spread of Alien Species in Japan and Education

Japan has preserved its native biodiversity by experiencing little species migration from other regions into Japan due to its geographical characteristics. After the 1800s, the Japanese native ecosystem saw an influx of alien species. Among them, the small Indian mongoose (*Herpestes auropunctatus*) and the largemouth bass (*Micropterus salmoides*) are two of the most well-known invasive alien species in Japan (Fig. 11.1).

Other well-known alien species that turned invasive in Japan include lance-leaf tickseed (*Coreopsis lanceolata*) and Senegal tea plant (*Gymnocoronis spilanthoides*) for ornamental purposes and raccoons (*Procyon lotor*) and red-eared sliders (*Trachemys scripta*) as pets. American crayfish (*Procambarus clarkii*) as a food source. Recent incidences involving alien species, such as the redback spider (*Latrodectus hasselti*), Asian hornet (*Vespa mandarinia*), and fire ant (*Solenopsis* sp.) hurting humans brought the public's attention to this issue.

Many nonformal educational sources, such as museum exhibitions, workshops, and Internet articles, portray them negatively. However, some TV shows present alien species as both an "enemy" and an "ally," like in the case of largemouth bass. Moreover, not all alien species are harmful. Some are beneficial and integral to the Japanese lifestyle. As a result, alien species are frequently mentioned in elementary school textbooks, making their comprehension critical (Doi & Hayashi, 2015).

In schools, students and teachers tend to release *Egeria densa* into the river rather than discard it into the trash when it grows uncontrollably. *Egeria densa* is an invasive aquatic plant commonly used as an experiment material. Explaining the reason behind this action requires an understanding of the Japanese indigenous view of nature, life, and death. The following section summarizes the Japanese indigenous view of the same using the literature on the topic (e.g., Terada, 1939; Nishida, 1972; Ogawa, 1998; Tachikawa, 1998; Watanabe, 2003; Shimazono, 2009; Meli, 2017).

Japanese Indigenous View of Nature

Torahiko Terada (1939) was one of modern Japan's first great scientists who was also known for his literary works. He wrote, "nature is both a loving mother and a stern father. Submitting to our stern father and fawning on our loving mother are both necessary to guarantee our well-being" (Meli, 2017). This parent-child-like view of the relationship with nature resembles Takeo Doi's (1971) idea of *Amae no*

Fig. 11.1 Two of the most well-known invasive alien species in Japan (Left: The small Indian mongoose, initially introduced to exterminate harmful local species. Right: Largemouth bass, imported as a gamefish). (Courtesy of the Japanese Ministry of Environment)

kozo, translated as *the anatomy of dependence*. This unique view of the Japanese that sees nature as indifferent to humans contrasts the Western objective, structural, mathematical, and confrontational view of nature (Watanabe, 2003).

Ogawa (1998) discussed how the Japanese viewed nature during ancient times. People of the Jomon era (13,000–3000 years ago) believed that all things (animals, insects, birds, plants, natural phenomena, and weather) have their spirit (animism). People of the Yayoi era (3000–1700 years ago) viewed humans as an integral part of nature. This proenvironmental belief continues to affect the Japanese views in modern times.

For instance, before consuming a meal, the Japanese usually say "Itadakimasu," meaning "I humbly receive." It holds a deeper nuance than the translated verse. It expresses gratitude for living organisms that gave their lives to become part of the meal. Across Japan, *Mushidzuka* (mound of insects) (Taguchi et al., 2011) was built to respect the life of insects that were exterminated for agriculture, silkworms that were used for threading, and fishes and shellfishes that were sacrificed for food.

Tachikawa (1998) explained that the Japanese words *Kachofugetsu* and *Sansensomoku* symbolize their harmonious relationship with nature. The former translates to "flower, bird, wind, and moon," referring to the natural splendor of the world. The latter refers to all aspects of nature close to humans, translating to "mountain, river, grass, and trees." This sense of belonging has accustomed the Japanese to accept death as a natural part of the life cycle.

Shimazono (2009) added that the "Obon Festival" exemplifies the Japanese indigenous belief in the afterlife. During the festival, families traditionally sent their ancestors' spirits back to their permanent dwelling place under the guidance of fire in a ritual known as *Okuribi* (sending fire). Fire also marked the commencement (Mukaebi) and closing of the festival. The festival implied that life and death are so close that fire lit in the land of living can be seen by those in the afterlife.

We may conclude with the concept of "Mujokan" (Nishida, 1972). *Mujokan*, which translates as "impermanence," refers to the constant change in all things and phenomena, including natural disasters. In this context, it is intuitive for the Japanese to let "nature take its course" by releasing alien species they cannot cultivate or keep as pets into the wild rather than "disposing them properly."

11.3 Methods

11.3.1 Research Participants

This study recruited sixth-grade students ($N = 262$) aged 11–12 years old from four public elementary schools in the Western regions of Japan. Purposive sampling was administered for two reasons: (1) they study food chains involving alien species and the impact of these species on human life and the natural environment, and (2) other grades in the elementary school do not include lessons on biodiversity. Intervention lessons and data collection were conducted between March and June of 2019.

11.3.2 Instruments

A scaled response questionnaire was administered in both pre- and posttest (Box 11.1). The questionnaire was original to this study. Questions 1–5 are identical in the pre- and posttest. They assessed the interests, experiences, knowledge, and perception of students regarding alien species. Questions 6–7 were included posttest only and aimed to assess whether students desired additional information about alien species and understood appropriate actions in dealing with alien species following the intervention lessons.

Box 11.1: Questionnaire About Alien Species

Question 1. Are you interested in alien species? Please choose the closest option.

Interested   A little interest   Not much interest   No interest

Question 2. Have you ever seen alien species? Choose the one that is close to your idea.

Every day   Often   Not so much   Never seen   Unsure

Question 3. If you have seen alien species, please write where you have seen it (river, TV program, book, etc.; If you remember the title, please write it.) (Multiple answers)

Question 4. Do alien species inhabit your area? Choose the one that is close to your idea.

Strongly agree   Disagree   Strongly disagree   Unsure

Question 5. What do you ever think about alien species? Choose the one that is close to your idea.

Good   Not good   I can't say either   I am not decided

Please write the reason.

Question 6. Do you think you can do something to prevent the problem of alien species from worsening?

Please circle the one that is close to your idea.

Agree*   Disagree   Unsure

*For those who agree, what steps do you think you could take?

Question 7. What do you think about the current lessons, and what more would you want to learn?

Please write freely.

11.3.3 Data Analysis

The responses to the questionnaire were analyzed using the SCAT (steps for coding and theorization) method (Otani, 2008). This method is beneficial in small-scale, one-time data collection surveys like the one conducted in this study. Furthermore, it can be asserted that the possibility of falsifiability by non-analysts is ensured, thereby avoiding the risk that the analysis becomes arbitrarily determined.

11.3.4 The Intervention Lesson

Aim of the Lessons

The lessons were developed to facilitate students' understanding of alien species in a multifaceted, comprehensive, and appropriate manner. Furthermore, the lesson aimed to help students increase their ability to predict the future consequences of the present actions and deal appropriately with alien species.

Content of the Lessons

The lesson strategy involved (1) factoring compliance with the Japanese national curriculum, limitation of the lesson time (two consecutive 45-min classes totaling 90 min), and the previous study (Schönfelder & Bogner, 2017) that indicated how an indirect experience was equally effective as the direct experience of the lecture. Figure 11.2 illustrates the flow of the lessons, while Table 11.1 lists alien species that were included in the lesson.

11.4 Results and Discussion

11.4.1 Transcript of the Intervention Lessons

Box 11.2 provides a summarized transcription of the dialogues between teacher "T" and students "S" during the intervention lessons. Italicized sentences refer to a direct reference.

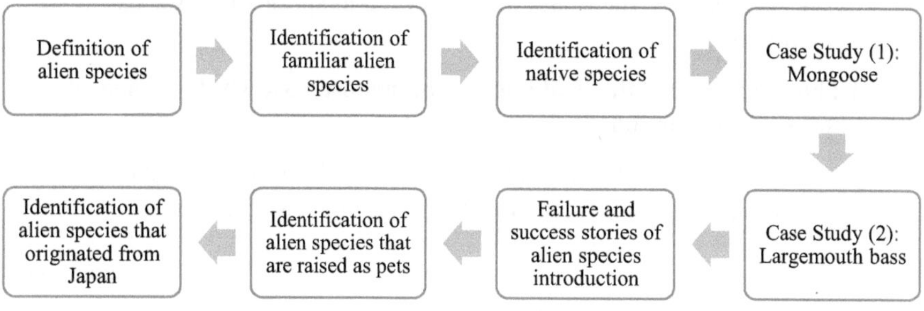

Fig. 11.2 Flow of the intervention lesson

Table 11.1 Alien species introduced during the lessons

Type of alien species	Example
Introduced 10,000–13,000 years ago	Cucumber, eggplant, radish, rice, morning glory, cat, dog, pigeon, sparrow, cabbage, white butterfly
Introduced 400–150 years ago	Onions, potatoes, carrots, sweet potatoes, spinach, white clover, and daisy fleabane
Introduced 150 years ago–present time	Chinese cabbages, strawberries, peppers, king trumpet mushroom, pill bugs, and common field-speedwell
Alien species that became invasive	Small Indian mongoose, largemouth bass, crayfish, coreopsis, water hyacinth
Alien species that are kept as pets	Racoon, red-eared slider, ferret
Native Japanese species that turned invasive in other countries	Japanese knotweed and kudzu

Box 11.2: Dialogue During Intervention Lesson

(1) Teacher explained the definition of alien species

T: "Alien species are animal and plants who were displaced both intentionally or unintentionally from its' original habitat to other location."

(2) Students identified the alien species from familiar plants and animals

T: Showed pictures of plants and animals that were introduced in Japan in chronological order: (1) 10,000–1300 years ago (cucumber, eggplant, radish, rice, morning glory, cat, dog, pigeon, sparrow, and cabbage white butterfly); (2) 400–150 years ago (onions, potatoes, carrots, sweet potatoes, spinach, white clover, and daisy fleabane); and (3) 150 years ago–present time (Chinese cabbage, strawberries, peppers, cabbage, king trumpet mushrooms, pill bugs, and common field-speedwell).
S: Exclaimed "Oh! No Way!" in surprise that plants and animals that they know are part of their daily lives are alien species, especially pill bugs and common field-speedwell introduced recently.

(3) Teacher gave examples of native plants and animals

T: Showed pictures of saw-tooth stag beetle, rhinoceros beetle, butterbur, and horseradish.
S: Asked, "Are there any native vegetable species that we eat in our daily life?"

(4) The teacher explained about the mongoose case study (failed alien species introduction)

T: "There is a highly venomous snake in Okinawa called Habu."
"It has been causing many incidents since the last 100 years."

(continued)

Box 11.2 (continued)

"Around that time, a prominent Japanese zoologist proposed the idea of introducing mongoose from India, after seeing the animal won a battle with Cobra snake in a show, then in 1910 17 mongooses were introduced to Okinawa."

"But even after it, the number of Habu snake did not decrease."

T: After narrating the story, the teacher asked the students to think individually for 3 min to "predict about the future of this situation" and then share their thoughts with the group.

T: After the group sharing, the teacher continued the story, saying "unfortunately most of the mongoose did not attack the snake."

S: The students laughed when the teacher explained that "mongoose was diurnal while the Habu is nocturnal. This means that in the first place, the opportunity for both of them to face each other was rare."

T: The teacher continued to explain that "since predating on the venomous Habu will be too risky, the mongooses chose to predate on weak prey including rare native species such as the flightless Okinawa rail, and Okinawa tree lizard."

S: Then many expressed their understanding of the phenomenon.

T: The teacher continued to explain, "not until 1980, 70 years after the mongooses were introduced, people start to notice that they were eating rare species instead of the venomous Habu."

S: Many expressed surprise at the explanation.

T: Showed a graph that illustrated that the "mongoose population had grown from 17 originally introduced to an estimated 30,000 in 1980 70 years later. But after finding out that the mongoose had radically reduced the native species population, it started to become a big deal, and the government set a plan to exterminate the mongoose population, resulting it to reduce into 1200 in 2001, 1000 in 2006 and 200 in 2010." the teacher then asked, "is it okay if the mongoose population is reduced like this?"

S + T: Discussed: "Since we know that the population can grow from 17 to 30,000, if we stop after the population reached 200, there are still a possibility that the population can increase again."

T: The teacher questioned the students: "What should have been done to prevent this from happening?" after explaining that every year 300 million yen was spent for the extermination.

S: Answered: "They should have study them properly before introduced the mongoose at the first place."

(continued)

Box 11.2 (continued)

(5) The teacher explained about the largemouth bass case (failed alien species introduction)

T: "I will tell you another failed example of alien species introduction."
"In 1925 after business trip to the USA, a businessman thought that it will be idea to bring largemouth bass because it is interesting to catch and delicious to eat" (proceeded to show a video about fishing). "Initially the fish only introduced to Lake Ashi, but now it is distributed all over Japan including the islands." (Proceeded to ask a question) "how can it be possible when the fishes can only swim in the water?"
S: (Many answered) "They must have been carried around by people who liked to fish!"
T: Showed video of the largemouth bass predating on small fishes and waterfowl.
S: (Surprised) "They even try to swallow something as big as that!"
T: Explained a graph that showed the population change of fishes in a swamp. "Just two years after the introduction of largemouth bass, harvested population of tamoroko and motsugo, small fishes with a total length of 5 cm, dropped from 10,000 tons to 0 tons. Five years after, harvested population of bigger fish such as carp dropped from 7000 tons to 3000 tons."
S + T: Continued to explain: "After the phenomenon was found to be a big deal, the country worked together and manage to reduce its population temporarily." after being asked for ideas to solve this issue, some students mentioned: "Do the same thing as the mongoose again" (with a grin).
T: Explained: "Largemouth bass is rare type of fish, because the male parent protects young fishes from hatching until it grows big enough, ensuring them to reach adulthood. Moreover, since catch-and-release is a standard in largemouth bass fishing, reducing its' population through fishing is not an option." "in USA, beaver and alligator was introduced to naturally reduce the largemouth bass population," and then proceeded to ask the students: "What if we do the same in Japan?"
S: Responded: "Absolutely no! Remember what happened with the mongoose!?"

(6) Students compared failure and success stories of alien species introduction

T: Explained another case of alien species introduction thought to bring "something good" but turned out to be the opposite (including raccoon, red-eared slider, crayfish, coreopsis, water hyacinth). The teacher contrasted them with alien species (onion, rice, radish, green pepper, cab-

(continued)

Box 11.2 (continued)

bage, milk cow) successfully introduced and benefitting human life. The teacher then proceeded to ask: "What are their differences?"
S: "It is okay when human is raising them," "but it will be difficult when human stop taking care of them."
T: Followed up with another question: "What should have been done to prevent the failed cases?"
S: "We need to think carefully before importing/introducing them!"

(7) The teacher explained about alien species raised as pets

T: Proceeded to talk about red-eared sliders that are usually sold during festivals and pet shops when they are small and around 3 cm in size and then explained: "Their lifespan is 20–30 years, and they can reach up to 30 cm in size, in which they color will change from beautiful green to blackish tone."
S + T: Discussed ferrets, another popular animal species raised as pets. Their behavior, size, diet, and excrements were discussed.
S + T: Discussed *Egeria densa*, a popular waterweed and an alien species that is usually used in medaka fish experiments.
T: Instructed students to think about the proper way of dealing with the alien species and asked them to write their ideas on the answer sheet.

(8) The teacher explained about invasive alien species that originated from Japan

T: Showed a picture of how Japanese knotweed and kudzu were introduced intentionally and were becoming invasive in other countries.
S: "I thought Japan was the victim here, but actually it is also a perpetrator!"

11.4.2 Pre- and Post-intervention Questionnaires

Prior to the intervention (Figs. 11.3, 11.4, and 11.5), students were disinterested in alien species, unaware that they existed about them, and viewed them as a distant existence. After the intervention lesson, 87% of the students answered that they were a little interested in alien species (Fig. 11.3). Sixty-five percent responded that they saw them often or every day (Fig. 11.4), and 60% agreed or strongly agreed that they lived around them (Fig. 11.5). The χ^2 difference result ($P = 0.01$) indicated a significant increase in students' knowledge and understanding of the topic because of the intervention lessons.

In response to the question in Table 11.2 about where students had seen alien species, the students elaborated on the preperception that alien species lived far

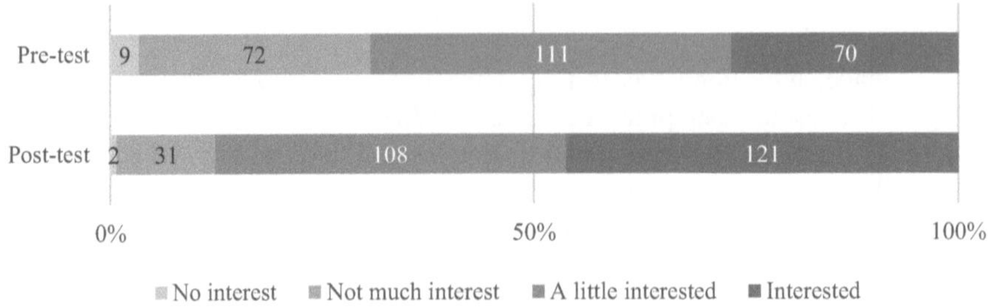

Fig. 11.3 Are you interested in alien species? ($N = 262$)

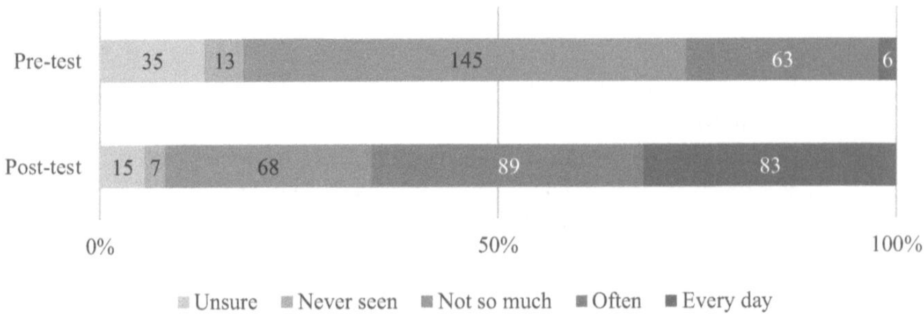

Fig. 11.4 Have you ever seen an alien species? ($N = 262$)

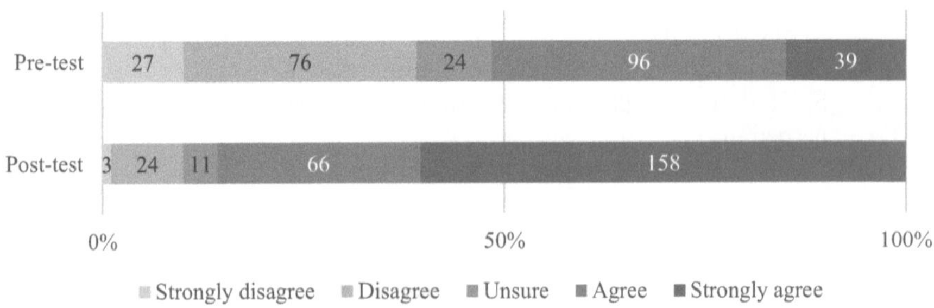

Fig. 11.5 Do alien species live near you? ($N = 262$)

Table 11.2 Where have you seen alien species?

	No answer	Others	Museum	School, nursery school	Super market	Dining table	Books	Home, friend, relative house	Out door	TV
Pre	31	6	13	8	1	0	40	6	56	175
Post	23	12	11	24	28	29	35	43	88	160

away from them and could only be seen on television shows. However, following the intervention, there was a statistically significant increase in the number of students who responded that the alien species were found in "home," "books," "dining table," "supermarket," and "school." Their responses demonstrated the effectiveness of the intervention lessons.

Figure 11.6 reveals that before the intervention lessons, students' preperceptions of alien species were negatively skewed. This could be attributed to the constant portrayal of alien species as "villains" in two popular TV shows. In the show "Emergency SOS! Drain the Pond!" entertainers caught alien species after draining the pond and filling it with clean water. In another show, "Grill Trouble," entertainers caught and cooked invasive alien species.

A deep dive into the reasons for their answers (Table 11.3) revealed that students who thought positively about alien species mentioned: "I thought alien species are very funny when I saw them on TV," "The existence of alien species allows us to know many things," "they are beneficial to human," and "human is the problem and not the alien species themselves," Those who viewed alien species negatively reasoned, "All native species connect with each other under their food chain, but alien species don't connect with native ecosystem. Therefore, the number of them increases rapidly, and their food (availability) decreases. It makes bad influences on the native food chain."

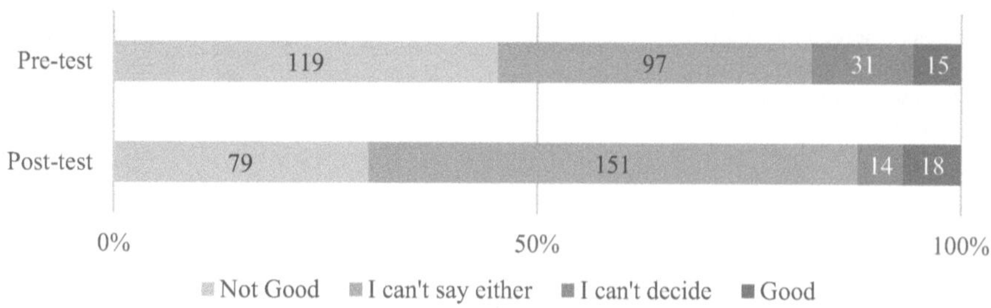

Fig. 11.6 What do you think about alien species? ($N = 262$)

Table 11.3 Why do you think so? ($N = 262$)

Answer	Pretest	Posttest
They are harmful to native species	86	65
They are both harmful and beneficial	54	127
I cannot answer	45	14
They are harmful to human	22	19
They are beneficial to human	18	12
Humans are the problem not them	16	20
No answer	11	3
They are from foreign countries	10	2

Students who answered "I can't say either" to the question on whether they were "good" or "bad" before the intervention mentioned, "There is a possibility of harm by an alien (e.g., red imported fire ant). However, there are also good things (such as dogs) that people feel happy or cute. Then, I cannot say either." After the intervention, there was an increase in the number of students who answered neutrally, as is apparent in the statement, "Alien species can be both harmful and beneficial."

Tables 11.4 and 11.5 show responses to post-test-only questions. The students who believed that they could help solve the alien species problem outnumber those who could not or did not want (Table 11.4). Among students' ideas for dealing with invasive species (Table 11.5), predicting the consequences of the current action related to alien species came out on top. Another response was to educate their friends and family about the dangers of releasing alien species into the wild.

In light of the findings presented above, the intervention lessons successfully met the learning objectives. First, the students thoroughly understood alien species from a multidimensional perspective. Second, students were able to predict the future consequences of the present actions. Third, students were able to establish appropriate relationships with alien species.

Following the intervention lesson, half of the students desired to learn more information about alien species (Table 11.6), which can be classified into three categories. The first category is concrete information, such as the name, population, and ecology of alien species; their introduction and establishment in Japan; and

Table 11.4 Do you think you can do something so that the problem of alien species will not worsen further? ($N = 262$) (Posttest only)

No	I cannot answer	Yes
5	71	186

Table 11.5 What do you think you can do so that the problem of alien species will not worsen further? ($N = 191$) (Posttest only)

Group	(*N*)	Typical text data
Don't release	95	*When keeping alien species as pet, we need to be responsible throughout their lifespan and should not let them escape to the wild*
Anticipate	41	*Before we deciding to keep alien species as pet, we need to think about what we should do when we cannot keep them anymore*
Exterminate	10	*I want to participate in exterminating alien species*
Don't bring in	9	*I think it is important not to bring alien species from other country when it is not necessary*
Don't breed/keep as a pet	9	*We should not keep alien species as pet*
Don't spread	8	*For the largemouth bass we should return it to the original place (where it was intended to be introduced)*
Inform others	8	*After knowing about the danger of alien species, I will inform my family and friend about it*
Learn/appreciate more	6	*If we learn more about alien species, I think the problem will not becoming worse*

Table 11.6 What students want to know more about alien species ($N = 143$) ※No description (119)

(Concrete information on alien species)		(Influence by alien species)	
Alien species' name and population	52	Merits	17
Details of transfer	15	Demerits	5
General knowledge	14	(Measures against alien species)	
Alien species from Japan to places abroad	7	Distinction of dangerous species	7
How to identify alien species	7	After transfer	6
The ecology of alien species	4	The water's edge	3
The current state of alien species	2	(Nothing special)	4

other general knowledge. The second category concerned the merits and drawbacks of alien species and their influence. The third category is to take precautions against alien species invading the native ecosystem.

11.4.3 Biodiversity Conservation, Alien Species, and Sustainability

The developed curriculum exemplifies that the implementation of goal 4, "quality education," may facilitate the achievement of goal 15, "life on land," in the Sustainability Development Goals (SDGs) (UN, 2015). The curriculum especially contributed to targets 15.4, "conservation"; 15.5, "prevent the extinction of threatened species" (e.g., the Okinawa rail and Okinawa tree lizard in the mongoose case); and 15.9, "integrate ecosystem and biodiversity values" (e.g., how to properly deal with an invasion by alien species).

With regard to the framework of education for sustainable development (ESD), the curriculum increased students' knowledge of alien species (cognitive dimension) and their interest in protecting biodiversity (socioemotional dimension). Furthermore, it informed them about the appropriate actions (behavior learning dimension). It also facilitated students to acquire some key competencies of sustainability (UNESCO, 2017), especially systems, critical, and anticipatory thinking.

About systems thinking and critical thinking, introducing invasive and noninvasive alien species trained students to think about problems from a multidimensional perspective. For instance, before the intervention lessons, many students viewed alien species negatively, which is contributed by the portrayal of alien species as "villains" in television shows. After the intervention, most students understood that some alien species could be "allies" (e.g., vegetables in the supermarket), as apparent in their neutral responses.

Case studies on the unsuccessful introduction of the mongoose and largemouth bass, helped the students achieve anticipatory thinking. For instance, in response to the question about the kind of action they could take, many students listed anticipatory as one of them. Specifically, this refers to the process of predicting the future consequences of raising alien species as pets and committing to their lives while preventing their release into the wild.

The issue of invasive alien species is a problem directly impacting the public's thoughts and behaviors. Thus, the topic of invasive alien species would be easier to implement at the elementary school level. Furthermore, it provided a context on how a simple isolated action (like releasing pets in the wild) contributed to devastating effects in the long run. This curriculum may enable an understanding of complex sustainability issues that are often controversial, like climate change.

Elementary schools should serve as a foundation for children to learn to understand alien species and identify ways to deal with them in the future. The curriculum on this topic should include several topics: (1) an introduction to familiar alien species, (2) the history of their introduction, (3) understanding how alien species could become invasive, and the action to be taken to handle them.

The developed curriculum has shown its potential to transform the Japanese indigenous views on nature, life, and death (Table 11.5). After the lessons, most students responded that alien species kept as pets must be responsibly kept throughout their lives and should not be released into the wild. Although there is no guarantee that the students may continue to think this way after growing up, this is a big step toward conserving biodiversity.

11.5 Conclusion and Implication

The literature review revealed that there are four requirements for developing a curriculum about alien species. Firstly, students should interact with familiar and unfamiliar organisms. Secondly, they should have a firm grasp on the origins of alien species and the impacts they bring. Thirdly, they must evaluate the best course of action and base their decisions on it. Fourthly, the developed curriculum must reflect the context of each country.

The implementation of the developed curriculum demonstrated that 90-minute lessons could potentially change students' knowledge, awareness, interest, and behavior. Before the lessons, the students tend to hold unidimensional views about alien species as "bad," as portrayed in informal education sources (e.g., TV). Upon receiving the lessons, their perceptions transformed into a neutral and multidimensional view that alien species could be "bad" and "good," depending on how humans dealt with them.

The curriculum introduced in this chapter was developed based on Japan's sociocultural context of biodiversity and alien species. Hence, the intervention lessons must be tailored to each country's unique characteristics, which may help improve their situations. The program could be enhanced by incorporating lessons learned from each country's challenges. Science education is critical in educating students about invasive alien species and preparing them for future action. Education may eventually change their minds and behaviors, promoting biodiversity conservation.

Acknowledgments Part of this research was supported by JSPS KAKENHI Grant Numbers JP18K02660 and JP 17H02700. I am grateful to Mr. M Sakamoto (Hiroshima City Insectarium) for triggering this study and for his suggestions. I would also like to express my gratitude to Dr. T Kishimoto (Museum of Natural and Environmental History, Shizuoka, Japan) for all the helpful discussions on curriculum development.

References

Alexandar, R., & Gopalsamy, P. M. (2011). Biodiversity conservation through environmental education for sustainable development – A case study from Puducherry, India. *International Electronic Journal of Environmental Education, 1*(2), 97–111.

Doi, T. (1971). *The anatomy of dependence: The key analysis of Japanese behavior* (English trans. John Bester). Kodansha International.

Doi, T., & Hayashi, T. (2015). Analysis of textbooks describing alien species and changes in students' recognition through lesson: Aiming to further the development of environmental education in elementary school. *Journal of Science Education in Japan, 39*(3), 212–224. (in Japanese).

Farmer, J., Knapp, D., & Benton, G. M. (2007). An elementary school environmental education field trip: Long-term effects on ecological and environmental knowledge and attitude development. *The Journal of Environmental Education, 38*(3), 33–42.

Lindemann-Matthies, P. (2002). The influence of an educational program on children's perception of biodiversity. *The Journal of Environmental Education, 33*(2), 22–31.

Meli, M. (2017). Terada Torahiko's "The view of nature of the Japanese." Translation with Critical Essay (Part 1 of 2). *Kansai University Literary Theory Collection, 67*(3), 1–38.

Millennium Ecosystem Assessment. (2005). *Ecosystems and human well-being* (Vol. 5). Island press.

Ministry of Environment. (2012). *The national biodiversity strategy of Japan 2012–2020*. https://www.env.go.jp/press/files/en/528.pdf

Nishida, M. (1972). *Genealogy of impermanence*. Ofusha.

Ogawa, M. (1998). *Rediscovery of "science" – Western science as a different culture*. Nouson-Gyoson-Bunka-Kyokai. (in Japanese).

Otani, T. (2008). "SCAT" a quantitative data analysis method by four-step coding: Easy stable and small scale data-applicable process of theorization. *Bulletin of the Graduate School of Education and Human Development (Educational Sciences), Nagoya University, 54*(2), 27–44. (in Japanese).

Schönfelder, M. L., & Bogner, F. X. (2017). Two ways of acquiring environmental knowledge: By encountering living animals at a beehive and by observing bees via digital tools. *International Journal of Science Education, 39*(6), 723–741.

Schreck, R. C., Marchante, H., Freitas, H., & Marchante, E. (2013). Public perception of invasive plant species: Assessing the impact of workshop activities to promote young students' awareness. *International Journal of Science Education, 35*(4), 690–712.

Secretary of the Convention on Biological Diversity. (2010). *Year in review 2010*. Secretary of the Convention on Biological Diversity.

Shimazono, S. (2009). Views of death and life in modern Japan: A historical prospect (ethics major lecture abstract in 2009). *Kokushikan Philosophy, 14*, 1–14. (in Japanese).

Skukan, R., Borrell, Y., & Ordás, J. (2020). Find invasive seaweed: An outdoor game to engage children in science activities that detect marine biological invasion. *The Journal of Environmental Education, 51*(5), 335–346.

Tachikawa, S. (1998). *Japanese view of life and death*. Chikuma-shobo. (in Japanese).

Taguchi, R., Seki, I., & Kato, N. (2011). A study of monuments to pray for life in water. *Bulletin Institute of Oceanic Research and Development, Tokai University, 32*, 53–97. (in Japanese).
Terada, T. (1939). *The Torahiko Terada essay collection* (Vol. 5). Iwanami Publishers. (in Japanese).
Umezu, T., Kawai, K., Fujii, H., & Doi, T. (2017). Science lessons in elementary school focused on immigration of alien species. *Abstract of a paper presented at the 12th conference of the European Science Education Research Association.*
UNESCO. (2017). *Education for sustainable development goals: Learning objectives*. UNESCO.
United Nations. (2015). *Transforming our world: The 2030 agenda for sustainable development.* United Nations.
Waliczek, T. M., Parsley, K. M., Williamson, P. S., & Oxley, F. M. (2018). Curricula influence college student knowledge and attitudes regarding invasive species. *Hort Technology, 28*(4), 548–556.
Watanabe, M. (2003). Japanese view of nature: Comparison with western countries. *Annual Report of Institute of Humanities and Social Science, Keiwa College, 1*, 11–34. (in Japanese).

Part III
Informal Science Education for Sustainable Development

Chapter 12
The Value of Zoos as Educational Resources for Sharing the Sustainable Development Goals with the Public: A Japanese Perspective

Shiho Miyake

12.1 Introduction

Currently, zoos are receiving significant academic and political attention as notable stakeholders in the United Nation's (UN's) Sustainable Development Goals (SDGs). One major global task of sustainable development is biodiversity conservation, and wildlife-oriented facilities such as zoos can promote this goal (Convention on Biological Diversity, n.d.). Patrick and Tunniclife (2013) and Rennie (2014) argued for the educational effectiveness of zoos in science education as learning programs and exhibitions that target primary and secondary school students are known to raise children's interests in living creatures and the natural environment. This suggests that zoo visits can lead to children acquiring environmental and scientific literacy. However, few studies have examined how and whether zoos can educate older age groups, such as undergraduate students. To address this gap in the field, I conducted a longitudinal study with undergraduate university students with the aim of developing educational methods for zoos that increase public awareness of biodiversity and environmental issues, especially in terms of the relationship between wild animals and humans. In this chapter, I introduce some research examples from the educational design based on our collective experiences during this empirical study. These demonstrate the value of zoos as educational resources for sharing the SDGs with the public.

In Japan, education for sustainable development (ESD) is aimed at informing the public about the SDGs. The Ministry of Education, Culture, Sports, Science and Technology created four ESD frameworks: curriculum development, teacher education, promotion of educational effects, and activities for youth (MEXT, 2020). At

S. Miyake (✉)
Kobe College, Nishinomiya, Japan
e-mail: miyake@mail.kobe-c.ac.jp

H. Fujii, S.-K. Lee (eds.), *Science Education for Sustainable Development in Asia*, Education Innovation Series, https://doi.org/10.1007/978-981-99-8711-5_12

the same time, the government sets a goal document, Society 5.0, which aims to foster richer sensibility and intellect concerning the SDGs in the next generation of young people. Society 5.0 envisions:

> A human-centered society that balances economic advancement with the resolution of social problems by a system that highly integrates cyberspace and physical space. … Social reform (innovation) in Society 5.0 will achieve a forward-looking society that breaks down the existing sense of stagnation, a society whose members have mutual respect for each other, transcending the generations, and a society in which each and every person can lead an active and enjoyable life (Cabinet Office Government of Japan, n.d.).

One issue in higher education is how to educate students in ways that encourage lifelong learning. While academic disciplines in higher education traditionally emphasized what had been taught by scholars, today there is a shift to "what can be learned by students" (Kurokawa, 2012). Cartrette and Melroe-Lehrman (2012) noted that "students without formal experience in authentic research activities often have misconceptions about the process of research, its goals and outcomes, and the nature of the research enterprise" (p. 1093). Students are required not only to absorb knowledge and skills but also to embody and express their own motivations, attitudes, and values. In other words, students must learn how to live based on what they have learned. While not all students become academic researchers, all of them become active citizens in society. Thus, we must ask ourselves what kind of educational research activity is suitable for the SDG era that imparts the meaning of sustainability and encourages participation. Considering how to educate young people, specifically university students around 20 years old, is essential to help them understand and integrate the SDGs. This chapter then examines two questions:

1. How can we make an impact on the SDGs' environmental and educational topics in the context of zoos?
2. How can undergraduate students conduct their undergraduate research projects in a zoo, and what does this process consist of?

12.2 Zoo Designs That Reflect SDGs

Zoos are the ideal place to deliver scientific literacy to a wide range of people. They are recognized as informal public institutions for science education in which both children and adults can gain knowledge and skills by utilizing zoo resources (Falk & Dirking, 1992; Patrick & Tunniclife, 2013; Rennie, 2014). Zoo experiences act on both cognitive and emotional aspects to promote long-term memory and environmental awareness for children (Falk & Dirking, 2000); also, the experiences raise people's awareness of environmental conservation and responsibility (Bruni et al., 2008). Researchers have examined the educational functions of zoos and how established programs provide educational goals and content. Zoos have also worked to change exhibition designs to better reflect the animals' habitats and biographies.

Therefore, we must consider how zoos are visually designed to serve education and research from the viewpoint of environmental and scientific literacy.

Japanese zoos are currently expected to incorporate aspects such as biology, ecology, biodiversity, and human society (Maki & Watanuki, 2011; Ministry of the Environment, 2016), all of which can be used to examine the issues targeted in the SDGs. Zoos are also useful for teaching environmental ethics, including decision-making related to sustainable societies, sciences, and environments that center on human-wildlife relationships (Hayama et al., 2012). The concept of environmental ethics is emphasized as a part of citizenship education (Levinson et al., 2020) or as an educational policy that aims to develop consumer awareness in striving for a sustainable environment and society (Okano & Sasabuchi, 2017).

For the purpose of the study, I visited 12 zoos in Japan and overseas between 2017 and 2020 to find out which SDGs perspectives were reflected in zoo design (Table 12.1). Instead of using the zoo's original name, the author provided each visited zoo with a "zoo code."

12.2.1 Environmental Ethics Topics in Zoo Designs

Based on photographs taken at the zoos, I extracted the environmental ethics topics in the SDGs from the zoo designs, as reflected in their visual elements. I found the following five topics that I correlated with the numbered SDGs: (11) Sustainable Cities and Communities, (12) Responsible Consumption and Production, (13) Climate Action, (14) Life Below Water, and (15) Life on Land.

Table 12.1 Zoos visited with "zoo codes"

Zoo code	Country	Year established	Area (ha)	Species	Entrance fee/adult
AHJ	Japan	1967	15.2	101	1000JPY (May, 2020)
DZI	Ireland	1831	28	400	17.5EUR (Sept, 2017)
KZJ	Japan	1975	47.8	60	580JPY (May, 2020)
OHJ	Japan	1951	8.1	130	600JPY (May, 2020)
PZC	Czech Republic	1931	60	676	200CZK (Sept, 2018)
SNU	U.S.A	1889	66	390	Free (Feb, 2018)
SSS	Sweden	1891	30	32	120SEK (Nov, 2017)
SVA	Austria	1752	17	700	20EUR (Nov, 2019)
TOJ	Japan	1915	11	200	500JPY (May, 2020)
TSA	Australia	1916	21	350	55AUD (July, 2017 Ferry fee included[a])
WSA	Australia	2006	0.7	N.A.	60AUD (July, 2017 Aquariums entrée included[a])
LZU	U.K	1826	14	755	24.3GPB online (Sept, 2020)

[a]This information was collected from April to September, 2020

Goal 11, Sustainable Cities and Communities, refers to making cities and human settlements inclusive, safe, resilient, and sustainable (United Nations Development Programme, 2020). This goal may suggest that it is limited to cities and humans, but some zoo exhibitions involve animals within these ideal communities. For example, the Japanese KZJ zoo treated a red-crowned crane that had lost a foot in an accident by developing an artificial leg to help its mobility. On the cage front was a signboard explaining the treatment and describing how a local veterinarian and staff members of the zoo developed the artificial leg. Additionally, the WSA in Australia displayed a model of a kangaroo that was struck by a car on the road. These exhibitions demonstrate that zoos are exhibiting how wild animals are impacted and harmed by human activities.

Goal 12, Responsible Consumption and Production, is to ensure sustainable consumption and production patterns (United Nations Development Programme, 2020). At first glance, this goal also does not seem to apply to animals, but zoos are also attempting to address this goal. For example, the CPZ in the Czech Republic sells beverages in cups made of biodegradable plastic and offers appropriate trash cans for their disposal. In this regard, the zoo encourages visitors to use eco-friendly products and dispose of them appropriately. Similarly, the Australian TSA introduced the sea eco-label at their sea lion shows to raise visitors' awareness of their ability to select sustainable fishery products. The eco-label is published by the Marine Stewardship Council (MSC) and the Aquaculture Stewardship Council (ASC). These standards can be applied to wild fish or seafood from fisheries that have been certified by the MSC/ASC Fisheries Standard, a science-based set of requirements for sustainable fishing (Marine Stewardship Council, n.d.; Stichting Aquaculture Stewardship Council Foundation, 2012). Therefore, the sealion show informs zoo visitors that purchasing these labeled products help conserve the ocean. This exhibition is also closely linked with SDG 14, Life Below Water, which aims to conserve and sustainably use the oceans, seas, and marine resources (United Nations Development Programme, 2020).

Goal 13, Climate Action, means taking urgent action to combat climate change and its impacts (United Nations Development Programme, 2020). The SVA in Austria offers a commentary on climate change in the polar bear exhibition building under the "endangered arctic habitat" label. While the exhibition building is devised to seem naturalistic for the animals, the commentaries are related to climate change and biodiversity conservation. In addition, the LZU in the UK has a position statement on climate change (Zoological Society of London, 2011), and its exhibitions now reflect how climate change affects the habitats of the animals.

In terms of Goal 15, Life on Land, the goal is to protect, restore, and promote the sustainable use of the terrestrial ecosystem through the sustainable management of forests, combating desertification, and halting or reversing land degradation and biodiversity loss (United Nations Development Programme, 2020). Several zoos prominently express this goal in their exhibitions by presenting the International Union for Conservation of Nature's (IUCN) Red List ranking on their animal commentary boards. The IUCN Red List of Threatened Species is the world's most comprehensive inventory of the global conservation status of plant and animal species, and it is the most authoritative guide on the status of biological diversity (IUCN, 2020). The SNU in the USA displays the cheetah's threat level on its commentary board with the conservation status of the animal. The zoo has also developed extensive research on the cheetah's ecosystem and on conservation for future population protection. In another example, the Czech Republic's PZC posted the animal's Red List ranking, along with its species' information (length, weight, and habitat), on Bongo's commentary board. The IUCN's Red List rankings are presented differently depending on the species: some species' rankings are prominently displayed, while others are shown inconspicuously because they are included within the animal information content. Nevertheless, presenting an animal's endangerment levels is becoming the norm in today's zoos.

Furthermore, zoos reproduce landscapes, habitats, and ecosystems for animal housing and paddocks. For example, the DZI in Ireland has 13 hectares in an African area that includes the African savannah enclosure and contains multiple species, such as scimitar-horned oryxes, that are extinct in the wild, as well as giraffes, rhinos, ostriches, and zebras (Dublin Zoo, 2020). Austria's SVA has recreated a rainforest jungle in which there are plants and frogs (amphibians) lining the areas where people pass through. Visitors feel as if they are entering the rainforest, and because the habitat is enclosed, the zoo is able to control its appearance, humidity, and temperature and create a rainforest ecosystem. As it is difficult for most people to actually visit the places where this type of wildlife lives, zoos create landscape designs that reproduce these environments and help people get a sense of the atmosphere of these habitats.

Another significant factor in a zoo's Life on Land design is the attempt to show the natural behaviors and lives of animals. Although landscape recreation can reveal an animal's physical environment, an exhibition design that leads animals to behave naturally when eating, sleeping, and playing can also be developed to enrich the experience. For example, the SNU in the USA has developed an enrichment program that considers animal habits, cognition, sensory experiences, food, and toys (Smithsonian's National Zoo & Conservation Biology Institution, n.d.). According to Swaisgood and Shepherdson (2005), this type of enrichment in zoos is an important research focus that reduces the zoo animals' stereotypical behaviors, which refer to repetitive and invariant behaviors (which lack apparent immediate functions) that the animals engage in. Today, enrichment is also thought to contribute to animal welfare.

In the zoos that I visited, I found the following enrichment exhibits. In the Czech Republic's PZC, the Asian elephants were kept in a family unit. When I visited, I saw two young elephants playing with their trunks entwined. In addition, the zoo installed several large artificial rocks in the enclosure to allow the elephants to rub their bodies when they are itchy or to hide themselves to rest. Additionally, Sweden's SSS installed ropes between trees in the Colobus guereza enclosure to promote jumping, an enrichment design that reproduces the monkey's living environment and revives their natural and lively movements. In other words, a zoo's enrichment design can reproduce an animal's living environment and encourage natural behaviors, revealing to the visitors how these animals actually live in nature.

As I have mentioned, I found that the recent zoo exhibition design corresponds to five topics in the SDGs. Zoo visits provide people with a first-hand opportunity to understand the SDGs' global goals and raise awareness of the relationship between humans and wildlife, specifically with regard to endangered species, landscapes, habitats, ecosystems, impacts of human activity, and required behaviors for humans (e.g., consumption and recycling). These five topics include elements that convey the current state of wildlife and more sustainable product options while encouraging people to make better decisions as members of society. In fact, these zoo designs inform people about the present, past, and future of environmental problems, reflecting a fact, reason, and prediction approach, which is the essence of scientific literacy.

12.3 The Promotion of the SDGs at a Zoo: An Empirical Study with Undergraduate Students

I started conducting an empirical study in 2015 to develop environmental education materials for the public's understanding of biodiversity conservation. The study was conducted with my students who were participating in the environmental sociology and environmental education laboratory and in cooperation with the Tennoji Zoo in Osaka city. We developed a *kamishibai* (theatrical show with the use of pictures) with two kinds of animals that were residing in the zoo (Asian elephant and Oriental stork) and presented the projects to visitors. In this section, I will illustrate how the undergraduate students and I created and developed the zoo animal picture-story show project developed in 2015, 2017, and 2018. The project demonstrates how undergraduate students were actively involved in examining perspectives of the SDGs and how they reflected on authentic research issues.

12.3.1 The Kamishibai Project on Zoo Animals

The Asian Elephant (2015)

This particular project started when my undergraduate students and I went to the zoo looking for a graduation research topic in 2015. After coming across an Asian elephant named Rani-Hiroko ("Princess Hiroko" in Hindi), we were immediately interested in her personality, which was full of character, as described in the exhibition commentary. A few days into our trip, one student had the opportunity to talk to the elephant's keeper, who shared Rani-Hiroko's interesting story, and as a result, we all revisited the zoo to hear about her background and personality, deeply empathizing with her circumstances. Every week, we held discussions on the environmental issues related to Asian elephants, finding that palm oil plantations (raw material for food and detergents that we use in our daily lives) have progressively taken over the elephants' habitats. These types of impacts (deforestation and habitat deprivation) caused by human activity are seriously concerning and require a problem-solving idea in order to achieve coexistence (WWF, 2015). Therefore, we agreed that an elephant-motif story would be an appropriate topic for raising Japanese people's environmental awareness as most of them are unaware that palm oil products are regularly used and that Asian elephants' habitats are negatively impacted by the materials.

Creating a *Kamishibai* as an Environmental Education Tool

Kamishibai is a traditional storytelling tool for children, and nowadays, it is known globally as a new interactive medium for teaching (McGowan, 2015). Several education researchers have suggested the value of picture stories as an educational medium for children as they help them transform their experiences into knowledge and acquire community values (Schank & Berman, 2002; Avraamidou & Osborn, 2009). Using the *kamishibai* medium, we began recreating Rani-Hiroko's story by referring to the Tennoji Zoo's staff blog for information on her daily life at the zoo. We also tried to incorporate Hiroko's personality and peculiar behavior. For example, one keeper felt anxious and was concerned about Hiroko because she often violently shook her trunk and trampled her droppings. Thus, we mainly organized the story with this keeper's perspective in mind, attempting to understand why Rani-Hiroko behaves in such a way every day. Then we composed a story with Rani-Hiroko as the main character, speaking in the first person "I." Another character was the zoo keeper, who talked about Rani-Hiroko, and the story commentary roles (narrators) were left to the readers/picture-story theater performers. During the performance, one of the narrators explained the environmental issues related to oil palm plantations by showing the visitors that many of our familiar products, such as instant noodles and chocolates, contain this oil. For example, the label "vegetable oil" we see printed on an instant noodle bag is, in fact, referring to palm oil.

We used cutwork pictures with several different colors of paper and pasted them on a sketchbook (B4 size), so the pictures were visually much clearer than if they had been painted. My students verified the compositions many times to ensure that the pictures and the story properly matched. Furthermore, we presented the project in advance to four elephant keepers at the Tennoji Zoo and received positive feedback (easy to understand, accurate story, etc.). In addition, they advised us to make the words simpler for children to understand and to improve speaking speed and tone. Based on this feedback, the final *kamishibai* was 8 min long with ten pictures.

As a result of these trial-and-error activities, we proceeded to create the picture story; we realized that we should propose various arrangements to deliver the story. These arrangements centered on (1) animal selection to create the story, (2) environmental problems to convey, (3) content design to introduce an episode in the animal's life, (4) story design of drawing and development, and (5) demonstration design of proper casting.

The Asian Elephant's Revival (2017): Comparisons with 2015 Through a Design Thinking Methodology

In 2017, three students joined the project as we continued the collaboration with the Tennoji Zoo. Based on my own experiences with the 2015 version, we had two objectives for improving the *kamishibai* project: (1) enhancing the environmental problem message and (2) upgrading the pictures. To analyze whether the 2017 version had progressed from the 2015 version, I used Stanford University's design-thinking methodology's five steps: emphasize, define, ideate, prototype, and test (Hasso Plattner Institute of Design, 2012). This method "melds an end-user focus with multidisciplinary collaboration and iterative improvement and is a powerful tool for achieving desirable, user-friendly, and economically viable design solutions and innovative products and services" (Plattner, 2010). Indeed, scholars such as Kurokawa (2012) have found that it leads to reasonable solutions for any problem and guarantees concrete proposals and feasibility. Recently, Japanese higher education has begun focusing on this method to construct curricula and educational contexts for university students. Thus, I organized the students' efforts in regard to their 2015 and 2017 project activities in terms of these five steps.

In 2015, the students were interested in the elephant's strange behaviors and knew that her mother was killed in their habitat in India (emphasize step). Based on what the keeper told them, they created a story that linked a conflict between elephants and humans to Rani-Hiroko's background. In terms of the define step, they set the theme of the picture story as "Why do humans kill elephants?" Then as the ideate step, they established two characters (elephant and keeper) and used a 19.6 W (50 cm) × 14.1H (36 cm) sketchbook for the pictures and inserted eight pages of cutouts made with colored paper into it (Fig. 12.1a). The students presented the theater play to their colleague undergraduate students, who were taking an environmental education course, and they evaluated the correspondence between the pictures and story and the story's comprehensibility (prototype step). Finally, as the test step, we performed the picture story for the zoo visitors, with several presentations in front of rest areas and the zoo's elephant enclosure (Fig. 12.1a).

In 2017, the emphasize step involved the surroundings of the elephant enclosure as these related to the oil palm plantations. The zoo reproduced a tropical forest, a

Fig. 12.1 Picture size difference. Left: 2015 (**a**); Right: 2017 (**b**)

water pond large enough for an elephant to soak in, and a watchtower that monitors elephants approaching a human village in India. The students also realized that the oil palm plantations – expanding due to the growing global demand for vegetable oils – reduced the area of the elephants' habitat. As large quantities of palm oil are consumed daily in Japan, the students wanted to emphasize that we are also involved in the Asian elephants' environmental problems. As a result, the students defined the story's message as "environmental problems have no borders" (define step). Regarding the *ideate* step, in addition to the elephant and the keeper, the students added a pigeon named Mamesuke, who happened to fly over the zoo as a new character playing the narrator. This choice improved the story's fluency because the characters developed the entire story themselves. They also expanded the size of the pictures to 29.9 W (76 cm) × 21.6H (55 cm), and the drawing paper was laminated for waterproofing in the case of rain (Fig. 12.1b). In 2015, undergraduate students evaluated the prototype, focusing on whether the picture story conveyed the environmental problems occurring in elephant habitats and related them to Japanese people and other palm oil consumers. To test the project, the students performed the story for the zoo visitors during the picture-book storytelling event held every month for parents and children and on other occasions in front of the elephant enclosure. However, in 2017, one notable problem occurred in that few people gathered for the picture-story show. We surmised that this happened because the performances were held on weekdays and afternoons.

12.3.2 Dealing with Environmental Issues in Japan

An Oriental Stork in (2018)

In the 2018 project, the students had to choose another animal for their *kamishibai* because Rani-Hiroko had died. They asked the zoo staff members for suggestions on which animals to focus. They proposed several animals, including sika deer (Sika: *Cervus nippon*), red-crowned cranes (Tancho: *Grus japonensis*), raccoon dogs (Tanuki: *Nyctereutes procyonoides*), Asian black bears (Tsukinowaguma: *Ursus thibetanus*), and an Oriental stork (Kounotori: *Ciconia boyciana*). The students chose storks as these birds are unique and relevant to the environmental biodiversity issue in Japan. Specifically, *Ciconia boyciana* is a large, carnivorous, and currently endangered bird that inhabits only the East Asian region (IUCN, 2018). The Oriental stork's uniqueness stems from its extinction in the wild in Japan and its restoration in Japan in the 1970s. Due to the national restoration project's success (Osako, 2014), the present population is over 200, including those in captivity and in the wild (Hyogo Park of the Oriental White Stork, 2019a, b). Thus, this animal, whose wild extinction and restoration took place in a short period, was an ideal topic for biodiversity conservation education.

Meeting Rico-chan, an Oriental Stork at the Tennoji Zoo

As I learned how to create a picture story, I led my students to collect biographical and personality information on the stork at the Tennoji Zoo. They interviewed the keepers and obtained factual material about the bird. The bird was a female, and her name was Rico. She was subsequently nicknamed Rico-chan as "chan" is a diminutive suffix for a female and expresses affection. The following five steps summarize the students' efforts in the 2018 project.

For the emphasize step, the students wanted to focus on the bird's story. Rico-chan, an old female stork who lived quietly and alone in an old rusty cage, had very few visitors, even though she lived across from the popular penguins. According to a keeper, Riko-chan had raised 27 children since coming to the zoo in 1991. Some of her offspring continue to breed in the Hyogo Park of the Oriental White Stork, and others have moved to numerous zoos nationwide, contributing to species conservation. The story theme was "The Great Mother Rico-chan," and we focused on her contribution to population recovery based on the past extinction of wild storks in Japan (define step).

The ideate step involved resolving the 2017 problem of low audience attendance. As a solution, we conducted quizzes before and after the main story to capture the audience's attention. For example, the prequiz posed yes/no questions such as "Does a stork sing?" and "Do Japanese storks live in the wild?" The postquiz questions were multiple choice: "Q1. How many children did Rico raise? 3, 16 or 27?" and "Q2. Why did the Oriental Storks disappear? (1) They were overhunted for their feathers, (2) Agricultural chemicals on fields harmed their prey, or (3) Too many of the tall trees in which the birds build their nests were cut down." The students designed the pictures using paper cutouts for a total of 17 pages with 15.7 W (40 cm) × 21.6H (55 cm) dimensions. As one keeper noted that the laminate made it difficult to see the pictures in the sunlight, we did not take this additional step. As before, the students introduced the prototype to other undergraduate students, who assessed the pictures' visibility, the story's comprehensibility, and the topic's ability to raise the audience's awareness of environmental issues. For the final test step, we performed the picture book for the zoo visitors in front of the Oriental Stork enclosure and during the picture-book storytelling events held for parents and children.

12.4 Impressions of the Kamishibai Project

12.4.1 Audience Evaluations

The Story of Rani-Hiroko

We asked the audience for our *kamishibai* to describe their impressions. For the 2015 Rani-Hiroko story, children under 10 responded with "poor elephant mom"; "I felt sorry for Hiroko, she came from afar, and I feel sorry for her mom"; "I first learned that elephant tusks were expensive and sell at high prices"; and "good pictures." Young adults in their 20 s considered that "While the pictures are for young children, the vocabulary used was more suitable for middle to upper grades of elementary school." Others within this age group were fascinated by the animal itself, with comments like "It is amazing that elephants can grab a peanut with their nose." People in their 30 s also found the content informative, mentioning "I did not know how the animals came to the zoo," "This is the first time I heard about the elephant's story," "I learned about why the elephant was brought here. I did not know that it was related to vegetable oil," and "I will start to learn about the environmental problems of elephants myself. The performance went well." Others reflected on the content's correspondence with children's comprehension levels: "The content is interesting, but some words were difficult for young children, such as deforestation"; "It was easy to understand, and both adults and children can enjoy it"; and "It was easy for children to understand. You should spread this story to more people."

Older adults, in their 40s and 50s, reflected on the educational materials and the environmental aspects of the story with comments like: "The picture-story explained the environmental problems in a very easy-to-understand manner"; "It was a good story for children"; "The story is educational"; "Most of us live without thinking about elephants, but I think we have to consider what our lives are supported by and what we sacrifice"; and "I didn't know about vegetable oil, so it was amazing to listen to your story." Other people within this age group reflected on the content in a more personal manner: "When I watched a video that was broadcast by the zoo, this elephant was playing with poop all the time, I was wondering why. Now I understand, you taught me" and "It was a shocking story, but it was nice to hear it."

For the 2017 presentation, people in their 20s mainly reflected on the environmental message, aside from two individuals who commented on the content's relevance for children: "My two-year-old son listened very carefully" and "The story is expressed in simple words, and it would be easy for children to understand, but it may be better to make it a little more colorful." We obtained responses like "I thought this was someone else's problem, but even if robbing or poaching animals is occurring overseas, because we consume products related to wildlife habitats, then we have to review and improve our behavior." Another person similarly noted: "When I hear about the over-hunting of elephants, it seems to be unrelated to me, but when I hear about chocolate and potato chips, I feel part of the problem." One individual in their 20 s reflected:

> We Japanese have no chance to see elephants unless we go to the zoo, but I understand that the palm oil contained in what we usually use is one of the causes of the environmental issues related to elephants. I learned that elephants have various emotions, both angry or sadness. I agree that there are environmental issues across the border, but it turns out that our lives are also affected by far away wildlife.

Furthermore, people in their 30s and 40s mostly reflected on the story's impact on children: "It is difficult for a two-year-old child to understand, but the idea of talking about an elephant in front of an elephant is good." "It was easy to understand since the story involved the interaction between only three characters. I listened to it attentively with my child" and "The picture-story explained that the number of elephants is decreasing. I thought it would be easy for children to understand." In terms of environmental issues, one elderly individual also noted: "We have to think of wildlife issues."

The Great Mother Rico-chan
We only received one child's response for this story, who commented: "Very good! The quizzes are fun. Your talk is amazing!" People in their 20s had similar responses: "After listening to this picture-story, the image of the animal may change … I want to hear other animals' story as well" and "The picture-story show was good." The participants in their 30s mostly reflected on the story's educational capacities: "Adults and children can learn from the easy-to-understand picture-story show"; "The children listened to the story properly and I learned a lot as well"; "I was surprised that Rico-chan hatched 27 children"; "My children were happy with the quizzes"; "I am happy to have the opportunity to see the picture-story with my child"; "My children, 3 and 6 years old, enjoyed the story and quizzes"; "I learned that the stork lays 2 to 5 eggs at a time"; "I knew that Toyooka has storks, but not that the species is in the Tennoji Zoo"; "I didn't know there is such a deep story behind the stork"; "I am glad to hear this story, because no other media teaches us such information"; "I am glad to hear the story with my children"; and "I would like to hear other animals' stories." Lastly, people in their 40 s had similar comments related to the quiz and Rico-chan herself: "Rico-chan had 27 children, which is amazing … It is a great success to have offspring all over Japan"; "It is very good and the quizzes were also well developed"; and "I am surprised to learn that Rico-chan's children are all over the country."

Based on the above participants' impressions, we discovered that the use of *kamishibai* helped attract and keep people's interest as they evaluated it as an effective tool that facilitates people's understanding of these complicated issues. We also noted that the participants were interested in animals and in environmental issues. In general, the audience offered positive feedback on the animal picture story. From their responses, it is evident that they were interested in both the quizzes and the story and had favorable impressions of the show's visual features. This suggests that not only the content but also its appearance is important to consider when creating a *kamishibai* exhibition.

12.4.2 Students' Evaluations of Their Efforts

The students participating in the 2015 and 2017 projects who created and performed "The story of Rani-Hiroko" reflected on their efforts and their impact. Some of them felt that they positively contributed to the zoo and to the visitors' knowledge. For example, Student A noted: "We provided zoo visitors with an educational resource that helps to raise awareness of environmental issues through the graduation research project. There are still few picture-stories for environmental education, so it is important that children should have more access to such environmental problems through child media." Similarly, Student B was happy that they received positive feedback from many age groups and that they were able to inform the visitors on matters that are not well known: "It turns out that we live without knowing that the foods and detergents we often use in our daily lives are likely to be depriving elephants of their habitat." Student C reflected not only on the teaching content but also on the fact that knowledge can lead to better results:

> Most of the zoo visitors were unaware of the oil palm plantation issue affecting wildlife habitats, so it was good to be able to tell the story. It is impossible not to use palm oil products, but I think that knowing the facts will lead to further steps for solving environmental problems.

Other students reflected on the information that they themselves learned: "I learned that human beings are creating a world where wildlife is having a difficult time living. I also learned various things from Rani-Hiroko's personality and behavior regarding the impacts of environmental issues" (Student D). Student E mentioned that they learned that "All animals kept in the zoo have various backgrounds and history" and that "if parents and children observe the zoo animals and talk about the history and habitat issues, children will have more opportunities to think about environmental issues."

The 2018 students, who created and performed "The Great Mother Rico-chan," also discussed their impressions, mostly reflecting on the educational opportunity that the project provided. For example, Student F considered that "Investigating how the consciousness of the audience has changed is also an issue for the future. Knowledge of environmental problems based on environmental education materials is a step toward solving environmental problems in the future." Student G similarly noted:

> Through the project, I met Riko-chan, a Japanese stork, who was not popular in the Tennoji Zoo. We learned that she was a great mother and that she contributed to the recovery of the population of Oriental storks in Japan. Furthermore, we learned about the human efforts involved in returning the Japanese stork to the wild. If we go to the zoo, observe animals, and understand their background, we will have an educational opportunity to think about environmental issues, rather than just watch the animals. According to the audience, they learned about storks for the first time. We would like to give story-telling to more people in the future.

Student H also reflected on the learning process as well as the children's attentiveness:

> Until I joined the project, I didn't know anything about storks and was not very interested in them. By actually talking to the keeper, I was able to get a deeper understanding of Rico-chan's legendary story. It was really amazing that one stork, Roco-chan, raised 27 children. Knowing about Rico-chan in detail, I felt a familiarity and empathy toward her. The children also listened to the picture-story show enthusiastically and positively. There were some difficult words in our *kamishibai*, but I thought they listened attentively. I was very impressed.

Through the *kamishibai* projects, the students investigated the environmental issues themselves and found appropriate themes and topics for the public. They were also aware that the animals in the zoo had several different backgrounds. Although the picture-story productions went smoothly, I believe that the students gained a lot from the struggles they encountered. While the students collected different types of information from their interviews with the zoo keepers, some interesting episodes did not appear in the contents of the picture stories because they did not match the environmental theme. For example, in the 2015 project, the keeper told us about husbandry training (training for the purpose of ensuring that animals accept health care work smoothly). After more than a month of discussion, we decided to drop the episode because it did not match the environmental issues related to habitats. Through this task, the students learned that when providing informative content to others, they need to obtain much more than what they end up presenting and that usually not all the information they obtain is included. The students also might have learned that even outcomes that take hard work sometimes need to be discarded to meet one's goals. In line with this, they gained a learning attitude that centers on the type of work that is required rather than the simple completion of a given task. This kind of process may sometimes cause students to have frustrating and unsuccessful experiences, but such negative incidents also provide a real-world perspective, reflecting what happens in real learning, that engenders lifelong learning skills.

12.5 Conclusion

This chapter focused on the following two questions:

1. How can we make an impact on the SDGs' environmental and educational topics in the context of zoos?
2. How can undergraduate students conduct their undergraduate research projects in a zoo, and what does this process consist of?

To examine the above questions, this study examined the SDGs' perspectives, as reflected in zoo exhibitions, and introduced specific examples of undergraduate students using zoos to complete their research activities. Based on evaluations of the activities, this study confirmed that zoos have a meaningful educational function even for undergraduate students and older populations as they can stimulate people to acquire environmental values, awareness, and learning behaviors. The results suggest that when individuals observe real animals in this environment, it increases their awareness of the fact that humans should protect the earth, animals, and their habitats. In other words, zoos can stimulate an empathetic response when individuals are in the presence of animals and learn about what happens to them as their environmental issues. As a result, people are reminded of the necessity to protect the earth.

Further research should assess the future of zoos and how to improve them to respond to the challenges of the SDGs in terms of human resource development. Although zoos are an emotionally valuable place for interesting, fun, healing, and happy experiences, they also offer a philosophical perspective that informs people of what can be done for the future of the planet. Therefore, it is important to further consider how zoos can be utilized, perhaps through trial and error.

Acknowledgment and Notes Although some of the study contents have been previously published and/or presented at academic conferences, this article provides additional descriptions. I would like to express my sincere gratitude to the students, the zoo visitors, and the staff members of the Tennoji Zoo who participated in the research project. The research project was funded by the JSPS KAKENHI [Grant Number 17H04490].

References

Avraamidou, L., & Osborn, J. (2009). The role of narrative in communicating science. *International Journal of Science Education, 31*(12), 1683–1707.

Bruni, C. M., Fraser, J., & Schultz, P. W. (2008). The value of zoo experiences for connecting people with nature. *Visitor Studies, 11*(2), 139–150.

Cabinet Office, Government of Japan. (n.d.). *Society 5.0, science and technology policy*. Council for Science, Technology and Innovation. https://www8.cao.go.jp/cstp/english/society5_0/index.html

Cartrette, D. P., & Melroe-Lehrman, B. M. (2012). Describing changes in undergraduate students' preconceptions of research activities. *Research in Science Education, 42*, 1073–1100.

Convention on Biological Diversity. (n.d.). *Biodiversity and the 2030 agenda for sustainable development.* https://www.cbd.int/development/doc/biodiversity-2030-agenda-technical-note-en.pdf

Dublin Zoo. (2020). *African Savannah.* https://www.dublinzoo.ie/animals/animal-webcams/african-savanna/

Falk, J. H., & Dirking, L. D. (1992). *The museum experience*. Whalesback Books.

Falk, J. H., & Dirking, L. D. (2000). *Learning from museums*. Altamira Press.

Hasso Plattner Institute of Design. (2012). *Stanford design guide. Five stages in design thinking* (T. Kashino, Ed. & T. Nakamura, Trans.). Eirene Management School. (in Japanese).

Hayama, S., Doi, T., & Narushima, E. (2012). *Coexistence with the wildlife*. Chijin Shokan. (in Japanese).

Hyogo Park of the Oriental White Stork. (2019a). *Outdoor stork field population as of August 20, 2019.* http://www.stork.u-hyogo.ac.jp/ex_situ/captive_ows_num/

Hyogo Park of the Oriental White Stork. (2019b). *Breeding stork population as of September 12, 2019.* http://www.stork.u-hyogo.ac.jp/ex_situ/captive_ows_num/

International Union for Conservation of Nature (IUCN). (2018). *Oriental stork. The IUCN red list of threatened species.* https://www.iucnredlist.org/species/22697695/131942061

International Union for Conservation of Nature (IUCN). (2020). *The IUCN red list of threatened species*. https://www.iucn.org/resources/conservation-tools/iucn-red-list-threatened-species

Kurokawa, T. (2012). Design thinking at undergraduate and post graduate levels. *Science & Technology Trends, 131*, 10–23. (in Japanese).

Levinson, R., Paraskeva-Hadjichambi, D., Bedsted, B., Manov, B., & Hadjichambis, A. C. (2020). Political dimensions of environmental citizenship. In A. C. Hadjichambis, P. Reis, D. Paraskeva-Hadjichambi, J. Činčera, J. Boeve-de Pauw, N. Gericke, & M. C. Knippels (Eds.), *Conceptualizing environmental citizenship for 21st century education* (pp. 17–28). Springer.

Maki, S., & Watanuki, K. (2011). Zoos from the perspective of science and technology policy. *Journal of Industry-Academia-Government Collaboration, 7*(5), 12–14. (in Japanese).

Marine Stewardship Council. (n.d.). *What does the blue MSC label mean?* https://www.msc.org/what-we-are-doing/our-approach/what-does-the-blue-msc-label-mean

McGowan, T. (2015). *Performing kamishibai: An emerging new literacy for a global audience*. Routledge.

Ministry of the Environment. (2016). *Ideal public function measures of zoos and botanical gardens in 2015*. https://www.env.go.jp/press/files/jp/102671.pdf. (in Japanese).

Okano, T., & Sasabuchi, K. (2017). Policy trends towards mainstreaming biodiversity in socioeconomics. *Japanese Journal of Ecology, 67*(2), 205–215. (in Japanese).

Osako, Y. (2014). Oriental stork (Kohnotori). In *Red data book* (pp. 36–37). Ministry of the Environment. (in Japanese).

Patrick, P. G., & Tunniclife, S. D. (2013). *Zoo talk*. Springer.

Plattner, H. (2010). Forward. In H. Plattner, C. Meinel, & L. Leifer (Eds.), *Design thinking: Understand-improve-apply* (pp. v–vi). Springer.

Rennie, L. J. (2014). Learning science outside of school. In N. G. Lederman & S. K. Abell (Eds.), *Handbook of research on science education. Volume 2* (pp. 120–144). Routledge.

Schank, R. C., & Berman, T. R. (2002). The pervasive role of stories in knowledge and action. In M. C. Green, J. J. Strange, & T. C. Brock (Eds.), *Narrative impact: Social and cognitive foundations* (pp. 287–314). Lawrence Erlbaum Associates.

Smithsonian's National Zoo, & Conservation Biology Institution. (n.d.). *Animal enrichment*. https://nationalzoo.si.edu/animals/animal-enrichment

Stichting Aquaculture Stewardship Council Foundation. (2012). *ASC's new consumer label for responsibly farmed fish*. https://www.asc-aqua.org/news/latest-news/ascs-new-consumer-label-for-responsibly-farmed-fish/

Swaisgood, R. R., & Shepherdson, D. J. (2005). Scientific approaches to enrichment and stereotypies in zoo animals: What's been done and where should we go next? *Zoo Biology, 24*(6), 499–518.

The Ministry of Education, Culture, Sports, Science and Technology (MEXT). (2020). *2020 UNESCO SDGs leader training (ESD) promotion project open call for participants*. https://www.mext.go.jp/unesco/001/2019/1413440_00001.htm on 30 Aug 2020. (in Japanese).

United Nations Development Programme. (2020). Sustainable development goals. https://www.undp.org/content/undp/en/home/sustainable-development-goals.html

World Wildlife Fund (WWF). (2015). *Palm oil & biodiversity loss*. http://wwf.panda.org/what_we_do/footprint/agriculture/palm_oil/environmental_impacts/biodversity_loss/

Zoological Society of London. (2011). *Climate change position statement by the Zoological Society of London*. ZSL Living Conservation. https://www.zsl.org/sites/default/files/media/2014-03/zsl-climate-change-position-statement-220411.pdf

Chapter 13
ESD Implementation Using a Whole-Community Approach in Okayama, Japan: Potential and Challenges from the Teacher Education Perspective

Hiroko Shibakawa

13.1 Why Is the Whole-Community Approach Necessary for ESD?

The importance of education has been re-emphasised in response to the various unsustainability challenges in modern society. Environmental, economic, social, and cultural issues are intricately intertwined and difficult to resolve without cross-sectoral and cross-border collaboration. At the root of all these challenges are the problems of our values and lifestyle. If these do not transform, we cannot hope for a solution. Thus, education for sustainable development (ESD) was born from the logic that education is important in changing our values and lifestyles.

However, educators should be aware that, as Jucker (2002) argued, education is not only a solution; it is also a problem, and we must simultaneously transform education. The size of the carbon footprint is proportional to the high rate of advancement to higher education. This shows that current education has not developed the knowledge and skills necessary to address sustainability challenges. Orr (1996) wrote: 'It is worth noting that this is not the work of innocent people. Rather, it is very much the results of work by people with Bas, BSs, LLBs, MBAs, and PhDs. The modern educational system does not guarantee good sense, prudence, and wisdom'. In ESD, such a critical view of the modern educational system is essential, as opposed to simply applying it to conventional education or updating it slightly. What is being asked in ESD is what education is *for*. There is no such thing as a sustainable society. However, it can also be said that ESD aims to address the

H. Shibakawa (✉)
Okayama University, Okayama, Japan
e-mail: hirokoshibakawa@okayama-u.ac.jp

H. Fujii, S.-K. Lee (eds.), *Science Education for Sustainable Development in Asia*, Education Innovation Series, https://doi.org/10.1007/978-981-99-8711-5_13

question of the kind of education that can enable the creation of a sustainable society.

What kind of education is 'education' in ESD? Many different things come to mind based on the country, region, and person. For example, a teacher standing with a textbook at the podium asks dozens of students at their desks, 'What happened to our county in CE 645?' Students respond by raising their hands. This scenario is very common in Japan even today. Conversely, some may feel no sense of incongruity at all. Others may feel a sense of incongruity about the situation where an older person who knows the answer may be averse to asking young 'unskilled' pupils about the results of cramming. In this modern education system, teachers who are highly adapted and successful at school tend to repeat the education they received and communicate the values they acquired. What Freire (1972) criticised as 'banking education' continues in today's public education in Japan. Instead of creating new value, this education functions to 'reproduce' and strengthen the values of the powerful and the majority, as Bourdieu and Passeron (1990) clarified. Freire said that education does not form a society but that a river of society builds education in a way that is in the interests of those in power and, therefore, does not allow them to resist power (Freire & Illich, 1975).

'Sustainable society' is not an answer, but ESD is an activity that can create new value by all people while facing the worldwide 'strain' that has occurred in past developments. ESD needs new pedagogy, as creating new societies requires new pedagogy. To make steady progress on the ground, it is necessary to transform existing education from within, using ESD. This internal transformation, as Gadotti (2010) described, aims at 'rehumanising' education. To think of humane education, educators will have to face up to real problems and think critically about the origin of the inhumanity or unsustainability in modern society. When asked what makes us human beings, we may get 100 different answers from 100 different people. However, there is a theory of learning that foresees the future to come, confronts the problems of modern education, and proposes a new direction for education, which influences public education in Japan: the theory of lifelong learning. The image of a human who adheres to the philosophy of lifelong learning is the image of thinking, learning, acting, living with others, and trying to be human so that we can be guided by 'the four pillars of education' (Delors et al., 1996). Lifelong learning can be said to create democracy. As Biesta (2011) pointed out, there is a bias towards school-based learning, as well as much reflection and criticism of modern education. Lifelong learning does not simply refer to self-contained learning, in which children and adults continue to learn but is rather a radical idea.

Dewey (1916) noted that societies comprise small communities and that democratic education does not merely promote participation but is education in which those communities interact freely and fully with other groups. Significant ESD practices already exist in the 'communities of practice (CoP)' (Lave & Wenger, 1991) that form communities among people who think, act, and collaborate with those facing difficulties in society and who are at the forefront of sustainability issues. ESD empowers all people to work together to create a sustainable society and will emerge as a process of legitimate peripheral participation in CoP for the creation of a sustainable society. However, as long as they remain in a group pursuing a single value, true 'sustainable development' cannot be achieved. Relying on Dewey, the criteria for democratic education are the amount and diversity of consciously shared interests and the extent of interaction and freedom with heterogeneous groups. If a community lacks any of these, the result is a skewed education (Lave & Wenger, 1991). In creating a sustainable society and promoting and deepening ESD, it is important to revitalise CoP to include ESD. The whole-community approach does not merely mean involving all local communities, such as school districts and towns. Conscious of the potential existence of ESD practices in the local community, it is an awareness that diverse CoP, including schools, social education institutions, and civic organisations, will interact more freely and fully and create new activities. This does not mean that the whole-community approach is universally effective, but that it is an effective approach in light of the status of ESD in Japan and other countries, where school education leans toward ESD and where there are almost no teachers who have already experienced ESD from their early stages.

13.2 History and Status of Community-Based Educational Approaches in Japan

Nearly 30 years have passed since ESD first appeared at the Rio Summit in 1992. At the Johannesburg Summit in 2002, the 'United Nations Decade of ESD' (UN-DESD), proposed by Japan, was unanimously adopted. In Japan, the Ministry of Education, Culture, Sports, Science and Technology (MEXT) proposed that UNESCO Associated Schools (ASPnet schools) be expected to serve as a base for ESD promotion.[1] Since then, the enrolment rate of ASPnet schools in Japan has increased dramatically. In 2014, when the world conference of the UN-DESD was held in Aichi, Nagoya, and Okayama, the number increased from 16 in 2005 to 913. Okayama City was also one of the first seven cities in the world to receive Regional Centre of Expertise (RCE) certification from the United Nations University in 2004. Of 163 public elementary and junior high schools (JHSs) in the city, 51 joined the ASPnet schools. Each JHS district has a community learning centre (CLC), a facility for social education and lifelong learning. CLCs have a history of playing an important role in building local communities in Okayama. The ideal of CLCs – 'gather, learn, and connect' – is widely shared by the general public, and all CLCs have been developing efforts to solve the problems of the local community. There are many non-profit organisations (NPOs) and non-governmental organisations (NGOs) in Okayama that make international contributions and have been engaged for many decades in various grassroots educational projects like environmental and human rights education and education for international understanding. With

[1] https://www.mext.go.jp/unesco/004/1339976.htm (in Japanese).

favourable environmental conditions, Okayama City, the Board of Education, and the Graduate School of Education of Okayama University have worked together to increase the number of ASPnet schools and to support ESD projects in the local schools and communities.

After 2014, the city board of education took the lead in applying for the ESD consortium project promoted by MEXT and commissioned several ESD coordinators, including the author of this chapter, with the aim of promoting ESD through the cooperation of various entities, not only with the 51 ASPnet schools but also with others. Even at that time, the city's basic plan for the promotion of education already clearly stated a model in which it promoted school development in cooperation with the community under the name of 'Okayama-style integrated education' and provided integrated education in JHS districts. The city also stated that ESD is applied to all forms of education and that all children should be nurtured as bearers of the creation of a sustainable society.

However, such community-based education is not limited to this area. The concept of community-based learning in Japan – where schools and communities work together to promote education and children go out and learn about the community – is not new. Compulsory education has accumulated the practice of 'community-based learning' in various ways, such as 'local community study' since the 1960s, 'collaboration/integration between formal education and social education' since the 1980s, 'community school' since around 2000, and 'localism' or 'regional science' introduced in the Period for Integrated Studies. Although there are varieties in definitions and approaches to 'localism,' it was born out of the experience that the Minamata region in Kumamoto. This region suffered from Minamata disease, caused by the release of methylmercury in the industrial wastewater from the major corporation's factory. However, the region has recovered gradually through the learning in which the residents cooperated, faced with the environment, and took action. Considering that Minamata is also the origin of ESD, the origin of 'localism' can be said to be ESD itself.

13.3 Potential and Challenges of Community-Based Educational Approach

Learning about local communities includes various sustainability issues, such as the environment, human rights, welfare, and disaster prevention, as well as issues rooted in real life. Therefore, the ESD learning method of 'Think Globally, Act Locally' (UNESCO, 2009) is neither too new nor too old, and it seems that the existing school curriculum of Japan can be achieved without radical reform. MEXT documents for schools on ESD suggest that ESD is not new and that existing efforts will be strengthened further by reviewing existing efforts from the perspective of ESD, by providing a common objective of 'Building a Sustainable Society', and by linking them together.[2] 'The Period for Integrated Studies', established under the 1998 Courses of Study, the broad standards for all schools in Japan, is based on the philosophy of lifelong learning and clearly states the policy of valuing experience, independently exploring issues, and fully utilising the resources of the local community. The Courses of Study, which have been in full implementation since 2020, share the philosophy of 'creating a better society through better school education' between schools and society and emphasise the realisation of an 'open-to-the-society curriculum' which nurtures the qualities and abilities necessary to become a future creator in cooperation with society. The preface, which was added to the Courses of Study for the first time, clearly states the idea that 'foster leaders for the creation of a sustainable society'. It notes: 'To realise this philosophy, it is very important for communities and schools to collaborate systematically and continuously', and as a concrete effort, 'Integrated promotion of community schools (School Management Council System) and community school collaborative activities is important'.[3] This appears to be a tailwind for ESD.

It was also recommended in the Course of Study that 'localism' and 'regional science' introduced into school education be practised, especially in the Period for Integrated Studies. However, as public education in Japan values neutrality and fairness, and the presence of various physical and institutional restrictions, these practices tend to be mild and conservative practices, far from their origin. There are various limitations[4] to school education in addressing discrimination, poverty, and unfair structures in local communities; connecting with those involved; taking action to resolve these issues; and engaging in learning to promote full understanding. Therefore, there is a tendency wherein learning is limited to 'discovering the charm of one's local area' (which is 'localism') and 'searching for something'.

Most ASPnet schools in Okayama City recognise this 'community learning' as ESD and practise it. However, 'community learning' includes, for example, 'Local treasure discovery', 'Learning from local masters', and 'Learning about the facilities that support the community' at the primary education level. At the secondary education level, they talk to local war veterans, visit local companies, and learn about local disaster prevention. These are regional learning. The mainstream aims at 'Fostering children's attachment to and pride in their hometown'. The students are 'learners' outside school, and the local community and people are learning

[2] Guidelines for the UNESCO Associated Schools in Japan: https://www.mext.go.jp/en/unesco/title04/detail04/1373242.htm

[3] https://www.mext.go.jp/b_menu/shingi/chukyo/chukyo3/siryo/attach/1365161.htm (in Japanese).

[4] Many previous studies in Japan, including Wakatsuki (2005), also have pointed out the tendencies and limitations of the "political neutrality" of public education, and the problems they cause. For example, if a school tries to explore more specific and familiar local issues while tackling the problems of "poverty" and "discrimination," there is a high possibility that they will face problems related to a certain family that the student at the school does not want to be exposed to. When these problems arise, schools tend to be as vague as possible because it is difficult to ensure that they can adequately discuss, understand, and resolve them within the context of the school curriculum, properly considering the human rights of the students themselves.

resources and 'collaborators' but are not co-explorers. School education has spread from the school to the community and aims to contribute uncritically to patriotic education by adapting to existing society and nurturing those who love the place they belong to or, more specifically, to have a mind to serve the place. The process of learning and empowerment has not become one where each person, as originally planned in ESD, voluntarily becomes a social creator.

However, it is not easy to face the challenges of the local community. The results of a survey of Okayama's ASPnet schools also suggest that teachers do not know or have the opportunity to know much about local issues. Even if they see some challenges, for example, they are not sure about who to connect with and how to tackle them. For teachers who are not necessarily local residents, it is difficult for them to understand the local issues and take action. This means it is necessary to work with people in a particular area who are cooperative with the school in a way that is consistent with their schedule. Under such circumstances, teachers who have been assigned to ASPnet schools and have learned about ESD have raised many questions. The most typical questions are 'Our school conducts community learning every year, but I'm not sure if it's really ESD' and 'Is learning about the local community ESD? If so, what is the difference between ESD and the Period for Integrated Studies?' These symbolise the challenges of ESD, 'community-based learning', and the Period for Integrated Studies in school education in Japan.

13.4 Transformational Learning of ESD and Challenges in School Education in Japan

The scope of ESD covers all global issues. These have led to new and complex difficulties in promoting ESD in school education in Japan. One reason is that ESD is not clearly different from the existing approaches of the Period for Integrated Studies. It is broad, complex, and ambiguous and would make it difficult for teachers to find new meanings to tackle ESD. Moreover, in the context of busy classroom work,[5] community-based or experiential learning, which requires a considerable effort but does not necessarily produce clear outcomes, tends to exhaust and deplete teachers and their motivation for new initiatives. ESD cannot be promoted in Japan unless it is practised based on the accumulation of conventional school education and systems.

However, sustainable development seeks fundamental 'transformation' from the root of our lifestyles and values; ESD is an effort to create new values, and education needs to be reoriented based on the criticism of conventional modern education. The four pillars of learning have also had a significant impact on ESD in terms of Japan's education policy, particularly the creation of the Period for Integrated Studies. The fifth pillar of learning – 'Learning to transform oneself and society',

[5] According to TALIS 2018, Japanese teachers have the highest amount of overtime in the world.

added to the UN-DESD Midyear Report – is considered since it is the core; that it is most important is sufficiently established (Combes, 2005). ESD has a major mission to transform our society, lifestyle, and education through its implementation, which is the most significant and toughest challenge of all. For example, at some ASPnet schools in Japan, students learn about environmental issues in classes at school, visit local recycling plants in general classes, and experiment with the reuse of milk cartons at school, but there is no change in the way they live at home. After a somewhat special extracurricular lesson, students open and copy the textbooks, and teachers unilaterally repeat questions and explanations. This is a limitation of making schools alone shoulder ESD.

Even if ESD began in schools or in a top-down manner, it will never be learning to transform oneself and society, as its philosophy dictates, unless, in the end, a wide variety of people are engaged in it on a daily basis. Again, based on the concept of CoP in Lave and Wenger (1991), the multi-layered and multidimensional engagement of a wide variety of people to ESD can be called ESD CoP. In ESD, the process of cooperation among various entities is emphasised. Based on Mezirow's (1994) transformational learning theory, it is thought that people learn most and are encouraged to change when they encounter different values and cultures. An ESD implementation community is not a single, strong, and committed group that aims to create a sustainable society and is fully aware of ESD but rather an aggregate of diverse and small communities that are overlapping and inspired, albeit with different levels of awareness and approaches to ESD or sustainability. Participants have free access to multiple communities. This is important because such multidimensional participation invigorates each community. For the community of practice not to fall into a disjointed group of hobby clubs, as Lave and Wenger (1991) mentioned, ESD must have a mechanism in which heterogeneous and diverse values, including contradictions and conflicts, fully interact. This brings us closer to the original ESD objective of creating new values for a sustainable society. If the root cause of the unsustainability of our world is our values and worldviews, then real transformation will not occur unless those values and worldviews change. The transformation is not triggered by calling from the outside, nor is it completed within ourselves. The radical transformation will be most possible in small-group interactions.

Transformative learning at the core of ESD emerges in the formation process of ESD-CoPs. The reasons are as follows. According to Lave and Wenger, CoP is not easy to build. It does not comprise roles or positions but is born steadily in activities and practices where people have an identity. Teachers who hold the key to the transformation of education toward sustainability must be legitimate peripheral participants in ESD-CoP in any field.

In the next section, how ESD-CoP has been formed based on a field survey of local cases is analysed. It focuses on how teachers can realise essential ESD as a leader of change from the inside while facing various issues in education and clarifies the possibility and challenges of the whole-community approach.

13.5 Requirements for a Whole-Community Approach from the ESD Model Case

Based on an analysis of a specific case, the requirements for a whole-community approach to ESD are clarified. The author has overseen supporting ASPnet schools in Okayama City since 2012 as a member of the ESD Promotion Center of Okayama University. In the Minan school district of Okayama City, which is taken up as a case study, since 2012, all three municipal schools in the district associated the ASPnet. To clarify the process by which ESD emerges in the local community, interviews with teachers and other related parties, participating in classes and activities, and collecting data have been done since 2019.

13.5.1 Outline of the Minan School District and Its History to the Present

The district, with a population of about 31,000, is located about 5 km southwest of the prefectural capital, the centre of Okayama City. The Sasagase River flows through the district. This area used to be a purely rural area where rice, wheat, and rushes were produced, but it has rapidly changed to an urban area during a period of high economic growth. In 1968, a wholesale centre of the prefecture opened its doors in the same district, rezoning in the area steadily progressed, and small and medium business industrial parks and housing complexes were built there, taking on the role of a major wholesale centre in the prefecture. With the opening of the new train station in 2005 and of a bypass through the area in 2011, the convenience of traffic within the area has greatly improved. There has been a remarkable increase in population and mobility. The younger population and the number of student schools in the district have risen to the largest level in the city.

Now that population growth and mobility have become intense, juvenile delinquency has become conspicuous in JHS. According to interviews with residents, they were very worried about the JHS students sitting in parks and smoking. Several students who were unable to enter the classroom loitered around the school, throwing garbage or school lunches. The situation was that the rate of juvenile guidance was the highest in the jurisdiction of the police station, including this area.

Schools and communities with these educational challenges are now popular areas where many child-rearing generations want to live. Not only that, but local and volunteer activities that children can enjoy have become so popular that they draw lots to decide participants every time. Today, there are various activities by local residents aimed at creating a sustainable society throughout the year, such as environmental conservation, disaster prevention, health promotion, and cultural and artistic activities. Behind the acceleration of these changes is the growing awareness of sustainability and ESD and the sharing of ideas not only by educators and institutions but also by local leaders. What made this awareness and sharing possible is not so much the efforts of a single school but rather the gears of CoP, already existing in the community, which are beginning to resonate as they engage each other based on sustainability.

13.5.2 Building Cooperative Relationships with Local Communities

The turning point for the district was that local people started gardening in the courtyard. The Minan JHS asked those who were willing in the community for their cooperation in improving education. Those who agreed were registered as school partners, which led to various forms of assistance. At that time, the JHS wanted to provide security and guidance to students who could not attend classes through patrols, but such ideas were rejected by the residents because the local people knew such relationships with children would not improve from their experience. Instead of taking such security and guidance positions, the school partners planted flowers in the garden in the courtyard and offered to take care of them with the students.

Ms. A, who had been promoting this initiative, says she spoke steadily to JHS students who were not able to enter the classroom at that time and asked them for help. When a student helped her, she always expressed her gratitude and, at the same time, spoke from her heart about 'small lives', such as flowers and tiny creatures. Gradually, the students opened up and started to talk to her about various things. They came to her home to study and sometimes eat together. The students, who seemed to be in a state of disorder, each had their own scars and loneliness, and they opened their hearts by receiving positive and kind words and acknowledging their own existence, which they had never been exposed to in the classroom before. Over time, they began to study together with the local people, and they succeeded in entering high school.

13.5.3 Community Schools as an Educational Policy of Okayama City

The collaborative relationship between local people and schools, as well as the fact that the 'school partner' system was made possible and effective at the same time, can be attributed to the efforts of local collaborative and community schools, which were promoted as part of Okayama City's educational policy. After several years of preparation, the school district officially launched the Community School system in April 2012. The educational institutions in the school district had a system in which they could cooperate relatively closely, and they knew each other's educational activities well, and the mindset of 'raising children in the local community' was gradually growing (Fujieda, 2016). According to Fujieda's analysis, visualising the results of theoretical and collaborative reflection on the community and schools has helped the district develop further collaborative relationships. At first, people in the community were critical of schoolteachers. However, through the activities of school partners, it has become clear that they have a deeper understanding of teachers' work, and they have started to offer more support for the school in various ways. The improvement of schools in cooperation with these school partners has accelerated the establishment of a system of cooperation to solve regional problems such as environmental conservation activities, health promotion, welfare (such as child-rearing support), and disaster prevention. In addition, exchanges between special needs schools, facilities for the disabled, nursery schools and kindergartens, elementary schools, and JHSs in the districts have been systematically implemented throughout the year. At the community learning centre, various people, parents, and residents had the opportunity to meet and learn about educational issues in the school district.

13.5.4 Joining ASPnet Schools in a JHS District

At the same time, with the support of the Okayama City Board of Education, based on these achievements, the three municipal schools in the Minan district became affiliated with ASPnet schools, and the district as a whole undertook the ESD project while valuing each school's unique efforts. At first, the activities focused on the awareness that people's interest in the community had been fading so that children could understand the characteristics of the community and become familiar with it. For example, there are 'clean operations' in the Sasagase River and in various traditional waterways in the school district. The students can experience old ways of life and industry by getting on a traditional wooden boat and being guided by the local people. There are also exchange programs at welfare facilities. These initiatives were not particularly advanced as local studies or ESD in Japan. It shows that they had a tendency to value existing activities as ESD. In other words, rather than making extra efforts as a school, it added the values of ESD to the classes conducted in cooperation with local communities in such subjects as the Period for Integrated Studies or Living Environment Studies and began to recognise them as school characteristics.

The policy of the Okayama City Board of Education was also not to start something new but to reaffirm what was being practised now as an approach to the 'creation of a sustainable society'. On the other hand, Okayama University's ESD Promotion Center, a member of the network of universities supporting ASPnet schools, gave concrete, professional advice to schoolteachers on how to formulate a curriculum and review the learning objectives themselves to enhance the quality of ESD practice. The author visited schools to give lectures on ESD and held workshops with teachers to discuss why the current activities of cleaning rivers, experiencing waterways, and collaboration among schools could be potential ESD activities and what kind of measures are needed to further enhance ESD. Although curriculums and teaching contents did not change significantly because of this, such steady activities have gradually unearthed rich educational resources in local communities through an awareness of sustainability, and these have been linked to form the basis for the creation of a sustainable society.

13.5.5 The Teacher Who Made the Knot

In addition to these, there are teachers who have been steadily knot-working on existing sustainability activities in the region. Teacher A of Minan JHS realised that the activities of gardening by local people have turned the school into a place where local residents can enjoy and interact with students and that this has had a positive effect on the education of the entire school. She felt the need and joy as an adult living in the community to meet, connect with, and work to create a better future. As the experience broadened her own perspective, she thought it was necessary for students to have more opportunities to come into contact with the various local people who are involved in community building. Accordingly, she tried to develop vegetable gardening activities from the perspective of ESD, not from the perspective of volunteer activities.

Beginning in 2015, she started a 'study in Okayama' program that encourages students to go outside and helps them explore their interests without fear of failure, using the Period for Integrated Studies. For example, students explore the local community, find its challenges, meet the expert to ask questions or to understand the current situation, and suggest a solution of their own to the public at the end of the year. The aim is to nurture students' problem-solving abilities, not only by simply carrying out activities together or conducting research and study but also by placing importance on the process in which students are actively involved in community development and make their own proposals. The important point is that not only the teachers involved but also the people in the community learn from and talk with each other and share a common understanding of local issues and education.

13.5.6 Regional ESD Centres: Private Companies and Community Centres

Okayama City registers organisations that implement ESD projects every year as RCEs and provides financial support for their activities. More than 300 organisations have been certified as Okayama ESD projects. The young president of the construction company in the Minan district mentioned above is more interested in sustainability than before and has been involved in community development activities in various ways since 2010. They rebuilt their own office building in a sustainable way, integrating the concept of permaculture. The company started a project called *Kurashi no Tane (Seeds of Life)* to create a place where various people can gather, learn about the environment, and seriously discuss community development, which was certified as one of the Okayama ESD project. In Okayama City, ESD activities were more facilitated in CLCs than in the past, and around the time of the 2014 UNESCO World Conference held in Okayama City, many CLCs conducted training to understand the international context of ESD, review their activities from an ESD perspective, and support local people's ESD-related activities. As a result of the conference, these CLCs have further enhanced their ESD perspective and recognised their role as an important base. The CLC in the Minan school district is also one of the active centres of ESD. It serves as a hub for connecting these people and has long been involved in various social activities as an ESD, such as holding workshops to deepen the understanding of people with disabilities and discuss local issues and their solutions, in cooperation with special needs schools.

It did not take long for the teachers and these people to connect and work together in various situations.

13.5.7 ESD to Blur the Boundary of Formal Education, Non-formal Education, Informal Education: The Local Event Called "Soup"

Today, with the aforementioned president of the construction company and the retired principal of Minan JHS leading the group to create a park where citizens can create on the unused land in the community, the Park Management Association has been established and continues to be active with various actors involved, from children to adults. With the goal of building a sustainable city, they are engaged in serious discussions with local people about what kind of park and what kind of town to create by learning from experts and studying overseas cases. For example, in the regularly held forums, discussions on the themes of 'regional disaster prevention', 'the light trading – food and economic cycle', 'Satomachi life – water and green network', and 'energy future' are held with professionals and practitioners in each field, along with local residents, key persons, and city officials.[6] In addition, small-scale events such as 'SDGs Thinking Experiment' and 'Voluntary Society for Community Development' are being held, and community-building practices based on sustainability are being carried out together with residents.

Among them, the event called 'Soup', held a few times a year, deserves special mention as an advanced ESD case.[7] Citizens have created a vegetable garden on the planned site so that everyone can enjoy growing vegetables daily. In addition, by

[6] For more information on these reports, see the *Kurashi no Tane* website: http://kurashinotane.jp/tanemaki_top/ (in Japanese).

[7] The detailed annual report of the activities of the Park Management Association and the Soup event is available on their website: https://www.kitanagase-park.com/ (in Japanese).

holding 'play parks'[8] that make full use of the natural environment there, it is also a place where children and parents can gather naturally. On the day of the event, everyone is supposed to harvest the vegetables, cook them, and enjoy the boiled soup. JHS students are the most active volunteers. They not only learn about growing and harvesting vegetables while communicating with people of different generations on the spot but also make soup by cooperating with each other with less-than-usual amounts of fire, water, and electricity outside. This provides an opportunity for disaster drills in a very natural manner in this area as the park is supposed to be a base for disaster prevention in times of crisis. They also care for and support babies and young children so they can play safely and happily in nature. They learn about human growth and experience the value of nurturing life through communicating with parents and young children. In addition to organic vegetable plantations, several local small businesses and civic groups that are pursuing a sustainable environment, health, and welfare have opened stores during this event, and they are cooperating in various ways. For JHS students, experiencing the activities of local people with aspirations and alternative businesses is an important experience in drawing a concrete picture of a sustainable society. As an unexpected result, not only students from Minan JHS but also students from neighbouring JHS have been inspired to participate recently. Some have started to make on-the-spot presentations on sustainability, and others to plan and implement science classes for children.

Above all, creating an ideal local park for everyone with other people will naturally cultivate a perspective not only on the natural environment and disaster prevention but also on various issues related to local sustainability, such as welfare and human rights. In other words, it is a field for various people to practise ESD, where they are tested on how to devise, contribute, and exert their power to create a sustainable society in a real and familiar environment – which they cannot learn in the classroom.

One of the JHS students participating in these activities said:

> We met many adults of the school supporters who are involved in the JHS, and people who are active in the CLC, and we came to realise that we want to be involved in making the community better like that.

The activity expanded from adults who enjoyed lively activities to JHS students, spread through the entire JHS, and, at the same time, expanded to the community. 'It is as if there is a "a good circle of voluntarism" in Minan', she said. Seeing these JHS students, primary school students started saying, 'We want to do the same thing ourselves,' and started volunteering.

At the same time, how have adults in the community transformed? The organisers of the event 'Soup' are now considering more intensive collaboration with the school for the independence of the students. Collaboration with school curriculums

---

[8] Okayama City has been working on the Okayama City Playpark Promotion Project to enrich the mental development of children through outdoor activities and increase the momentum for raising children throughout the community. https://www.city.okayama.jp/kurashi/0000013007.html (in Japanese).

has been promoted in earnest, and with the cooperation of organisers, disaster prevention education at JHS has been jointly implemented beforehand, so that the lessons learned there can be used for the events. In addition, for example, Teacher A mentioned above is making efforts to incorporate a 'soup' event into the curriculum as an ESD, rather than as a one-off event or volunteer activity. These are still in the exploratory stage; however, it is not easy – it is a process that takes time to penetrate. However, as JHS students increasingly participate, adults are becoming more aware of the educational functions of the place. It can be said that this is a movement to restore education to a more humane state and that the formation of an ESD community is promoted when these communities stimulate the creation of a sustainable society and the education for it to create new activities.

There is a saying in the area: 'Minan for Everyone'. The words speak for the atmosphere of this community. It is not that they have created ESD in order to promote ESD nor that they practise ESD out of a sense of obligation. It is a state of foreseeing a sustainable society beyond the fact that each of them is tackling it daily with a reason for living. Moreover, based on the idea of a sustainable society, educators in both social education and school are knot-working on a wide variety of sustainability-related CoPs, and children also function as knots. Knot-working has played a role in supporting the activation of each. In other words, the requirements of the whole-community approach are, first and foremost, that it be a bottom-up process and a place where small numbers of diverse people learn from each other to nurture the soil and the earth. As mentioned at the beginning, ESD is a broad, holistic concept that embraces a variety of values and has a number of points. Therefore, it is also a problem in that activities proceed without the interaction of different values and do not lead to social transformation. However, through ESD-related learning programs that are clearly aware of the need to create a sustainable society either in the form of volunteers or service learning, the diverse people who have come into contact with the opportunities for children to grow will find a common goal in nurturing the next generation, connecting with each other, and stimulating each other to activate their CoP. Unlike what Lave and Wenger (1991) call a CoP, this is a situation in which not just one participant but several CoPs are coming and going in a multi-layered manner, and each CoP is creating mutual activities, including contradictions and conflicts.[9] If these are referred as an ESD-CoP, there should be a variety of entrances for participants. These are the effective learning approaches of ESD that blur the boundaries of formal, non-formal, and informal learning described by Wals (2011). This is the second requirement of the whole-community approach. Furthermore, as a new development of ESD, it is important and essential to be able to connect the diverse CoPs related to sustainability and education, which are inherent in communities, with different values. Catalysing such diverse CoPs is essential for teachers, schools, and, above all, students with a will. Sharing these devices and perspectives with people is the third requirement of the whole-community approach.

[9] For a detailed analysis of the process by which school-based service and volunteer learning connect local community practitioners and promote the formation of an ESD-CoP, see Shibakawa (2016).

13.6 Implications for Teacher Education

The requirements for the whole-community approach as a new development of ESD were considered, taking up the cases of ESD being developed in the entire region. In Japan, where the school education system has been established and many children receive public education, the idea that school plays a central role in education is widespread. Like any other developed country today, however, the model course of being able to do well in school, go to a good school, get a good job, and live a good life is no longer an illusion.[10] On the contrary, there is still a tendency to demand school education as a solution to social problems, and the burden on schools and teachers is increasing. In light of the emergence of the concept of sustainable development, it is necessary to move away from this illusion of education and transform it into a more diverse and open form that creates a sustainable society in which all people can live well, connected and nurtued. New education is needed to create new values. Such new education is not a mere addition to existing education or a penetration of the goals set by international organisations. Rather, it needs to be transformed from the inside as diverse values interact through repeated cooperation and opposition. It should be noted that ESD refers to such transformative education and cannot be implemented only within the framework of school education.

Where, then, should it be started? As can be seen from the case of Minan, the transformation always starts with a small group and spreads over time, sometimes producing unimaginable results. That does not mean we should wait for the chance. It is an important suggestion that the problem-conscious schoolteacher did not try to deal with the problem alone and put the problem to the community as an SOS, and the school was opened to the outside, which was an opportunity for improvement. When the teachers of Minan JHS were asked about the status of the present students, the realisation of a calm educational environment, and the results of their good relations with the local community, they all said, 'It is not that we have been aiming for ESD, but that the people of this region have already done great work and have been helped by it'. One teacher said, 'How can you tell a child to go out into the community without having adults really connected to the community and having fun living there? I'm just doing this because it's fun.' Being involved with the community means teachers will have more time to go out and do activities on weekends and after school. On this point, the teachers said, 'It is true that I'm busy, but I don't think it's a burden because it's my job to see our student's lively appearance and growth'. When asked about the state of ESD, a local person answered, 'I didn't think we were practicing such an amazing thing [ESD] because it was something we just love and do. It became an unimaginable dream school 10 years ago. I'm so happy about that.' Today, many local residents are involved not only in gardening but also in teaching, after-school study support, and various hobbies and learning courses on Saturdays. In the community, JHS students are always active in

[10] Changes in the situation and difficulties of young people in the transition period in developed countries including Japan have been clarified through a large-scale survey by Inui et al. (2017).

any event.[11] These statements also suggest that ESD in this region is not something that is special or must be done but something that is rooted in daily life and in which not only children but also adults are learning and empowered. That is the essence of the whole-community approach.

Conversely, the fact that teachers are still working overtime and no concrete and effective measures have been taken against it is a problem to be solved. Furthermore, in Okayama City, the number of school violence cases at elementary schools is higher than the national average, and the number of truant children continues to increase.[12] Thus, it cannot be said that the recognition of ESD and the number of project participants are linked to relative improvements in school education. Another problem is that in public schools in Japan, teachers are transferred every few years, so if a particular teacher is leading ESD activities, it will not be sustained after the transfer. One of the major challenges is that local communities and schools need to actively analyse, accumulate, and share the process of building, maintaining, and developing essential collaborative structures, rather than as the artisanship of one progressive teacher or local frontrunner. Accordingly, there is a need for political action to clarify how much in terms of human resources, time, and support for teachers and schools are specifically needed to make such efforts truly effective and sustainable and to appeal to educational administrators. It can be said that there is room for teacher education institutions and researchers to make a great contribution in this respect as well. Teacher education institutions should also provide opportunities for future teachers to learn about their local communities through experience. This is because it is impossible for teachers just to theoretically advance ESD with the whole-community approach, but it is necessary for them to have substantial experience in their own lives, e.g. at social education institutions or in the community, and gain effectiveness and self-usefulness. This is the first and the most important requirement for teachers to advance ESD, fully engaged with the local communies.

References

Biesta, G. (2011). *Learning democracy in school and society*. Brill | Sense.

Bourdieu, P., & Passeron, J.-C. (1990). *Reproduction in education, society and culture* (R. Nice, Trans.). Sage Publications.

Combes, B. P. Y. (2005). The United Nations decade of education for sustainable development (2005–2014): Learning to live together sustainably. *Applied Environmental Education & Communication, 4*(3), 215–219.

[11] ESD Promotion Center of Okayama University, where the author of this chapter is assigned, focused on this whole-community approach in the Minan area and produced downloadable video learning materials for teacher education and the further development of ESD. It is uploaded on the website – https://edu.okayama-u.ac.jp/promotion_center/ (in Japanese). For the English subtitled version - https://edu.okayama-u.ac.jp/promotion_center/?lang=english

[12] Based on the report dated 22 October 2019, 'FY 2019 Survey on Problematic Behavior and Truancy among Schoolchildren' by the Guidance Division of Board of Education of Okayama City.

Delors, J., Mufti, I. A., Amagi, I., Carneiro, R., Chung, F., Geremek, B., Gorham, W., Kornhauser, A., Manley, M., Quero, M. P., Savane, M.-A., Singh, K., Stavenhagen, R., Suhr, M. W., & Nanzhao, Z. (1996). *Learning: The treasure within. Report to UNESCO of the International Commission on Education for the twenty-first century*. UNESCO.

Dewey, J. (1916). *Democracy and education: An introduction to the philosophy of education*. Macmillan.

Freire, P. (1972). *Pedagogy of the oppressed*. Herder and Herder.

Freire, P., & Illich, I. (1975). *Dialogo Paulo Freire-Ivan Illich*. Ediciones Busqueda. Japanese edition: Freire, P., & Illich, I. (1980). *Taiwa – kyouiku wo koete – I. Illich & P. Freire* (trans: Sunami, K. et al.). Yasosha. (in Japanese).

Fujieda, S. (2016). A discussion of the collaboration between community schools and community at large Minan JHS as a contact point and asset of the greater district community. *Bulletin of Center for Teacher Education and Development, Okayama University, 6*, 150–159. (in Japanese).

Gadotti, M. (2010). Reorienting education practices towards sustainability. *Journal of Education for Sustainable Development, 4*(2), 203–211.

Inui, A., Honda, Y., & Nakamura, T. (Eds.). (2017). *Japanese youth in crisis – A cohort study of young people, their education and careers, 2007–2011*. University of Tokyo Press. (in Japanese).

Jucker, R. (2002). "Sustainability? Never heard of it!" some basics we shouldn't ignore when engaging in education for sustainability. *International Journal of Sustainability in Higher Education, 3*(1), 8–18.

Lave, J., & Wenger, E. (1991). *Situated learning: Legitimate peripheral participation (learning in doing: Social, cognitive and computational perspectives)*. Cambridge University Press.

Mezirow, J. (1994). Understanding transformation theory. *Adult Education Quarterly, Summer, 44*(4), 222–232.

Orr, D. W. (1996). *Earth in mind: On education, environment, and the human prospect*. Island Press.

Shibakawa, H. (2016). Critical reflection on the pedagogical theories of service-learning: Current challenges and new possibilities in educational practice for sustainable development. *Journal of Japan Academic Association of Socio-Learning and Service Learning, 29*(2), 124–135. (in Japanese).

UNESCO. (2009). *UNESCO world conference on education for sustainable development: Bonn declaration*. https://unesdoc.unesco.org/ark:/48223/pf0000188799.locale=en

Wakatsuki, K. (2005). Service learning for citizenship education. *Buraku Liberation Research, 163*, 67–79. (in Japanese).

Wals, A. E. J. (2011). Learning our way to sustainability. *Journal of Education for Sustainable Development, 5*(2), 177–186.

Chapter 14
Informal Education as a Strategic University–School Partnership for Enhancing the Quality of Science Education for Sustainable Development: An Example from Malaysia

Munirah Ghazali, Zainun Mustafa, Mohd Zohir Ahmad, and Shahrom Mahmud

14.1 Introduction

Rapid globalization and technological advancement have led the World Economic Forum (Baller et al., 2016) to release, in 2016, a list of the most valued skills in the twenty-first century, which includes creativity; curiosity; communication; collaboration; confidence; critical thinking, along with grit leadership; and adaptability. At the heart of these twenty-first-century skills are not just student grades but also their

M. Ghazali (✉)
RCE Penang, Universiti Sains Malaysia, Albukhary International University, Penang and Alor Star, Malaysia
e-mail: munirah@usm.my

Z. Mustafa
RCE Penang, Universiti Sains Malaysia, Albukhary International University, Penang and Alor Star, Malaysia

Universiti Pendidikan Sultan Idris, Tanjung Malim, Malaysia

M. Z. Ahmad
RCE Penang, Universiti Sains Malaysia, Albukhary International University, Penang and Alor Star, Malaysia

RCE Penang, Universiti Sains Malaysia, Penang, Malaysia

Universiti Pendidikan Sultan Idris, Tanjung Malim, Malaysia

S. Mahmud
RCE Penang, Universiti Sains Malaysia, Albukhary International University, Penang and Alor Star, Malaysia

RCE Penang, Universiti Sains Malaysia, Penang, Malaysia

Universiti Sains Malaysia, Penang, Malaysia

H. Fujii, S.-K. Lee (eds.), *Science Education for Sustainable Development in Asia*, Education Innovation Series, https://doi.org/10.1007/978-981-99-8711-5_14

ability to solve problems, and the inspiration underlying the importance of these skills is briefly summarized in the United Nations Educational, Scientific and Cultural Organization's (UNESCO's) four pillars of learning: learning to know, do, be, and live together. Furthermore, the successful implementation of university–school partnerships for achieving the goals of Education 2030, Education for Sustainable Development (ESD), and the Sustainable Development Goals (SDGs) requires close engagement and collaboration between stakeholders to promote multifaceted informal learning programs (Fig. 14.1).

Schools in Malaysia, like in many other countries worldwide, face challenges, including ESD elements in their formal curriculum (McKeown & Hopkins, 2005; Ramzy & Wahieb, 2012). These challenges, in turn, create a barrier to the meaningful commitment of universities and schools to the development of formal university–school partnerships. Therefore, partnerships between these two parties are foreseen to provide unique opportunities for and become key drivers of change and innovation in and for the societies in which they operate. By considering the potential pivotal role of these partnerships and acknowledging the barriers to their implementation, the Regional Center of Expertise (RCE) Penang has been and continues to be searching for ways to promote the formation of these partnerships through both informal and formal educational routes.

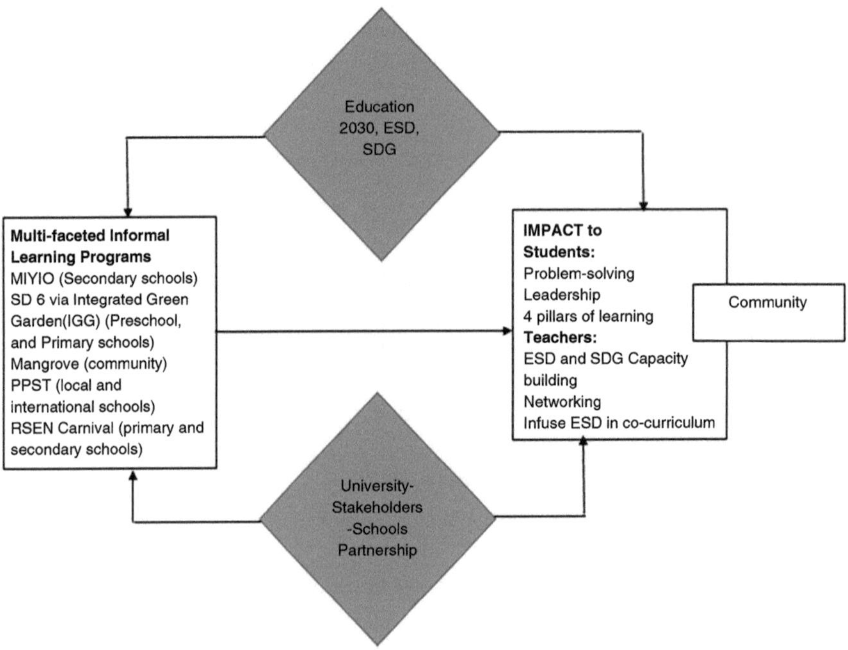

Fig. 14.1 Multifaceted informal learning programs involving strategic university–school partnerships

RCE Penang has successfully implemented the six multifaceted informal learning programs highlighted in Fig. 14.1, allowing for the inclusion of ESD goals, together with the formal curriculum through cocurricular activities and informal education. These informal learning programs are the following: (1) the Malaysian International Young Inventors Olympiad (MIYIO), (2) the Integrated Green Garden (IGG) project, (3) Rehabilitating mangrove forests in cooperation with local communities at Merbok Mangrove Reserve, (4) Program for Encouragement of Science and Technology Penang (PPST), (5) collaboration in the local ecosphere via school carnival and exhibition, and (6) Empowering Students' Leadership in ESD through Networking at the Asia-Pacific region level. All six programs were formed based on dynamic and flexible university–school partnerships, which, in turn, are related to an increase in the students' problem-solving skills, leadership skills, and the four pillars of learning proposed by UNESCO. For teachers, the target variables of this dynamic partnership are capacity building, networking, and the ability to make students aware of the ESD goals through informal education and cocurriculum activities. Figure 14.1 summarizes the university–school partnerships through the six multifaceted informal learning programs developed by the RCE Penang.

The discussion in this chapter is mostly focused on the university–school partnerships evoked by the MIYIO, which is an excellent example of an informal education project integrated into the curriculum through social learning, for five consecutive years. The MIYIO provides an avenue for teachers and students to realize SDG 4 in their daily lives as well as empowers them with twenty-first-century skills by embracing the aforementioned four pillars of learning. The following questions provide the framework that guided the explorations for this study, and they consider the strategic university–school partnerships that emerge through informal education projects and the role of these partnerships in enhancing the quality of science education for sustainable development:

1. How do university and school partners collaborate to advance the Education 2030 agenda, specifically ESD goals, through informal science educational projects?
2. How do universities, as institutions of higher learning, engage the school community in learning processes, especially regarding ESD goals, through problem-solving?
3. How can we support teachers and students in acquiring twenty-first-century skills through an informal curriculum and social learning?

This chapter will deliberate on these three questions by discussing the case of the MIYIO, which is one of the various university-school partnership projects initiated by RCE Penang. Herein we focus on the objectives, target groups, implementation process, resource use, opportunities, challenges, lessons learned, and implications of the MIYIO for science education for sustainable development.

14.2 Background

14.2.1 Informal Science Education in Malaysia

Generally, science education can be taught in both formal and informal settings. Formal education has systematic characteristics, structured curriculum outlines, clear goals, organized assessments, certified educators, and formal premises (e.g., schools and institutions; Saleh et al., 2020). The formal science curriculum in Malaysia comprises four components: scientific knowledge, skills, thinking, and values. Researchers have argued that classroom-based science lessons leave behind knowledge gaps regarding how to apply scientific knowledge in real-life contexts. By comparing informal education activities and conventional formal education activities at school, Giannakos (2020) asserted that children could develop their science literacy and learning experience from the diverse contexts and resources provided by informal educational settings. Indeed, informal education can be applied in various contexts and by using various resources, including through educational events (Vologzhina et al., 2020), interdisciplinary projects (Ryazanova et al., 2020), free-choice learning activities (Dunlop et al., 2019), multistakeholder involvement in extracurricular activities (Cambers et al., 2019), and science exhibitions (Héctor et al., 2016). The literature also shows that the integration of both formal and informal learning experiences improved children's scientific reasoning abilities (Gerber et al., 2001; Falk et al., 2012), content knowledge (Fenichel & Schweingruber, 2010), and motivation to learn science (Goto et al., 2018).

In Malaysia, informal education in school settings mostly involves cocurricular activities that are managed by the school in coordination with the Sports, Co-Curricular, and Arts Division, Ministry of Education (*Bahagian Sukan, Kokurikulum Dan Kesenian, Kementerian Pelajaran Malaysia*). Examples of common informal education activities in the country include the visitation of places presenting educational exhibits, listening to radio broadcasts, watching specific TV programs, reading journals and magazines, and participating in contests. Several studies have shown the benefits of participating in competitions to enhance science education toward sustainable development (e.g., Corbett & Marshall, 2013; Issa et al., 2014; Shukla et al., 2017).

14.2.2 Education for Sustainable Development

The ESD is a program promoted by the United Nations (UN) that encourages changes in knowledge, skills, values, and attitudes to promote sustainable development and a more just society through education. The SDG 4 was first announced in the UN's Agenda 21 action plan, which recognized the role of education in improving humans' capability to address the environment and development (MacNaughton & Koutsioumpas, 2017). While respecting cultural diversity, the SDG 4 empowers present and future generations to be responsible for and make informed decisions about environmental integrity, economic viability, and a just society. Hence, instead of fractional, time-bound learning, the SDG 4 proposes lifelong learning as an integral part of wholesome quality education. Achieving a holistic and transformational society requires viewing the learning environment as going far beyond learning specifications and outcomes, outmoded pedagogical paradigms, and individualized learning settings. Furthermore, transforming society is not a one-person job, meaning that the SDG 4 agenda to stimulate learning, core competencies (e.g., critical and systemic thinking), collaborative decision-making, and responsibility requires strategic coordination and management. Higher learning institutions can undoubtedly provide significant contributions to this agenda by becoming key partners of global development through proliferating educational networks and partnerships.

14.2.3 Strategic University–School Partnerships for Sustainable Development

At the core of good partnerships lies the ability to optimize and integrate the use of limited and diversified resources (e.g., financial, natural, or human resources) for achieving greater impacts and producing increased value. This core ability is described in the partnership paradigm put forward by the 2030 SDGs, which became known as the "Partnership Delta" (Stibbe et al., 2018). This form of resource integration requires a high degree of planning, attention to procedures, sensitivity to cultural differences, and commitment to building mutual trust. In this article, these partnerships are defined as follows:

> *Partnerships for sustainable development are multi-stakeholder initiatives voluntarily undertaken by Governments, intergovernmental organizations, major groups, and other stakeholders whose efforts are contributing to the implementation of inter-governmentally agreed development goals and commitments.* (Stibbe et al., 2018; p. 6)

Scholars have described that the value of university–school partnerships lies in their potential to be mutually beneficial for all those involved. These partnerships are especially critical in enriching the experiences related to and the means of building social capital (Ramaley, 2001). Kim et al. (2013) further emphasized this pivotal role by claiming that university–school partnerships could promote enriched experiential learning opportunities by providing shared goals, funding availability, interimpact mechanisms, interdisciplinary collaboration, distinct roles, and expert resources. Such partnerships can also be viewed as endeavors to recognize the contributions of universities to metamorphosing societies toward achieving sustainability at both the local (Dusenbery et al., 2008) and international levels (Kuzhanova & Klets, 2017). Thus, strategic university–school partnerships can be regarded as benchmarks and a requirement for achieving sustainable development.

These partnerships require the involvement of all relevant stakeholders and individuals, as well as the collaboration of different sectors and institutions. Regarding this topic of collaboration and involvement, one study remarked that people and organizations can work together and use various strategic mechanisms to share knowledge, advocate causes, and take action (Khelghat-Doost et al., 2011). The precursors of active participation of relevant actors in such partnerships are ensuring that they have common goals and share the vision of the path toward sustainability and community development. Willard and Creech (2008) claimed that the combination of efforts, networks, and different levels of partnerships would result in more significant impacts on sustainability issues than single, one-sided actions. Khelghat-Doost et al. (2011) further discoursed that higher education institutions should be championing the cause of ESD by promoting strategic university–school partnerships that also get the government, the community, nongovernmental organizations, and the private sector involved.

14.2.4 Regional Center of Expertise Penang on Education for Sustainable Development

In response to the lack of a shared understanding among stakeholders regarding the operationalization of the 17 SDGs (Sachs et al., 2019), and as a part of the UN's Decade of ESD, the concept of RCE was launched. The RCE is meant to be a social platform for multisectorial and integrated learning, disseminating information, dialogue, and collaboration to promote ESD among regional and local stakeholders. The network of RCEs has been increasing rapidly since the official recognition of the first seven RCEs in 2005 (Aipanjiguly et al., 2008). Since then, the RCE network has been working on defining strategic areas for collaboration that can bring together different RCEs interested in developing actions for these specific areas. In general, RCEs should serve as translators of the global objectives put forward in the SDG 4 for the context of the local communities in which they operate. Accordingly, various RCEs have been positioned across the globe and have official links to UN agencies, formal education institutions, and informal educators; this strategic placement enables these centers to bridge the communication between and promote efficient information sharing among relevant stakeholders.

Hosted by the Universiti Sains Malaysia (USM), the RCE Penang, in collaboration with its stakeholders, has been engaging annually with more than 200 schools and over 1000 students in Malaysia – as well as about 50 schools and more than 200 students in the Asia-Pacific region – through various projects and competitions focused on ESD. This RCE has the role of addressing the existing needs and challenges in employing an educational framework capable of ensuring sustainability in the region. As mentioned in its mission statement, educational programs form the backbone of the RCE Penang because they are the promoters of the integration and reorientation of existing platforms and programs toward the advancement of sustainable development. Initially, the concept of ESD was somewhat marginalized in the region wherein RCE Penang is localized, and there were several difficulties in explaining this concept to invested stakeholders and getting the notion to be understood; nonetheless, by collaborating with strategic partners and coordinated communication, RCE Penang has been disseminating a framework for the development of regional ESD projects and programs.

RCE Penang has undergone a dynamic developmental process, garnering recognition over time as the information and reference pool for ESD enlarges not only in Penang but also in the national arena. RCE Penang continuously attempts to consolidate the linkages between the various organizations, institutions, nongovernmental organizations, agencies, and concerned individuals invested in achieving the shared goals of ESD via the promotion of projects and activities that emphasize the vital role of partnerships within communities. RCE Penang's MIYIO is an example of an excellent collaborative project (involving the community, ESD stakeholders, and industry partners) for developing global linkages by expanding cooperation in innovation, education, environment, and sustainable development. After more than 5 years of organizing the MIYIO, RCE Penang has created global allies that help strengthen the activities related to SGD 4 via informal science education projects.

14.3 Malaysian International Young Inventors Olympiad

Year after year ever since the onset of this informal science education project, the partnerships created for organizing the MIYIO have played an indispensable role in spreading sustainable education to students, youths, and communities in Malaysia.

14.3.1 Rationale and Objectives

The MIYIO aims to be a platform for youth to share their research findings and developments regarding science and engineering. Its purpose is to encourage and channel students' scientific endeavors through the promotion of experimental scientific research. In the MIYIO, students participate by acting as school representatives and producing authentic innovations according to the themes and issues in ESD and the SDGs. It is an annual program that has been held since 2014 and will hopefully continue to thrive under the strategic partnership between RCE Penang and schools. This competition successfully reached over 1000 schools and more than 5000 students over the past 5 years. However, to abide by the national movement restriction order due to the COVID-19 pandemic, the MIYIO 2020 was canceled.

14.3.2 Implication

The MIYIO requires students to innovate using sustainable elements, undertake scientific research to sharpen their thinking skills, and continuously encourage participation across 17 transnational disciplines – hence sharing goals with SDG 4. Some examples of projects related to the 2019 MIYIO included biogas-food waste, heat shield made of *kapok* (*Ceiba pentandra*) fiber, toilet fragrance from *pandan* (*Pandanus amaryllifolius*) leaves, biodegradable plastic produced from corn (*Z. mays*) cob, sorbents extracted from walnut shells (*Juglans regia*), alternative medicine derived from olive stone (*Olea europaea*), diarrhea prevention for sheep using *apollinaris* herb, oat (*Avena sativa L*) oil-based biodiesel, and the use of *hypericum* (*Hypericum perforatum*), a natural herb, to fight tobacco addiction. According to the laws of the International Science and Engineering Fair, the scopes and rules of the research presentation were determined.

Considering its characteristics, the MIYIO has eventually become an interactive meeting arena and platform for future collaborations. Specifically, the competition requires participating students and teachers to be physically gathered in one place to share their inventions and ideas. Further, government agencies embraced the project by endorsing innovations, coordinating patent development and applications, and copyrighting novel ideas. The local newspaper, citing data from the Intellectual Property Corporation of Malaysia, reported that in 2017 alone, 30 innovative and creative products featured by students in the competition passed the initial screening (The Sun Daily, 2018). These descriptions show that the shared purpose and vision regarding the competition brought in allies and associates that empowered the program, consequently impacting the broader community. The "shared responsibility" networking paradigm of this competition allowed for shared goals indirectly to all actors involved: participants (school representatives), the school's hosting community, government agencies, and nongovernment organizations. Accordingly, the MIYIO has become an annual flagship program under the Residential School Division, Ministry of Education, and it even won the Acknowledged Flagship Project (Thematic Category: SDG 17 – Partnerships for the Goals).

14.3.3 Partnerships

The Ministry of Education of Malaysia has always encouraged schools nationwide to make efforts to ensure that children achieve their full potential. As the governing body, the Ministry of Education must affirm the school's overall direction while allowing school administrators and teachers to operationalize their ideas into a workable plan. Sekolah Menengah Sains Tun Syed Sheh Shahabudin and Sekolah Menengah Sains Sultan Mohamad Jiwa, two high schools in Malaysia, endeavored to package the ideas from their teachers and students to apply them on a large scale. The principals of these two high schools initially sought advice from the RCE

Penang on the organization of competition at the national level. Early discussions with the RCE Penang officials were strategically planned to allow for partnerships between the respective schools to be built. Besides consultation with academic experts, RCE Penang provided a platform for Sejahtera Schools Club and Regional Sejahtera ESD Network members to participate in the competition. Through the internal link among USM academicians, who got involved in the discussion by the intermediation of RCE Penang, it was decided that the juries – namely, faculty members from various disciplines with vast experience in judging innovation competitions – for the competition would stem from USM and related industries. This goes to show the uniqueness of the partnerships gained via the MIYIO, considering that the project itself started as an initiative from two high schools attempting to conduct an informal science learning platform using a school-based club.

The MIYIO competition is advantageous for the development of university–school partnerships, particularly for the purposes of the RCE Penang, by giving way for schools and universities to collaborate in the organization of an international event; this partnership, in turn, boosts capacity building for teachers and students as they experience the processes related to such organization. This leads to university–school partnerships that resonate geographically and attract more government bodies, nongovernment agencies, private organizations, and even international institutions to collaborate. As time went on, with the schools serving as active hosts and through expert consultation, the RCE Penang eventually started to consistently play the role of a central link between potential allies for the organization of the MIYIO, and this has been so for 6 years. This allowed for the university–school partnerships enabled by the project to evolve and generate sufficient revenue for the sustainable promotion of the projects.

In the year after a MIYIO, the organizing committee often endeavored to create more like-minded alliances by contacting the Residential School Division, Ministry of Education, Malaysia. While the schools involved in this process identified and utilized their available resources and outstanding boarding school structures to provide hospitality for the participants of the MIYIO, the Residential School Division primarily served as the line of authority, the funder, and the host of the competition nationwide. The Intellectual Property Corporation of Malaysia also acknowledged the authentic innovations created by the project and how they addressed issues related to ESD and SDG, which led this institution to actively respond to the MIYIO by promoting public awareness regarding the intellectual innovations and providing advisory services to the participating schools.

Then as the active involvement of a large number of government bodies started to take place, it became undeniable that the competition had drawn the attention of strategic allies from several key stakeholders and communities. For instance, in 2019, several organizations teamed up as co-organizers of the MIYIO, including the USM, various government bodies (Ministry of Education Malaysia, Fully Residential and Excellence Schools Management Division, Kedah State Education Office, Kuala Muda Yan District Education Office, the Centre for Global Sustainability Studies, Universiti Teknologi MARA, Universiti Malaysia Perlis, Standard and Industrial Research Institute of Malaysia, Malaysian Institute of

Pharmaceuticals and Nutraceuticals, Sungai Petani Municipal Council, and Kedah Educational Technology Division), and nongovernmental organizations (the Jerai Hotel and Parents Teachers Schools Association of participating schools). The MIYIO also garnered the attention of the local media, and all related allies actively posted promotional campaigns for the competition on their official websites.

The RCE Penang also played the role of providing an extended and smooth platform for members of other RCEs, allowing for international schools to get engaged on the MIYIO. The other RCEs promoted the MIYIO on their official websites through announcements. Then students from Malaysia and other countries in the Asia-Pacific region revealed their interest in the competition itself, thus serving as an example of how external links can bridge geographical challenges. Based on our own records, the MIYIO has been able to attract more than 800 teachers and students from Malaysia and Thailand, Iraq, the Philippines, Kyrgyzstan, Turkmenistan, Tajikistan, Indonesia, Nigeria, and the United States of America; more than 300 projects have been exhibited in the competition (143 international projects and 175 Malaysian projects); the competition also sometimes drew interest from the private sector, with examples being the participation of the Perda City Mall in 2015 and the Mydin Mall Bukit Mertajam in 2017.

The strategic partnerships among universities and schools that emerged through the MIYIO have led to a growing network of associates and allies. Some of the outcomes of the fruitful collaborations enabled by the project include receiving reciprocal consultancy, financial support, bureaucracy assistance, provision of juridical expertise, and the influx of participants from both local and global communities. As aforementioned, these strategic partners and allies also promote the MIYIO on their websites or through official communication channels, allowing the project to reach massive audiences that grew year after year. Table 14.1 shows the number of participants by year.

Table 14.1 shows the steady increase in the number of participating schools, although the total number of participants displays slight fluctuations. A summary of the partnerships involved in the organization of this informal science education project, which configures a strategic university–school partnership based on ESD shared goals, is shown in Fig. 14.2.

Table 14.1 The number of participants in the MIYIO by year

Year	Themes	Total number of schools	Total number of participants
2015	Innovation Core of Excellence	180	680
2016	Innovation Core of Excellence	180	498
2017	Innovation Core of Excellence	220	424
2018	Nurturing Ideas Culturing Innovations	230	518
2019	Towards Sustainability-Driven Innovations	230	318

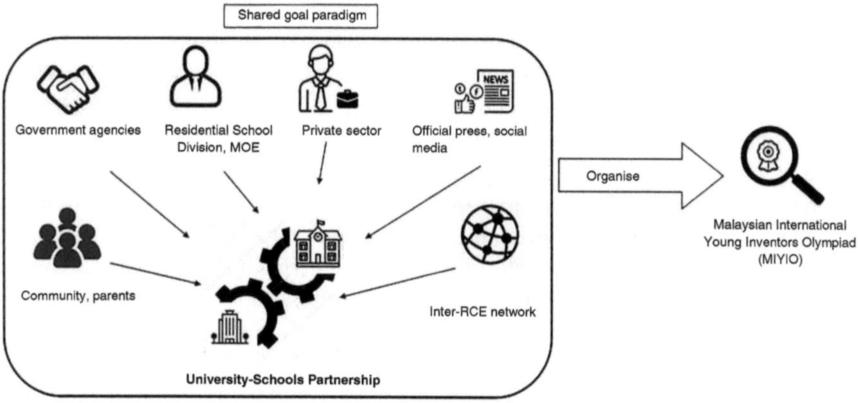

Fig. 14.2 Overview of the partnerships involved in organizing the MIYIO

14.3.4 The Shared Goal Paradigm in Partnerships

The case of the MIYIO fits the "Combine/Integrate" element of the "Partnership Delta" concept (Stibbe et al., 2018). From our experience, strategic university–school partnerships offer a comprehensive value chain for the affected students, who are encouraged to become active problem-solvers and present their ideas in recognized platforms. Particularly, the MIYIO required students to work collaboratively to identify issues in society and then propose solutions using scientific inquiry. To this end, students must plan, resource, design, experiment, analyze, conclude, reflect, improve, finalize the experiment, report, and present a case; this is clearly a mimic of real research work, which is likely to provide greater benefits for students than winning a trophy, considering that students must assimilate scientific content and contexts and then propose practical and product-based solutions. Hence, the MIYIO bridges the interdisciplinary characteristics of science learning and elevates the student's ability and knowledge.

The high flexibility of this informal learning mode is also advantageous for the cultivation of scientific thinking and manipulative skills. Compared with infusion of sustainability in the subject matter, classroom-based experiment, competitions such as the MIYIO provide more opportunities for students to embark on the scientific exploration based on own interests and creativity; the related preparation process for the competition provides more venues for the students' communication and presentation skills to be honed; deliver more pathways for them to train themselves to be more aware about existing problems; provide further opportunities for students to play an active role in solving these problems. Besides enabling the development of a scientific mindset, scientific thinking skills, and a scientific attitude, the MIYIO's final delivery mode (i.e., idea presentation) promotes the learning of important twenty-first-century skills, including confidence, collaboration, and creativity. The involved teachers also encourage the students' engagement in the latest innovations through partaking in the competition and prioritizing sustainability.

Thus, the strategic partnerships underlying the MIYIO functioned as "enablers" of the translation of existing competition practices in Malaysia at the school level to the national and international levels by providing the networks that helped magnify these practices. Particularly, these partnerships endeavored to juxtapose the essential resources and coordinate actions of multiple actors, with RCE Penang bringing together organizations with different mandates and cutting across interconnected the SDGs to deliver more integrated solutions. A synergic coordination was created by aligning action, resource sharing, collective learning, capacity building, access to private services, and preventing task duplication.

The pool of diverse resources accumulated through these partnerships creates a new and effective method of communication for all those involved. The various communication platforms promoted by the partnerships underlying the MIYIO increased the engagement of different actors and the dissemination of information. Partners could develop norms and standards for schools to use in the development of well-recognized, impactful inventions related to sustainability. The situational exchange of knowledge and experiences promoted by the MIYIO allowed partners to learn from each other, while students and teachers improved their scientific skills, thinking, and problem-solving for interests that stretch beyond the boundaries of the school curriculum. Thus, the leveraging of the knowledge, expertise, and capacity of the hosting schools and RCE Penang allowed for the MIYIO and its underlying partnerships to deliver more opportunities for school-going children to actively transform the social systems surrounding them and embed these systems with environmentally sustainable innovations.

14.3.5 Challenges

Both schools and universities tend to encounter funding constraints for championing ESD, the SDGs, and Education 2030 goals. In this context, the MIYIO and its related university–school partnerships may represent an attractive initiative for funders potentially interested in championing Education 2030s.

In the year 2020, the COVID-19 pandemic led the Malaysian government to take serious action to contain the related virus, leading to the restriction of travel to and from other countries. Students in Malaysian boarding schools were also mandated by the government to stay at home and engage in virtual learning activities. Additionally, the restrictions imposed by the government based on the Movement Restriction Order did not allow for mass gatherings (i.e., when they followed a specific set of requirements). Considering that the major target population of the MIYIO is school-aged children, their participation may require the accompaniment of those responsible for them; with all these hindrances on the table, the organization committee decided to postpone MIYIO 2020 until further notice, having not yet been conducted to the date of publishing of this article.

14.4 Conclusion

The MIYIO is the brainchild of the attempts of two high schools in Malaysia to elevate the quality of their cocurricular activities. Eventually, it became an annual event and a platform for young scientists to share their science, technology, engineering, and mathematic research findings and develop proenvironment inventions. It all started with the initiatives of two Malaysian schools to reach out to RCE Penang for expert advice on how to organize an international-level competition. The subsequent "reach" that the MIYIO achieved through the assistance of RCE Penang implies the recognition of the role of this RCE as a public reference for ESD and the SDGs in Malaysia. The RCE took this as an opportunity to promote the SDG 4 goals and celebrate the initiative of these schools via the development of an informal science education project. Through the MIYIO, RCE Penang established a close network that serves as a mechanism to nurture the outstanding ideas of principals, teachers, and students at the national and international levels. Further, the collaboration of USM scholars, who have thorough experience in organizing international seminars and colloquiums, helped the schools play their roles as executive parties. Meanwhile, RCE Penang provided schools with information, framework analysis, and expert guidance, as well as coordinated efforts to gather other allies toward the shared goal.

The key practical implication of exploring the case of MIYIO is that our society must open its eyes to this partnership format and start viewing universities as valuable idea generators, major influencers, and potential champions of the SDG 4 goals through the application of their expertise.

References

Aipanjiguly, S., Mochizuki, Y., & Fadeeva, Z. (2008). Emerging communities of practice: Regional centres of expertise discuss themes and evaluations. *Journal of Education for Sustainable Development, 2*(1), 17–20.

Baller, S., Dutta, S., & Lanvin, B. (2016). *Global information technology report 2016*. Ouranos. https://core.ac.uk/download/pdf/51343122.pdf. Accessed on 28 Sept 2020.

Cambers, G., Diamond, P., & Verkooy, M. (2019). The role of informal education in climate change resilience: The Sandwatch model. In L. Filho, W. Hemstock, & L. Sarah (Eds.), *Climate change and the role of education* (pp. 371–383). Springer.

Corbett, K. S., & Marshall, J. M. (2013). *STEM applications: Integrating informal learning with the formal learning environment*. Paper presented at the ASEE annual conference and exposition, conference proceedings. https://peer.asee.org/stem-applications-integrating-informal-learning-with-the-formal-learning-environment. Accessed on 28 Sept 2020.

Dunlop, L., Clarke, L., & McKelvey-Martin, V. (2019). Free-choice learning in school science: A model for collaboration between formal and informal science educators. *International Journal of Science Education, 9*(1), 13–28.

Dusenbery, P. B., Harold, J. B., McLain, B., & Curtis, L. (2008). Space weather outreach: An informal education perspective. *Advances in Space Research, 42*(11), 1837–1843.

Falk, J. H., Randol, S., & Dierking, L. D. (2012). Mapping the informal science education landscape: An exploratory study. *Public Understanding of Science, 21*(7), 865–874.

Fenichel, M., & Schweingruber, H. A. (2010). *Surrounded by science: Learning science in informal environments*. The National Academies Press.

Gerber, B. L., Anne, M. L. C., & Edmund, A. M. (2001). Relationships among informal learning environments, teaching procedures, and scientific reasoning ability. *International Journal of Science Education, 23*(5), 535–549.

Giannakos, M. N. (2020). An introduction to non-formal and informal science learning in the ICT era. In M. Giannakos (Ed.), *Non-formal and informal science learning in the ICT era*. Lecture notes in educational technology. Springer.

Goto, T., Nakanishi, K., & Kano, K. (2018). A large-scale longitudinal survey of participation in scientific events focusing on students' learning motivation for science: Antecedents and consequences. *Learning and Individual Differences, 61*, 181–187.

Héctor, E. G. H., Palomares, G. S., Alfonso, A. O. T., & Rosas, P. L. (2016). The trailer of science in informal education in Jalisco, Mexico. *Mathematics Education, 17*(2), 90–93.

Issa, G., Shakir, M. H., & Al-Bahadili, H. (2014). Competition-based learning: A model for the integration of competitions with project-based learning using open source LMS. *International Journal of Information and Communication Technology Education, 10*(1), 1–13.

Khelghat-Doost, H., Sanusi, Z. A., Fariddudin, T. F. F. D. T., & Jegatesen, G. (2011). Institutions of higher education and partnerships in education for sustainable development: A case study of the Regional Center of Expertise (RCE) Penang, Malaysia. *Journal of Sustainable Development, 4*(3), 108–117.

Kim, D. J., Park, Y. J., Cho, J. I., & Kim, D. (2013). General and specific characteristics of a university-schools partnership: Promoting learning opportunities for students with deafness or hearing impairments. *International Education Studies, 6*(1), 57–62.

Kuzhanova, N., & Klets, T. (2017). Methodological and psychological aspects of education for sustainable development in Russia with regard to international cooperation. *Proceedings of the 11th International Scientific and Practical Conference, I*(165), 168.

MacNaughton, G., & Koutsioumpas, K. (2017). Universal human rights education for the post-2015 development agenda. In J. Zajda & S. Ozdowski (Eds.), *Globalisation, human rights education and reforms* (pp. 15–33). Springer.

McKeown, R., & Hopkins, C. (2005). EE and ESD: Two paradigms, one crucial goal. *Applied Environmental Education and Communication, 4*(3), 221–224.

Ramaley, J. A. (2001). Why do we engage in engagement? *Metropolitan Universities, 12*(3), 13. https://eric.ed.gov/?id=EJ641308. Accessed on 28 Sept 2020.

Ramzy, O., & Wahieb, R. (2012). Branding the green education: Challenges facing the implementation of education for sustainable development in Egypt. *Discourse and Communication for Sustainable Education, 3*(1), 83–99.

Ryazanova, N., Naumov, V., & Kamennykh, N. (2020). Implement environmental education implementation trajectories for sustainable development in formal, non-formal, and informal education based on eduScrum project management methodology. *E3S Web of Conferences. EDP Sciences, 169*, 05002. https://doi.org/10.1051/e3sconf/202016905002

Sachs, J. D., Schmidt-Traub, G., Mazzucato, M., Messner, D., Nakicenovic, N., & Rockström, J. (2019). Six transformations to achieve the Sustainable Development Goals. *Nature Sustainability, 2*, 805–814.

Saleh, S., Muhammad, A., & Syed Abdullah, S. M. (2020). STEM project-based approach in enhancing conceptual understanding and inventive thinking skills among secondary school students. *Journal of Nusantara Studies, 5*(1), 234–254.

Shukla, S., Barkman, J., & Patel, K. (2017). Weaving indigenous agricultural knowledge with formal education to enhance community food security: School competition as a pedagogical space in rural Anchetty, India. *Pedagogy, Culture and Society, 25*(1), 87–103.

Stibbe, D. T., Reid, S., & Gilbert, J. (2018). *Maximising the impact of partnerships for the SDGs: A practical guide to partnership value creation*. The Partnering Initiative and United Nations.

The Sun Daily (2018, April 15). MIYIO 2018: 30 innovative products pass first screening for patent filing. *The Sun Daily*, p. 5. https://www.thesundaily.my/archive/miyio-2018-30-innovative-products-pass-first-screening-patent-filing-HUARCH540379. Accessed on 11 Nov 2020.

Vologzhina, S., Ryazanova, N., & Eroshenko, V. (2020). Regional case study of Sustainable Development Goals implementation: Informal education and third mission of universities. *E3S Web of Conferences, 169*, 05005. https://doi.org/10.1051/e3sconf/202016905005

Willard, T., & Creech, H. (2008). *Public policy influence of international development networks: Review of IDRC experience (1995–2005)*. https://idl-bnc-idrc.dspacedirect.org/bitstream/handle/10625/47632/IDL-47632.pdf?sequence=1

[illegible]

Chapter 15
Learning for Change as an Experiential and Transformative Learning Method for Sustainable Development Planning in Siam Reap, Cambodia

Irina Safitri Zen, Marilyn Mehlmann, and Prima Wahyu Titisari

15.1 Introduction

The human ability to take action and find solutions for problems has been emphasized in Chapter 36 of Agenda 21, which is the action blueprint from the Earth Summit conducted in Rio de Janeiro, Brazil, in 1972 (UNCED, 1992). Universities function as a place of research and learning to initiate (behavioral) change, improve leadership skills, and promote sustainable development (UNESCO, 2004). Further, higher education institutions (HEIs) are responsible for and should promote sustainable development. Prior research shows that universities can integrate teaching, learning, research, campus operation, and governance to accelerate changes (Velaquez et al., 2005; Zen et al., 2016; Balsiger et al., 2017; Mulà et al., 2017; Barth et al., 2020). However, efforts related to capacity building for lecturers and academicians to facilitate sustainable development changes remain scarce.

Research also shows that quality education and learning at all levels are urgently needed to stimulate changes and actions to achieve sustainable development. The Education for Sustainable Development (ESD) concept reflects the efforts of invested stakeholders in engaging and encouraging people to be constructive and creative when facing global challenges, as well as in creating resilient and sustainable communities (UNESCO, 2004; Kioupi & Voulvoulis, 2019). ESD also

I. S. Zen (✉)
International Islamic University Malaysia, Kuala Lumpur, Malaysia
e-mail: irinazen@iium.edu.my

M. Mehlmann
Legacy17 Cooperative Association & Union of International Association, Sollentuna, Sweden

P. W. Titisari
Universitas Islam Riau, Pekanbaru, Indonesia

H. Fujii, S.-K. Lee (eds.), *Science Education for Sustainable Development in Asia*, Education Innovation Series, https://doi.org/10.1007/978-981-99-8711-5_15

proposes to empower people of all ages to take responsibility for creating a sustainable future. The emphasis is not only on acquiring knowledge about sustainable development or simple knowledge transfer but also on the translation of this knowledge into new skills, which in turn may bring forth attitude changes that promote new norms and values concordant with positive societal transformations. On this topic, Bushell and Goto (2011) described that transforming people's perspectives about their relationships with and roles in making sustainability a reality is necessary for achieving a sustainable society.

Considering that the basic requirements for maintaining the existence of human life are protecting and preserving our natural, social, and cultural environments, there is importance in adopting the general framework of ESD to empower humans to manage natural resources. Accordingly, the ESD framework has been developed and applied in several countries to support the dealings with local problems that have global impacts. Complex problems related to the whole of Earth need to be resolved through multidisciplinary and multidimensional approaches. To this end, the ESD framework proposes the incorporation of key and complex problems pertaining to sustainable development into teaching and learning systems and curricula, for example, topics on climate change, disaster risk reduction, biodiversity, poverty reduction, and sustainable consumption (Pauw et al., 2015; Titisari et al., 2020; Zen et al., 2020). ESD can be carried out within the formal education system or through informal activities based on the participation of the community, government, and companies focused on environmental issues. As a future-oriented framework, ESD focuses on protecting and restoring the environment by promoting engagement in actions that preserve our current ecological and social systems, actions that are meant to support sustainability by considering environmental, social, cultural, and economic systems as being interrelated. This confirms the orientation of ESD goals toward the development of skills and values that enable humans to contribute to environmental conservation and sustainable society (Bushell & Goto, 2011).

Nonetheless, when these topics reach HEIs, the integration of ESD into the curriculum becomes an intellectual discourse, making efforts to train the existing academia to stimulate change and act on urgent matter. There have been some formal initiatives aimed at preparing the next generation to become change agents, such as the Cambridge Institute for Sustainability Leadership (https://www.cisl.cam.ac.uk/) and the Postdoc Academy, which strive for transformation through entrepreneurship promotion, hoping that future entrepreneurs can instill HEIs with innovation, lead to inspire positive changes, and strive for an academic leadership position (Barth et al., 2020). However, the need for capacity-building interventions aimed at the existing academia to accelerate changes toward sustainability is an issue that remains to be tackled, especially on the south side of the globe.

Our program, the ASEAN Sustainability Leadership Program, is a capacity-building leadership program for sustainability developed under the guidance of the Higher Education Leadership Academy (AKEPT), Ministry of Higher Education Malaysia (akept.mohe.gov.my/). The program was developed in 2015 as an attempt to use the ASEAN community's spirit, together with the strong aspirations of HEIs to support the achievement of the Sustainable Development Goals, Agenda 2030, which came right after the United Nations Educational, Scientific and Cultural Organization's (UNESCO's) ESD 2005–2014. In general, the program intends to empower leaders of HEIs in ASEAN member countries through participatory and action-based research, as well as experiential and transformative learning experiences comprising lecture inputs, workshops, and technical visits. The program considers the fact that rapid change requires rapid learning through experience; specifically, it proposes that what leaders need to know to act on transformative changes toward sustainable development cannot be derived only from books or even conventional research. Instead, such knowledge must be driven by experimentation and action research. Hence, the program adopts the experiential learning cycle proposed by Kolb (1984), which operationalized a methodology that focuses on learning for change (L4C), and the L4C method developed by Mehlmann and Pometun (2013).

15.2 Sustainability Leadership

Since the sustainable development paradigm was proposed based on the agreement of world leaders at the Earth Summit in Rio de Janeiro, Brazil, in 1972, the indexes of development started to stretch beyond economic growth, expanding to include many other dimensions of human and environmental well-being. Further, the importance of the role of the leader in managing sustainable development started to be emphasized around this time. Considering these delineations, one may argue that leadership capacity building is highly relevant for sustainable development, especially within the context of HEIs and their direct contact with the world's youth. While short-term frameworks used by decision-makers tend to focus on economic factors, the development of leadership needs to be holistic and presupposes prioritizing strongly the relationship between humans and nature in the long term.

The basic definition of sustainability leadership relates to the basic principles of environmental science, namely, interaction, interdependence, harmony, and diversity, as well as the recognition of sustainability, with the latter being defined as the enablement of society to meet the needs of the present generation without jeopardizing the needs of future generations. This definition emphasizes interaction between a person and a group of people sharing common interests while recognizing the needs of future generations (Ferdig, 2007; McCann & Holt, 2010).

This rationale is manifested in the conceptual definition of sustainability leadership in the context of HEIs. Hargreaves and Fink (2004) posited seven principles of sustainable leadership in HEIs, with the first being that sustainable leadership should create and preserve learning processes that nourish and engage students intellectually, socially, and emotionally; the focus here is on the learning process, not academic achievements. Hence, the primary focus of sustainable educational leaders should be to sustain learning (Stoll et al., 2002). Second, sustainable leadership should secure longitudinal success, meaning that it should tackle challenges related to letting go of the role of leader, moving on, and planning for one own's obsolescence. Third, sustainable leadership should upkeep others' leadership through developing shared leadership at various levels of HEIs' professional community. Fourth, sustainable leadership should address social justice issues, developing interconnections to ensure benefits for all students, schools, and the community. Fifth, sustainable leadership should develop human and material resources; through the provision of either intrinsic or extrinsic incentives, it should provide time and opportunity for other leaders to network, learn from and support each other, and coach and mentor their successors. Sixth, sustainable leadership should develop environmental diversity and the capacity to cultivate and recreate the environment, which in turn has the potential to stimulate continuous improvement through colearning efforts. Seventh, sustainable leadership should engage through activism in environmental issues and the community, promoting colearning processes toward innovation.

Moreover, Tideman et al. (2013) postulated that sustainable leadership comprises the following principles. *Context based* recognizes interdependence, complexity, ambiguity, interconnectedness, resource constraint, regulators, and mega-trends. *Consciousness mindsets* respect different worldviews, beliefs, mental models, and attitudes. *Continuity* promotes a focus on long-term goals, courage, strength, common purpose, centeredness, and change processes. *Connectedness* serves the needs of all stakeholders, has both long- and short-term influence, and promotes collaboration, trust, fairness, altruism, relatedness, and needs instead of wants. *Creativity* advances innovation for sustainable, shared value creation; sustainable business models; and new value measurement models. *Collectiveness* scales up for collective impact, embeds sustainability in business structures, and promotes sustainable consumption. These principles provide the foundation for the ESD method applied in the program explored in the current study: the ASEAN Sustainability Leadership Program.

15.2.1 Experiential and Transformative Learning Through Sustainability Leadership

Capacity building for transformational leadership in HEIs provides a platform for transdisciplinary, participatory, and action-based research characterized by experiential and transformative learning. The concept of L4C was adopted to integrate a transdisciplinary approach to the service-learning program for the promotion of sustainable consumption in HEIs (Barth et al., 2014). The concept of L4C is part of transformative learning for ESD (Balsiger et al., 2017) and social learning (Tilbury, 2007).

Balsiger et al. (2017) outlined some key conditions for accelerating the implementation of transformative learning: *normativity*, which empowers learners to perform autonomous critical acts; a *social context* characterized by transformative learning and a training environment that are concordant with the descriptions in the UNESCO Global Action Programme (2014); *liminality*, which notes the importance of a disorienting dilemma for transformative learning and implies that coping with liminality is a responsibility of such learning endeavors; *transformative teaching* with an emphasis on personal experience, interdisciplinarity, and transdisciplinarity, which serve to enable teachers to play the role of coaches who encourage colearning conditions; and *leadership for change* through training. A prior study noted the lack of training programs for developing institutional leadership and transformative education (Mulà et al., 2017). Hence, our program contributes to filling in this gap regarding the lack of sustainable leadership training programs.

According to O'Sullivan (1999, 2003), transformative learning involves *experiencing a deep, structural shift in the basic premises of thoughts, feelings, and actions*. This conscious shift either dramatically and/or irreversibly alters our understanding of ourselves and our self-locations; our relationships with other humans and with the natural world; our understanding of the relations of power in the interlocking structures of class, race, and gender; our body awareness; our visions of alternative approaches to living; and our awareness of possibilities for social justice, peace, and personal joy. This definition highlights the importance of providing learning platforms that promote transformative learning. For these opportunities to arise, Mezirow (1997) posits that the educator's role should be more of a coach who assists the learners, enables their practice to recognize frames of reference, and encourages the redefinition of problems from different perspectives. Following these conceptualizations, the ASEAN Sustainability Leadership program was designed to provide capacity building for the extant academicians in HEIs, enabling these adult learners to foster change toward sustainable development. The shift in HEIs toward transformative learning can be initiated by changing the perspectives of these leaders about the relationships with and the roles in making sustainability a reality.

According to Mezirow (2000), the mechanisms for facilitating transformative learning can be expressed through the following key phrases connected to the learning process: (1) a disorienting dilemma, (2) critical questioning and self-reflection, (3) sharing of one's transformation process with others, (4) exploring new roles and actions, (5) planning a course of action, (6) provisionally trying new roles, (7) building competence and self-confidence in the new roles and relationships, and (8) reintegrating them into own life based on a new perspective. Furthermore, the actualization of sustainability should be supported by interdisciplinary and transdisciplinary frameworks that allow various disciplines, actors, and local communities to strive toward solving the same issue (Barth et al., 2020). Accordingly, these two frameworks were used concomitantly in the design of the ASEAN Sustainability Leadership Program. The program also recognizes and is based on the major concepts of learning for the twenty-first century, as proposed by Monk (2011): *learning to know*, referring to the capability of making connections, adapting to changes, and knowing how to learn; *learning to do*, describing the need for student-centered learning and leadership performance tasks that demonstrate the ability to apply knowledge in a creative way; *learning to live together*, which refers to using the concept of collaboration on local and global scales; *learning to be*, describing that the lifelong journey of self-discovery is a part of the learning process.

Experiential learning is important to motivate people by providing a hands-on approach to learning, one that shifts away from classroom settings, the use of the teacher role, and welcomes various ways of learning through the environment (e.g., field trips, action learning, service learning, technical site visits, and face-to-face interviews; Itin, 1999). Well-planned, supervised, and assessed experiential learning programs can stimulate academic inquiry by promoting interdisciplinary learning, civic engagement, career development, cultural awareness, leadership, and other professional and intellectual skills. Experiential learning involves reflection, critical analysis, and synthesis. Considering these delineations on experiential learning, the four stages of learning proposed by Kolb (1984) were adopted in the ASEAN Sustainability Leadership program. These steps involve knowledge input covering concepts, facts, and information acquired through formal learning and past experiences; activities for the application of knowledge in a "real world" setting; reflection, which is the analysis and synthesis of knowledge; and actions to create new knowledge. Furthermore, the program is also based on the concept of interpretation through reflection outlined by Mezirow (Cranton & King, 2003): first, content reflection through pondering about the question of *What did I do that led to the outcome?* Second is process reflection, which includes problem-solving checks in the classroom by the teacher, including by asking the question *Do I understand the needs of my students?* Third is premise reflection through questioning oneself, *Why do I feel responsible for this situation?*

15.3 Research Methodology

15.3.1 The Program's Approach

In general, the ASEAN Sustainability Leadership Program comprises a series of input lectures (30%) and training and interactive workshops based on L4C methods and technical visits (which function as participatory and action research; 70%). The five-day and four-night program took place in a selected hotel in Siam Reap, Cambodia, with the participation of 40 leaders of HEIs in ASEAN countries. The program comprised a one full day of lectures with the tagline *Sustainability Leadership Program in ASEAN Community Higher Education: Experiential & Transformational Learning*, eight hours of transformational learning training based on L4C, and two days of on-site experiential learning at Angkor Wat and Tonlé Sap lake. These two sessions of experiential learning comprised multistakeholder discussions that were co-organized with local stakeholders: the *Sustainability Tourism of Angkor Wat* was facilitated by the APSARA Authority, which manages the Angkor Wat and community-based tourism; the *Sustainable Livelihood of Fishers at Tonlé Sap Lake* was arranged by the Fisheries Community Association and the Royal University of Phnom Penh, Cambodia. The stakeholders involved in these two sessions of experiential learning were representatives of the government, non-government organizations (NGOs), and academia; these two sessions emphasize the issues of community-based tourism and sustainable livelihoods and include the representatives' sharing of some of their life experiences within the context of implementing sustainable development in Cambodia. Researchers from the Royal University of Phnom Penh, Cambodia, linked us with the local community. Furthermore, the program promotes capacity building by providing in-depth, wide-spectrum discussions on sustainable development in HEIs and leadership. The current study focuses on data collected from the participants' experiences with the *Sustainable Livelihood of Fishers at Tonlé Sap Lake*.

The two keynote speeches of the program were titled *Higher Education for Future*, delivered by Prof. Dr. Mohamad Kamal Harun, Higher Educational Leadership Academy (AKEPT), and *Sustainability Leadership – UTM Experience*, delivered by Prof. Dr. Hamdan Ahmad, Institut Sultan Iskandar (ISI), Universiti Teknologi Malaysia. There were also eight input lectures from diverse representatives of the academia, government agencies, NGOs, and research institutes, as follows: (i) *Sustainable Indigenous Ecotourism in Cambodia* by Dr. Soriya Yin, Royal University of Phnom Penh; (ii) *Sustainable Livelihood Approaches* by Assoc. Prof. Dr. Salfarina Abdul Gapor, University College of Technology Sarawak; (iii) *Sustainable Development Goals (SDG) ASEAN Community – the UNESCO Experience* by Mr. Tchadie Alain Michel, UNESCO Office Jakarta, Indonesia; (iv) *Reframing the Leadership Roles by Enhancing the Interconnections in Sustainable Development Using Quintuple Helix Framework* by Dr. Irina Safitri Zen, Universiti Teknologi Malaysia (UTM); (v) *Navigation in the Sea of Sustainability: Education for Sustainable Development, Leadership and Sustainable Development Goals* by Ms. Marilyn Mehlmann, Global Action Plan (GAP) International; (vi) *Governing the Sustainability: Case Study of River Basin Organisation in Australia* by Dr. Tariq Rana, Murray Darling Basin Authority – Australia; (vii) *Multi levels Bottom up Advocacy: Linking the Community to the Government* by Ms. Mardha Tillah, Rimbawan Muda Indonesia (RMI); and (viii) *Community Based Tourism: Multipronged approach in Support the Sustainable Development Goals (SDG) and the relation with Goals 8, 9, 11 and 13* by Dr. Norhazliza Abd Halim, Centre for Innovative Planning and Development (CiPD), Faculty Built Environment (FBE), Universiti Teknologi Malaysia.

The program had the following objectives: introduce the interdisciplinary and interconnection approach for sustainability leadership; provide direct information on experiential and transformative learning through the L4C method and on-site visits; enhance the communication skills of leaders and their appreciation of consensual decision-making through promoting interpersonal active listening and group decision-making simulations; enhance the quality of leadership by promoting humanist sensitivity toward serving societal needs; allow leaders to learn about various education and social structures from other participants; and strengthen the leadership capabilities and international social skills of the involved leaders by promoting interactions with other participants from various countries.

15.3.2 Tonlé Sap lake

Tonlé Sap lake is a seasonally inundated freshwater lake in the Mekong River Basin. The lake functions as a natural reservoir for the Siem Reap people in Cambodia, with the flow of the Mekong River feeding the lake and bringing in fertile soil, making Tonlé Sap the world's most productive inland fishery. Among the largest freshwater lakes in Southeast Asia, it is also known as the biggest fisher community living in stilt houses, also known as floating villages. The villages are characterized by low availability of clean water and sanitation systems, and although they do contain schools, religious houses, clinics, and other institutions, these generally have poor infrastructure.

The lake is home to the biggest freshwater bird sanctuary in Southeast Asia, which contains some of the rarest birds native to the Asian continent. About 149 species of fish and 200 species of higher plants are recorded in the Tonlé Sap lake, and it provides habitat for 16 globally threatened vertebrate species (Campbell et al., 2006). The best way to explore Tonlé Sap is by using a boat or kayak because they allow for exploring the floating forests and dense mangroves, and one can find drinks and snacks at the local waterside markets. Some of the local fishers provide homestay services for tourists to experience an overnight stay with a local fisher family, allowing tourists to get a true taste of the village life.

Tonlé Sap is also a part of the UNESCO World Network of Biosphere Reserves, being established as such in 2001 under a Royal Decree of the Cambodian government. The Tonlé Sap Biosphere Reserve (TSBR) covers 42,300 ha and has three basic functions: conservation, development, and logistics. The concept is supportive of community-based development ecotourism, assisting in sustainable local development (Baromey, 2008).

15.3.3 Learning for Change

L4C is a generic process tailored for collaborative learning, knowledge creation, and project assessment, and it should be developed according to the needs of participants (Benaim & Mehlmann, 2012; Mehlmann & Pometun, 2013). L4C is a colearning method that facilitates collaboration, analysis, and synthesis of learning from past experiences and connects this learning to expectations for further improvement; in the case of our program, the expectation is for HEIs to accelerate sustainable development. The L4C framework for the ASEAN Sustainability Leadership Program was customized into five steps, which are preceded by a structured preparation step; specifically, before the workshop, participants were given a template wherein they were required to describe the personal experiences from which they wished to learn.

In the five-day program, participants from various HEIs in ASEAN countries were trained in the L4C method and deployed it during the on-site visit to the Tonlé Sap lake. The diversity of the experiences in the program serves to ensure that all can learn, enabling participants to gain new insights regarding the transformative learning process. This process, in turn, occurs at the individual level and enables sustainability leaders to better handle the future demands of our currently interconnected world and its fast-changing environment and to anticipate global climate change challenges. The participants were engaged in collaborative learning, analysis, and synthesis activities.

The developers of the L4C method of the ASEAN Sustainability Leadership Program have systematically developed and tested methods and tools that facilitate the transformative change process, in particular among leaders, making use of individual and organizational learning activities. It engages participants/learners in an intensive process of interaction and reflection, resulting in a form of social learning that empowers leaders to lead the change toward greater sustainability.

In the program, the L4C workshops included a ball session; case study presentation; nominal group technique (NGT); Aha! Why, why, why?; deep listening and reflection; and Fleck's synergy method, which is a way of conducting meetings that are both highly effective and inclusive/creative. These are in line with our program objectives, namely, to expose participants to the dynamic interactions and interconnections necessary for translating the SDGs into a sustainability science approach. Most activities were carried out in groups, so-called "home groups," and each had a facilitator.

The Ball Session

The ball session is an ice-breaking session wherein all participants introduce themselves and share their intentions behind attending the program (Fig. 15.1). The instruction for each participant was as follows: please explain *who you are*, *why you are here*, and *what brings you here* in a maximum of 90 seconds. After giving the answer, the participant throws an imaginary ball to the next person they would like to hear from; the session continues until all have spoken (Buber, 1937).

Fig. 15.1 The ball session

Case Study Presentation

The case presentations were made in two steps: first, the appreciative presentation technique was used, which is similar in approach to the appreciative inquiry technique, and, second, the NGT.

1. *Appreciative Presentation*

In the home groups, participants took turns to present their prepared presentations: the experience from which they wished to learn. This process comprises parts (a) and (b), with part (a) being a factual summary of the work in question (e.g., statistics on the number of people involved) and part (b) being a personal assessment of what was the best aspect of the experience, namely, a "boasting" session about success and satisfaction. For many participants, part (b) was the most difficult as few cultures encourage people to be clear about their successes; still, this is an important skill overall, especially for leaders.

The presentations of all six participants in the home group are summarized on a single flip chart. Normally, each presentation is documented by the previous presenter, and the flip charts are posted afterward to allow other groups to see the summary of the presentations.

2. *Nominal Group Technique*

After the appreciative presentation technique, participants are directed to make considerations about the possibilities for improvement regarding their cases. Although the cases and the presenter's satisfaction with the case are unique, there is often a convergence of possible improvements for the activities; the NGT allows for synthesizing the diverse ideas that come up regarding these potential improvements.

The NGT is a structured group method involving problem identification, solution generation, and decision-making (Fig. 15.2), and it encourages contributions from every participant. This technique aims to train the leaders in a simple, rapid process to gather and synthesize inputs, facts, opinions, and evaluations from a group of students, a community, or colleagues. This method is particularly useful for leaders to engage everyone when the group is part of an established hierarchy; the three steps of the NGT technique are listing the problem, sharing, and synthesizing.

The first stage is to list potential improvements. Each participant thinks silently and writes down brief answers to the following question: *What would you prefer to have done differently or better?* It provides participants with the possibility of thinking that the solution to a problem, a research issue, or an insight that is already good could be even better. Each response is written on a separate small piece of paper, and each participant keeps their own notes.

In *the second stage*, participants share their answers: they should stand around a table that has been cleaned of everything or use post-its and stand in front of a clean whiteboard. Each member, in turn, reads one of their answers aloud. During this process, no discussion is allowed, but others can make some clarification by asking the following question: *What do you mean by that?* And the person reading own answer should clarify the statement as needed until everyone understands it; after reaching a general understanding for all participants, the participant places the

Fig. 15.2 Nominal group technique

answer on the table or whiteboard. Once each participant has read one answer, all participants are free to contribute with new ones, one at a time. Answers that seem to be related are placed close to each other on the table or whiteboard.

The third stage is synthesis, wherein the preliminary groupings of notes from stage 2 are consolidated into clusters, each with a descriptive title. During this process, further clarification may be needed to find the most appropriate cluster for each note. The quality of the synthesis is dependent on the quality of the naming of the clusters; for instance, a passive title like "Administration" is not actionable, but "Ensuring salaries are paid on time" is actionable. This stage should be allowed a relevant amount of time for its completion. At the end, each group shares the names of their clusters with others, with the possibility of discussion based on the following question: *how can this contribute to action formulation and decision-making?*

Deep Listening

Deep listening is a method developed by Ziegler (1995) to help people access their inner wisdom or intuition. It presupposes that each of us has knowledge in our subconscious to which we have no access to or in which we do not trust. At a certain level, we acknowledge that we know things that we may express as our "gut feeling," as a form of intuition. The deep listening technique aims at making this intuition accessible. It is also highly empowering to deeply listen to other people, and this is a key competence in the role of the leader as a coach.

In the program, deep listening was introduced to improve the quality of leaders' listening as it encourages the speaker to talk freely and describe how one wants to be supported or coached. The process begins by paying close attention and listening *in silence*; it is nonjudgmental, empathic, and nurturing. Deep listening is an important skill for leading sustainable development processes because those involved tend to be highly diverse and contribute with different insights, problems, and solutions related to human interaction with the environment. Furthermore, knowledge itself cannot change the leaders' behaviors unless they understand the issue, empathize with those involved, and understand the local context. Thus, a leader must have good listening skills, which allow one to understand others' messages deeply. Deep listening also allows for issues to be detected, analyzed, and solved with wisdom.

Instruction for Deep Listening Combined with "Why, Why, Why" for Deep Coaching

- *Participants form pairs and sit next to one another. One is designated the problem owner, and the other is the coach (after a while, they change roles).*
- *Think about the problem or concern that you want to talk about. Take a few minutes in silence to decide and formulate on it.*
- *As the coach, your job is to make the problem easier to solve for your partner.*
- *After about three repetitions of the question "why" to reach deeper levels on the topic, the question turns to "how." The coach asks what the problem owner might do in regard to the deepest level of "why." Then the coach asks, "And how might you succeed in doing that?" Usually, the process is complete after three iterations of the questioning of "how." Upon completion, the problem owner can proceed to formulate an intention to take action.*
- *For this exercise, each person has 9 min to express a problem and to work on it with the help of the coach.*

The first rule of this coaching technique is to know that the coach will not solve the partner's problem; instead, the job of the coach is to make it easier for the partner to solve own problem. The problem owner needs to tell the coach a problem and should acknowledge that the coach does not have to give a direct answer to the problem; instead, the coach should ask the problem owner questions like the following: *Why/in what way is that a problem for you personally? How might you change that? Is there anything you can do?*

Synergy Meetings

Fleck's synergy meeting is potentially the most useful of the methods comprising the L4C methodology because it can be used in any context with strict time limits, is focused on both outcomes and relationships, and aims at establishing a climate for inclusion, collaborative learning, and creativity. It can also be applied to a regular project meeting, such as an envisioning session or a board meeting. A synergy meeting takes place in three steps:

1. Dynamic agenda setting: each person may contribute to topics, including an estimate of the time needed for each topic. The proposer becomes the "owner" of the topic.
2. Negotiation: if the total time needed for all topics exceeds the time available, negotiation ensues. Only the owner of a topic can change time allocation.

3. The topics: they are dealt with logically, not chronologically. Anyone may suggest a topic as the next item to be tackled; since maximum times are set for each topic, all topics are dealt with within the meeting period, without overruns.

Participants were taught the principles both of this meeting method and of facilitation, and they applied the method to discuss their findings related to an activity (e.g., after the on-site visit).

15.4 Results and Discussion

The skills learned from the L4C workshop were deployed during the stakeholder discussions amid the on-site visits to the Tonlé Sap lake. The application of the L4C method in the interactive workshops and the on-site visits allowed for experiential learning to be achieved. The results showed that the L4C methods were successfully used in the program as the list of identified problems clearly demonstrated improvement in the leaders' skills of colearning, empathy, and understanding of the local people's problems. The NGT was applied for the examination of solutions to the livelihood problems in the villages at Tonlé Sap lake.

15.4.1 The Ball Session

Various response dimensions were provided by participants from various disciplines and different levels of academic qualification, from lecturers to associate professors and professors. Since all learners were adults, participants were generally able to listen and appreciate what others said, taking the information relayed by others as a basic input in the learning process. The following responses summarize most results from the Ball session:

(i) *I want to learn about all aspects of sustainability, and this is the right place for me to get knowledge on and share what I have done in my research on sustainability.*
(ii) *I am involved with SDG-related workshops, and those workshops push the early-career researcher to be involved in sustainability issues.*
(iii) *I want to learn, relearn, and realign my role in sustainability.*
(iv) *I want to share something and get something from all of you here. My own definition on sustainability: no beginning, no ending.*
(v) *I am very happy to be here. I have been learning topics about sustainability since 2008; it is very interesting to learn something new from each of you.*
(vi) *I came here because my boss wants me to replace him. So, here I am. In the Plenary session workshop, I hope to gain insights and be able to learn more from the program.*

15.4.2 Case Studies

Each participant brought and presented their own "case study" in the home groups. In each group, participants performed the following steps: in Step 1, they described the case with a few key facts (identified ahead of the workshop) and provided an answer to the question *What worked really well? What are you the most proud of?* Other group participants listened without any type of critique (without the need to respond, analyze, or prioritize; they only listened and endeavored to understand the topic at hand); case descriptions were noted and the notes displayed. In Step 2, participants deployed NGT to identify potential improvements. The results from this method are shown in Table 15.1, and the responses to the question *What would you prefer to have done differently or better?* are listed in Table 15.2. The responses reflect the future expectations that participants had for teaching, learning, research, and community work in HEI. Through this reflection session, participants are guided in the identification of their achievements and the work about which they are the most proud, thus serving as a tool for participants to reflect on their past research experiences and learn from their individual projects (Table 15.3). The summary of the participants' responses to *What worked really well? What are you the most proud of?* is shown in Fig. 15.3.

15.4.3 Nominal Group Technique

From the individual brainstorming sessions in silence and the collective learning of leaders from HEIs in ASEAN countries, the participants were requested to reflect on the question, *What would you prefer to have done differently or better?* The synthesis from these activities led to six clusters: leadership, funding, research, orientation, networking, and community (Table 15.4).

The participants provided several comments about the NGT session, such as the following:

Participant 1: *It was very difficult to draw a specific topic because we come from different backgrounds, countries, and economic contexts.*

Table 15.1 Summary of the reflections on the case studies based on the question *What would you prefer to have done differently or better?*

"What would you prefer to have done differently or better?"	Synthesis of the results in the clusters
Kuala Sepetang mangrove forest Port well tin mining area, tourism attraction Changing of the area and land usage Mangrove usage for charcoal Provide society job market Ecotourism attraction impact Issues: water quality and fish quantity	Resilience strategies Changing landscape Resilience community Leadership

(continued)

Table 15.1 (continued)

"What would you prefer to have done differently or better?"	Synthesis of the results in the clusters
Disaster (flood) Abandon resources Accessibility for people who cannot reach food sources Not fully utilized Victim needs/volunteer offers Evacuation hit Ensure the endorsement of a welfare department Early warning emergency system How to handle volunteer issues	Natural disaster Adaptation and risk reduction
Finding feasibility Model for indigenous ecotourism Promoting the political visibility of indigenous communities Economic support for the indigenous communities Community engagement in the urban area Fishing village in island issue? Accessibility to basic needs: electricity and water High loss (drawback)	Basic needs
Ecotourism in mountainous areas Products for tourism in the traditional village are not suitable for many people The government wants to develop ecotourism in the area, but it is not suitable because the model would change the traditional culture (especially for women)	Ecofeminism for tourism in sustainability development
How to change the attitude and behavior among youth Concerning consumption levels in Indonesia Sustainability consumption	Self-learning
Empowering learners Self-directed learning The use of information and communications technology	

Table 15.2 Responses to the question *What would you prefer to have done differently or better?*

Question	Responses
What would you prefer to have done differently or better?	Wanted the government to recognize community engagement and action research Misconceptions of community engagement led the government to marginalize social science research Wanted the government to give more funds to research
	Change behaviors Regarding publication and to show output to the university Talking about ISI/Scopus Agenda Describe that a low number of publications is better Put a belief, not just support the university Ask for respect Adaptive challenge
	Project at the initial phase Get support from family Help children from hospitals in remote areas Expected to extend the project to other areas in Malaysia To challenge: as a new initiative, have to make it sustainable Build an example from developed countries

Table 15.3 Summary of the responses to the question of *What worked really well? What are you the most proud of?*

Question	Responses
What worked really well? What are you the most proud of?	Participant 1: "[My] Persistence. I am still working on not giving up. Success [of the project] is commercialized. Farmers plant the herb, process it, and provide to companies after 3 years. Farmers still keep in touch [with the participant]. Farmers still keep in touch."
	Participant 2: "Having met diverse people, faced new challenges, and new environments."
	Participant 3: "Adaptive challenge, mix and match [of different techniques or methods]."
	Participant 4: "It was the first initiative in this area in Malaysia, it gave people something to believe in, and promoted open discussions about sustainable development in formal and informal ways.."
	Participant 5: "I provided small contributions that helped the community."
	Participant 6: "I assisted in the development of a democratic decision making process, developed the democratic system in the community, and helped people become aware of conservation topics."
	Participant 7: "I helped people to better understand the concept behind clean and safe water."
	Participant 8: "I helped introduce the concept of sustainable development to non-environment focused teachers."
	Participant 9: "I provided people with hands-on experience on cleaning the river in the neighborhood, and children were empowered to be part of the solution. ."
	Participant 10: "The project provided a platform for the sharing of concerns of different stakeholders, for the development of volunteerism, and delivered a manual for engineers."
	Participant 11: "The project provided active learning strategies, led to changes, enabled for cooperative learning and participatory learning to be promoted."
	Participant 12: "Through the project, community participation was institutionalized, and it became an open process for all."
	Participant 13: "The promotion of stakeholder coordination, the development of a policy and implementation plan for sustainable community, as well as peace education."
	Participant 14: "The project allowed for research to be performed continuously, and there have been attempts to translate the scientific knowledge into rea-life settings and practices."

Participant 2: *I feel scared if I don't understand what the other person is trying to tell the group or if I misinterpret the point.*

Participant 3: *At the beginning, I doubted [the exercise] and what it was about; but at the end, I felt very happy and I think that I was able to do the exercise, and now I'm curious about what to do next.*

Participant 4: *At first, it's difficult to understand everybody's opinion or statements. But at the end, the technique helps to make things easier to understand, more meaningful, and turns out to be simpler.*

Fig. 15.3 Summary of the responses to the question *What worked really well? What are you the most proud of?*

Table 15.4 Reflection on the results of the nominal group technique

"What would you prefer to have done differently or better?"	Synthesis from collective learning outputs
To have a better understanding of the scenarios that happened in our own communities nowadays and how to improve them To have developed a collective leadership To have changed the mindset of students in my own program about the responsibility related to being good human beings; to have made it clearer that we should use our knowledge, reasons, and heart together when making any decision To have explained about and invited everyone to use green and eco-products more, which would then help lower green product prices	1. Leadership
To have tried to provide internship sponsorship for students to go to other universities To have started a research project on a smaller scale To have proposed more feasible projects To have gotten funds for the research endeavors from the government	2. Funding
To have made visual arts as a part of the research methodology Research and documentation Process and policy research should be added in research To have enabled for visual arts method-based research to be seen as equal to other research methodologies	3. Research

(continued)

Table 15.4 (continued)

"What would you prefer to have done differently or better?"	Synthesis from collective learning outputs
To have provided orientation (on perspectives, principles, and processes) on ecotourism, biodiversity, and community development To have implemented the idea of "learning with the heart" ever since preschool and upward To have gotten more inputs on how to create awareness actions in the community, especially regarding the appreciation of culture and heritage To have helped the local fisher people to be the official group To have probably presented better results by providing life experience education to students; to have further discussed the conditions, problems, and challenges pertaining to gender rights within the context of women, gender, the LGBT community, and ecotourism	4. Orientation
To have increased student participation in lecturing and research To have trained students more on the need to be concerned about the environment To have increased the collaborative activities with university students To have made more collaborative research with different research disciplines	5. Networking
To have seen visual arts be accepted as one of the factors that can boost ecotourism To have reduced morality issues and increased community awareness To have expanded the environmental conservation group to the international level and to have conducted more networking at the international level To have aimed to produce better and happier communities To have facilitated community development and organization through student and academic programs, such as field outstation programs To have further changed people's behaviors, as well as to have begun with changing myself and then to try to change others	6. Community

Thus, the NGT allowed participants to reflect on their past experiences, describe their future expectations, and gain different perspectives (Fig. 15.4). This process is based on the transformative learning theory by Mezirow (1997), which proposes that adult learners can construct, validate, and reformulate the meaning of their experiences. The NGT technique involved deep thinking in silence for some moments, thus allowing for the learners to engage in critical reflections on own experiences, which in turn involves processing specific beliefs, attitudes, and emotional reactions; this process eventually leads to perspective transformation (Mezirow, 1990).

15.4.4 Deep Listening

Overall, participants gave positive feedback about this method. Despite this technique being challenging, it teaches participants to be patient and good listeners. Some of the feedback was as follows: *This session is like a counselling session*;

Fig. 15.4 A group discussion session during the L4C method

From my point of view, this activity is challenging because we have to be patient when being the listener, instead of giving any advice; *I personally think that this method is very useful to become more professional*; and *I feel strange when my coaching partner also has the same problem as me; it is just like I myself have the solution to my own problem.*

The deep listening method was used to promote discussion among various stakeholders, and through it, participants reportedly learned to listen to the local people and stakeholder issues in Tonlé Sap lake. Moreover, participants went to the lake, took the local fishing boat, and visited the fisher village and its facilities (e.g., schools, floating fresh market, and a health clinic). Each boat carried a mix of workshop participants, local people, representatives from local NGOs, NGOs, and researchers from the Royal Cambodia University.

Further, as participants got back from the technical visit, they were asked to reflect on their lived experiences through the visit in writing and to come up with potential solutions to the problems they got to know about. The results from this process are described in Table 15.5. This reflection session conducted on return from the on-site visit enabled the participants to learn from this experience. These tools can be used to facilitate the reflection process, the gathering of tacit knowledge acquired from interaction with local people, insights from observations and experiences while making on-site visits, and the collection of local issues. *The aim of this session was to make their learning more explicit and articulated, contributing to their professional performance in their own working contexts.*

The results in Table 15.5 show the participants' knowledge derived from the observations and the deep listening technique, which show the convergence of local/tacit and scientific knowledge that stemmed from the local people, stakeholders, and participants.

Table 15.5 Reflection, synthesis, and proposed solutions for the problems in Tonlé Sap lake

Reflection from the on-site visit to Tonlé Sap lake	Cluster and synthesis results	Proposed solutions
Difficulty in the provision of basic needs Poor infrastructure Transportation problems Poor public amenities and community areas Poor access to roads in the area Poor jetty infrastructure Non-organized jetty Poor funding	Public facilities and infrastructure	Providing clean water through cooperation between NGOs and universities
Noise pollution caused by motor boats for transporting tourists	Noise pollution	Better technology for boat engines Smaller boats for smaller groups of people, which in turn produce less noise pollution
Hard-to-get clean water Municipal wastewater discharged to the lake Lack of water treatment and toilets Poor water and sanitation	Water and sanitation	Providing clean water through cooperation between NGOs and universities Recycling water facilities Improve water treatment and toilets
No solid waste management available	Waste	Research project to promote the separation of solid waste Show the community how to recycle Provide knowledge for invested stakeholders that some waste can be sold and serve as a source of income for the economically disadvantaged Promote awareness about sustainable consumption, production, health, energy, water, and food, among others
Minimal health service availability and limited access to clinics and other health care facilities and services Fever is a common health problem People experience difficulties in going to a hospital or clinic	Health	Provide air medical services Improve health facilities Provide mobile clinics Educate the community with basic nursing courses and awareness
Low budget from the government to the community Loan interests are high Lack of job opportunities for women Financial support is limited Diminished support from the government for building basic infrastructure Lack of public-private partnerships Lack of coordination	Finance	Need to provide more money with low interest Potential financial support for the economically disadvantaged Create microfinance programs Support to build basic infrastructure from the government More proper coordination

(continued)

Table 15.5 (continued)

Reflection from the on-site visit to Tonlé Sap lake	Cluster and synthesis results	Proposed solutions
Children face difficulties in going to school Those with tertiary education refuse to come back because of a lack of job opportunities Kids spend their time helping their parents and other people to help with the family's income Not enough teachers Education is not a priority in some families, especially for girls Various school children drop out of school There are problems with keeping students interested in education	Education and its quality	Encourage children to study subjects or engage in vocational training that will help them generate income and develop the community Educate parents about the importance of education for their kids Provide a more appropriate school system and education Kids should stay longer at school and should be able to enjoy it Improve the quality of primary and secondary education Support government policies to improve the quality of education in remote areas Provide capacity building for teachers Build the infrastructure of schools Provide life-long learning opportunities Students should be able to learn with their hearts Teachers should serve as motivators for students to learn Students should receive interventions to enhance their awareness about sustainable development Set additional value through visual educational methods Ensure the successful reconciliation of social science and visual educational methods Need to promote the perception in the community that education is a joyful endeavor Change the students' behaviors regarding the environment Develop a curriculum regarding the environment Prohibiting hazing and "violence" during the introduction week for novel university students may have a significant impact

(continued)

Table 15.5 (continued)

Reflection from the on-site visit to Tonlé Sap lake	Cluster and synthesis results	Proposed solutions
Poverty Lack of tourism-related income activities for the community Lack of attraction services to make tourists spend more and return Different mindset and lifestyle Illegal fishing Very limited resource persons in the community	Socioeconomic issues	Promote networking with NGOs and other countries Promote and introduce the local culture and its foods and products Clearly identify the main tourist attractions Promote the capacity of community members to adapt and change toward sustainable development Promote leadership with a clear structure that is based on the community's grassroots Explore the possibilities of fishery management Increase the number of enjoyable tourist attractions and programs Training community members in making nets and in the fishing farm culture since fishing activities are strong in the community
People are not very appreciative of forestry, especially community members in the upper-level strata The dam in the upper area of the Mekong River causes the community to have less water flow, affecting the water ecosystem The weather in the community is extreme due to climate change There are drastic changes in the environment of the community in different places, with some being covered by water, while others are not	Nature and physical environment challenges	Influence community members to be concerned about forestry issues Improve the community's adaptive capacity through capacity building and training programs
Community development takes time Democracy and development How time is defined: cultural time Concern with agenda: democratic time	Challenges to transform the community	Improve fisher people's awareness of conservation Think, plan, and solve problems together with community fisher people Influence and advice company involvement in the public and private sectors Work with business sectors

15.4.5 Reflection and Contextualization

This session, allocated at the end of the program, invited participants to consider the answer to the following question: *How can a university most effectively contribute to sustainable community development?* The participants provided various responses, including the introduction of a curriculum regarding community development at the local level, raising teacher awareness, improving the curriculum for teacher training, promoting the engagement of alumni and retired professors, promoting collaborative learning and research at the institutional level, teaching students the skills to engage with communities, encouraging students to reach out to their communities, transfer of technology, providing better social protection, recognizing community development practitioners and promoting their work in the community, and integrating community engagement as part of the academia extension services. All responses are recorded in Figs. 15.5 and 15.6. Hence, transformative learning for sustainable development was manifested through the participatory, integrative, and reflective characteristics of the L4C method (Singer-Brodowski, 2017). Hence, transformative learning for sustainable development was manifested through the participatory, integrative, and reflective characteristics of the L4C method (Singer-Brodowski, 2017).

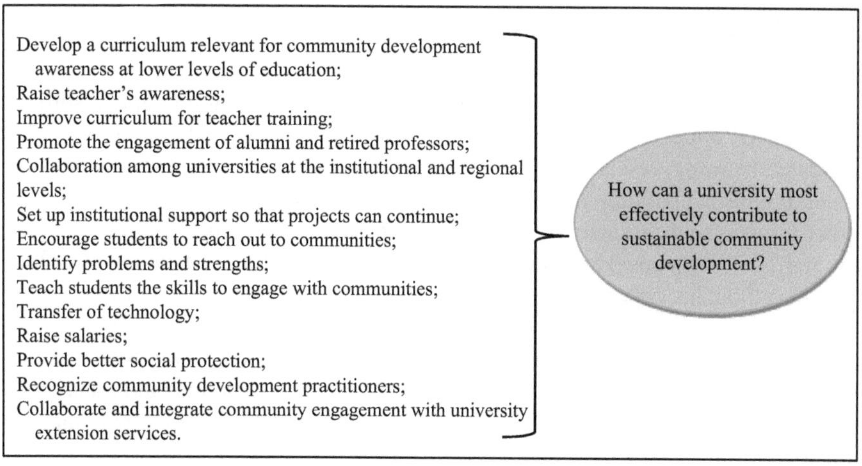

Fig. 15.5 Reflections on how universities can contribute to facilitate sustainable community development (Part 1)

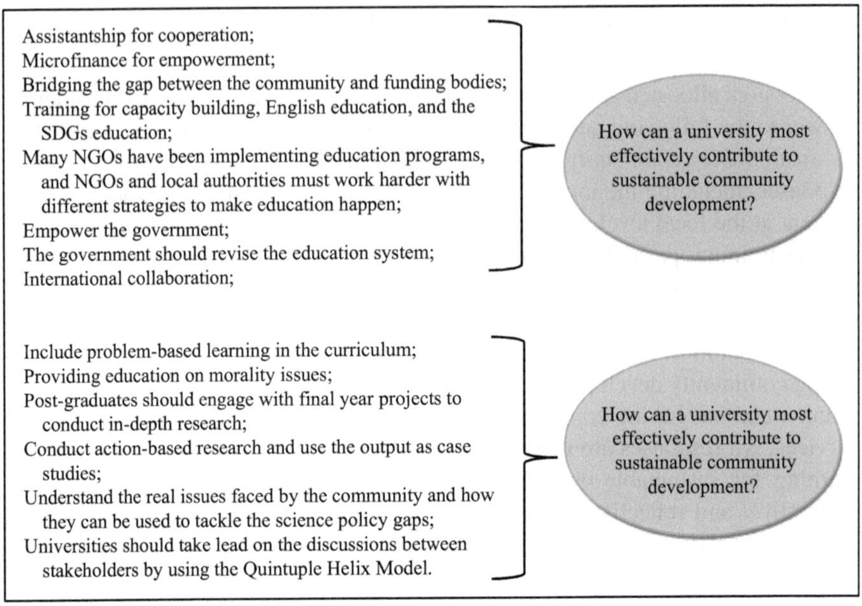

Fig. 15.6 Reflections on how universities can contribute to facilitating sustainable community development (Part 2)

15.5 Conclusion

Overall, the ASEAN Sustainability Leadership Program is an attempt to shift the participants' understanding of themselves and their roles in HEIs in the context of teaching, learning, research, and community development for sustainable development. It aims at enabling these stakeholders to reemphasize their relationships with other humans and the natural world. From an ESD perspective, the results of the program reflect the need to promote collective awareness for engagement in sustainable development, an issue identified as part of the reflection process from the participants' technical visit to the region and which represents the real livelihood problems of the fisher communities in the floating villages at Tonlé Sap lake. The implications of the program based on the participants' experiential and transformative learning and discussions, which were promoted by the L4C colearning method, are as follows:

- Most participants wanted to learn about sustainability, share their research experiences, improve sustainability action research, realign their roles, and gain knowledge from other participants. Further, various alliances were created through the program.
- The participants showed that they had diverse experiences and levels of knowledge and engagement in research and expressed that they had high expectations for future community-based research and high expectations for the support (e.g., for the promotion of social science research and better publicity of sustainable development research) from relevant ministries in the country.

- Participants highlighted six main areas that need to be enhanced to support sustainable development-based research: leadership, funding, research, orientation, networking, and community.
- Reflections from the on-site visits revealed several key factors that should be tackled for the area to achieve sustainable development, such as improvements in education, infrastructure, and health facilities.
- There is a strong potential for the improvement of teaching, learning, research, and community development for sustainable development in Malaysian HEIs and in the studied region.

Acknowledgment This study was funded by AKEPT, Leadership Academy, Ministry of Higher Education Malaysia for Program SD103: ASEAN Sustainability Leadership Programme, Siam Reap, Cambodia, in 2015. The program is co-organized with the Centre for Innovative Planning and Development (CIPD), Institute Sultan Iskandar, Universiti Teknologi Malaysia, UNESCO (United Nations Educational, Scientific and Cultural Organization) Regional Science Bureau for Asia and the Pacific, Jakarta Office, GAP (Global Action Plan) International, Sweden, and Royal University of Phnom Penh, Cambodia.

References

Balsiger, J., Förster, R., Mader, C., Nagel, U., Sironi, H., Wilhelm, S., & Zimmermann, A. B. (2017). Transformative learning and education for sustainable development. *GAIA-Ecological Perspectives for Science and Society, 26*(4), 357–359.

Baromey, N. (2008). *Ecotourism as a tool for sustainable rural community development and natural resources management in the Tonle Sap biosphere reserve*. Kassel University Press GmbH.

Barth, M., Adomßent, M., Fischer, D., Richter, S., & Rieckmann, M. (2014). Learning to change universities from within: A service-learning perspective on promoting sustainable consumption in higher education. *Journal of Cleaner Production, 62*, 72–81.

Barth, M., Bruhn, A., Lam, D. P., Bergmann, M., & Lang, D. J. (2020). Capacity building for transformational leadership and transdisciplinarity. *GAIA-Ecological Perspectives for Science and Society, 29*(3), 195–197.

Benaim, A., & Mehlmann, M. (2012). *Learning for change*. https://legacy17.org/Files/Article-related/L4C.pdf

Buber, M. (1937). *I and thou*. T & T Clark.

Bushell, B., & Goto, M. (2011). Leadership for sustainable society: A transformative learning approach. *Procedia-Social and Behavioral Sciences, 29*, 1244–1250.

Campbell, I. C., Poole, C., Giesen, W., & Valbo-Jorgensen, J. (2006). Species diversity and ecology of Tonle Sap Great Lake, Cambodia. *Aquatic Sciences, 68*(3), 355–373.

Cranton, P., & King, K. P. (2003). Transformative learning as a professional development goal. *New Directions for Adult and Continuing Education, 98*, 31–38.

Ferdig, M. (2007). Sustainability leadership: Co-creating a sustainable future. *Journal of Change Management, 7*(1), 25–35.

Hargreaves, A., & Fink, D. (2004). The seven principles of sustainable leadership. *Educational Leadership, 61*(7), 8–13.

Itin, C. M. (1999). Reasserting the philosophy of experiential education as a vehicle for change in the 21st century. *Journal of Physical Education, 22*(2), 91–98.

Kioupi, V., & Vaulvaulis, N. (2019). Education for sustainable development: A systemic framework for connecting the SDGs to educational outcomes. *Sustainability, 11*(21), 6104.

Kolb, D. A. (1984). *Experiential learning: Experience as the source of learning and development*. Prentice-Hall.
McCann, J. T., & Holt, R. A. (2010). Defining sustainable leadership. *International Journal of Sustainable Strategic Management, 2*(2), 204–210.
Mehlmann, M., & Pometun, O. (2013). *ESD dialogues: Practical approaches to education for Sustainable Development by and for educators*. Books on Demand.
Mezirow, J. (1990). How critical reflection triggers transformative learning. In J. Mezirow (Ed.), *Fostering critical reflection in adulthood* (pp. 1–20). Jossey-Bass.
Mezirow, J. (1997). Transformative learning: Theory to practice. *New Directions for Adult and Continuing Education, 74*, 5–12.
Mezirow, J. (2000). *Learning as transformation: Critical perspectives on a theory in progress* (The Jossey-Bass higher and adult education series). Jossey-Bass Publishers.
Monk, N. (2011). *Open-space learning: A study in interdisciplinary pedagogy*. Bloomsbury.
Mulà, I., Tilbury, D., Ryan, A., Mader, M., Dlouhá, J., Mader, C., Benayas, J., Dlouhý, J., & Alba, D. (2017). Catalysing change in higher education for sustainable development: A review of professional development initiatives for university educators. *International Journal of Sustainability in Higher Education, 18*(5), 798–820.
O'Sullivan, E. (1999). *Transformative learning: Educational vision for the 21st century*. University of Toronto Press Inc.
O'Sullivan, E. (2003). Bringing a perspective of transformative learning to globalized consumption. *International Journal of Consumer Studies, 27*(4), 326–330.
Pauw, J. B., Gericke, N., Olsson, D., & Berglund, T. (2015). The effectiveness of education for sustainable development. *Sustainability, 7*(11), 15693–15717.
Singer-Brodowski, M. (2017). Pedagogical content knowledge of sustainability: A missing piece in the puzzle of professional development of educators in higher education for sustainable development. *International Journal of Sustainability in Higher Education, 18*(6), 841–856.
Stoll, L., Fink, D., & Earl, L. (2002). *It's about learning (and It's about time)*. Routledge/Falmer.
Tideman, S. G., Arts, M. C., & Zandee, D. P. (2013). Sustainable leadership: Towards a workable definition. *Journal of Corporate Citizenship, 49*, 17–33.
Tilbury, D. (2007). Learning based change for sustainability: Perspectives and pathways. In A. E. J. Wals (Ed.), *Social learning towards a sustainable world* (pp. 117–132). Wageningen Academic Publishers.
Titisari, P. W., Elfis, Zen, I. S., Hendrayani, Y., Chahyana, I., Khairani, Janna, N., Suharni, N., & Sari, T. P. (2020). Students' perceptions of education for sustainable development (ESD) to achieve SDG 4 in Indonesia: A case study of Universitas Islam Riau. In N. Kaur & M. Ahmad (Eds.), *Charting a sustainable future of ASEAN in business and social sciences* (pp. 191–202). Springer.
UNCED. (1992). *Division for sustainable development, Agenda 21, Chapter 31*. http://www.un.org/esa/dsd/agenda21/res_agenda21_00.shtml
UNESCO. (2004). *Draft International Implementation Scheme*. UNESCO.
Velazquez, L., Munguia, N., & Sanchez, M. (2005). Deterring sustainability in higher education institutions. *International Journal of Sustainability in Higher Education, 6*(4), 383–391.
Zen, I. S., Subramaniam, D., Sulaiman, H., Omar, W., & Salim, M. R. (2016). Institutionalize waste minimization governance towards campus sustainability: A case study of green office initiatives in Universiti Teknologi Malaysia. *Journal of Cleaner Production, 135*, 1407–1422.
Zen, I. S., Saleh, M. N., Afrizal, T., Yaumidin, U. K., Titisari, P. W., & Hendrayani, Y. (2020). Quo vadis development: Assessing the livelihood of indigenous people's communities in Malaysia and the potential for community-based conservation effort. *Environment, Development and Sustainability, 23*, 6502–6523.
Ziegler, W. (1995). *Ways of enspiriting: Transformative practices for the twenty-first century*. Fia International LLC.

Part IV
Science Teacher Education for Sustainable Development

Chapter 16
Current Situation and Training of Chinese Elementary School Science Teachers in the Context of Education for Sustainable Development

Jing Zhang, Xinyi Shen, and Hong Liu

16.1 Introduction

The level of scientific literacy is directly related to comprehensive national strength. Students are a country's potential human resources. The level of scientific literacy of elementary school students will affect not only the quality of personal life in the future but also the country's scientific and technological capabilities as well as its future survival and development. Therefore, the Ministry of Education clearly lists the cultivation of primary school students' scientific literacy as the purpose of the curriculum. Scientific literacy should be cultivated at an early age.

16.2 Development of Science Education with China's Basic Education in the Twenty-First Century

In 2000, China launched a nationwide basic education curriculum reform. The Ministry of Education decided to move from the "Nature" curriculum in elementary schools to a "Science" curriculum, and it promulgated and implemented the "Full-Time Compulsory Education Science (Grade 3–6) Curriculum Standards (Experimental Draft)" in 2001 (Ministry of Education of the People's Republic of China, 2018). Since then, the reform of science education has been in line with international standards, and China's science education research entered a new stage.

J. Zhang (✉) · X. Shen
Beijing Academy of Educational Sciences, Beijing, China

H. Liu
Beixin'an Primary School, Beijing, China

H. Fujii, S.-K. Lee (eds.), *Science Education for Sustainable Development in Asia*, Education Innovation Series, https://doi.org/10.1007/978-981-99-8711-5_16

Since 2002, science education and education for sustainable development (ESD) have been integrated. At the beginning of 2017, the Ministry of Education issued a notice on the publication of compulsory education primary school science curriculum standards. The notice stipulated that "the new curriculum standards will be implemented from the fall of 2017, and elementary schools will offer science classes from the first grade, ensuring that students in grades one and two will have at least one class per week, and the science class hours of grades three to six would remain unchanged." At the same time, it proposed rationally allocating elementary science teachers based on actual conditions and gradually establishing a team comprising full-time and part-time teaching and research personnel, increasing the funding to ensure laboratory construction, equipment, and consumables.

Since the implementation of the curriculum reform, the primary science curriculum has continuously been improving in practical exploration, which plays a vital role in the cultivation of primary school students' scientific literacy. To ensure the smooth progress of national science literacy in the field of basic education and to promote the in-depth development of elementary school science education, the Ministry of Education revised and improved the elementary science curriculum standards and required their implementation in the fall of 2018. There are two main aspects.

One is the curriculum characteristics and ideas for curriculum design. First, the standard repositions the nature of the elementary science curriculum in terms of the basis, practice, and comprehensiveness. The standard asserts that early science education is indispensable to the formation of students' scientific literacy. In elementary school science courses, students acquire knowledge and skills to cultivate the ability to observe, compare, and analyze. They use scientific knowledge and skills to understand scientific phenomena around them and solve practical problems, which lays a solid foundation for future scientific knowledge, daily life, and even lifelong development. The practicality of elementary science courses can be manifested in inquiry activities as an important way for students to learn science, enhance their learning abilities in all aspects, and shape their scientific attitudes through practical activities, such as focusing on students' hands-on experience. It emphasizes the integration and interrelation of the four fields of material science, life science, earth science, and technology and engineering. The standard focuses on the combination of understanding nature and problem-solving to strengthen students' comprehensive abilities. At the same time, it emphasizes the integration of science courses, Chinese, mathematics, and other courses to promote the all-round development of students, which significantly reflects the comprehensive characteristics of elementary science courses. The redivision of science courses is not only conducive to integrating the new curriculum objectives of science, technology, society, and environment into the elementary school science curriculum system but is also more conducive to the integration and refinement of stage knowledge in accordance with the cognitive laws of primary school students so as to realize the curriculum purpose of gradually cultivating primary school students' scientific literacy, starting from basic education (Wang, 2015).

The second is the course goal. The standard gets rid of the concept of dividing the traditional three-dimensional target as the general target, and it emphasizes that the specific targets should be explained from four aspects: "scientific knowledge," "scientific inquiry," "scientific attitude," and "science, technology, society, and environment." The overall goals of each aspect are divided into four categories in detail according to different fields, elements, dimensions, and relationships. After the course objectives are updated, the main task that students need to accomplish in science courses is to understand the relationship between science, technology, society, and the environment. At the same time, it is necessary to strengthen the cultivation of primary school students' sense of innovation, environmental protection, and social responsibility. The fourth item in the general goal of scientific knowledge clearly states: "Understanding technology is an extension of human ability, technology is the power to change the world, and technology promotes the development of human society and the continuous progress of civilization." Two new curriculum standards have been added, and new technical and engineering content has been clarified. Based on current education, new responsibilities for society and the environment should be explicit. In addition to the application of science and technology in reality, new thinking about the impact of science and technology on ethics, the environment, and life has been added. All of this allows students to pay attention to the development of the surrounding technological world and experience the impact of science and technology on personal life and social development.

16.3 Current Situation and Training of Chinese Science Teachers in the Twenty-First Century

16.3.1 Overview of Chinese Elementary School Science Teachers

Teachers form the basis of the education plan. In recent years, the status of elementary science teachers in China has not been optimistic. We will bring to light the general situation of primary school science teachers mainly from three perspectives: number, gender structure, and educational background.

The Number and Gender Structure of Elementary School Science Teachers
The number of full-time science teachers from 2011 to 2015 increased annually, from 174,000 to 186,000. However, comparing the number of Chinese and math teachers, the number of science teachers still needs to increase. In 2011, the number of science teachers accounted for about 1/10 of the number of math teachers and about 1/12 of the total number of Chinese teachers. By 2015, the number of science teachers accounted for about 1/9 of the number of math teachers and about 1/11 of the total number of Chinese teachers. In 2001, the third to sixth grades of elementary school in China introduced science courses. According to the proportion of

Chinese and math teachers, the number of science teachers should be close to 2/3 of that of Chinese or math teachers.

The imbalance between male and female teachers in primary schools in our country is becoming increasingly serious, which will cause many problems, such as the disadvantage and feminization of boys. However, the ratio of male to female primary science teachers is relatively optimistic. The difference between the number of male and female science teachers was reduced from 36,000 to 26,000. In 2015, male teachers accounted for 57.0% of science teachers.

The Academic Structure of Elementary Science Teachers
The academic level of a teacher is an important indicator of the teacher's educational level. An academic structure is a basic indicator of the theoretical knowledge level, professional quality, and academic ability of teachers. It is also the most quantitative and authoritative indicator for measuring the overall quality of teaching staff. The academic structure of science teachers is an important indicator of the composition of science teachers. From the perspective of the educational structure of elementary science teachers, the number of teachers with a junior college degree is the highest, followed by teachers with a bachelor's degree, and teachers with a high school degree. Teachers with graduate diplomas accounted for a small portion. In 2015, science teachers who graduated from junior college accounted for 51.6%, science teachers who graduated from undergraduate programs accounted for 35.6%, teachers who graduated from high school accounted for 11.9%, teachers who graduated from graduate school accounted for 0.7%, and teachers who graduated below high school accounted for only 0.1%. From 2011 to 2015, the number of science teachers who were postgraduates and undergraduates increased each year. Among them, the number of teachers who were postgraduates increased from 359 to 1259, an increase of 260.7%. The proportion of primary science teachers with a bachelor's degree rose from 21.1% in 2011 to 35.6% in 2015, an increase of 14.5%, and the number of teachers with a junior college degree and below decreased each year. In 2015, the proportion of teachers who graduated from junior college, high school, and below high school decreased by 2.5%, 12.4%, and 0.2%, respectively, compared to 2011.

16.3.2 Infrastructure of Science Education

In this section, we analyze the status of the infrastructure of elementary science education in China based on three aspects: the compliance rate of experimental equipment in primary schools, the laboratory area per pupil, and the asset value of experimental instruments and equipment per pupil in all regions of the country. The regions included urban, town, and rural areas.

Compliance Rate of Primary School Experimental Equipment
The up-to-standard status of laboratory equipment refers to whether the school's equipment meets the standards set by various provinces, autonomous regions, and

municipalities. The compliance rate of experimental equipment is the percentage of the number of schools that have the required standard of experimental equipment and the total number of primary schools. It is a basic indicator of the availability of basic teaching equipment in schools in various regions.

The compliance rate of experimental equipment in each primary school region is calculated according to the total number of schools that have the required standard of experimental equipment and the total number of primary schools in each region across the country. From 2010 to 2015, the compliance rate of laboratory instruments in primary schools in our country increased yearly, from 54.6% in 2010 to 69.0% in 2015, an increase of 14.4% and an increase rate of 26.4%. The compliance rate of experimental instruments in primary schools in urban, township, and rural areas has increased annually. At the same time, the percentage of experimental instrument compliance in each year was as follows: urban area > town district > rural area. The conditions of urban areas are relatively good, so the instrument compliance rate is high, but the rural conditions are relatively poor.

Laboratory Use Area per Pupil

The per capita laboratory area of primary school students refers to the ratio of the total area of a certain grade of education laboratories to the total number of students in that grade. The larger the laboratory area used per pupil, the more resources available to teachers and students, which is more conducive to the development of science teaching. Based on the total laboratory area of primary schools in each region and the total number of students in primary schools, the average laboratory area used by students in each region was calculated. From 2000 to 2015, the laboratory area used by primary school students across the country increased significantly, from 0.14 square meters in 2010 to 0.21 square meters in 2015, an increase of 50.0%. There is little change in the area of laboratory use per capita in urban areas and towns. The difference between rural and urban areas has increased from 0.02 square meters to 0.10 square meters, and the difference has increased by four times. Every year, the laboratory area used by students in rural areas is the largest, which is far higher than the national level.

16.4 Training of Science Teachers and ESD

16.4.1 Enriching Teams of Science Teachers at Multiple Levels and Integrating ESD with Science Courses Through Relevant Training for Teachers

Guide Extra Science-Related Course Teachers in Middle and High Schools to Join the Elementary Science Teacher Team

Schools lacking science teachers guide teachers from other disciplines in their school to teach science, but because these teachers lack a professional background in science, the actual effect of science teaching remains unchanged. Currently, the number of physics, chemistry, and biology teachers in some middle and high schools may be relatively adequate. Elementary schools can guide these science-related course teachers to join the elementary science teacher team, which will not only solve the lack of science teachers but also to a certain extent improve the professional level of science teachers. Mastering the relevant concepts, principles, and methods of science is a prerequisite for engaging in elementary science education. For elementary science teachers, mastering professional subject knowledge is one of the requirements, and they need to understand the characteristics of related subjects and their related knowledge of people's production and life. In the process of practicing subject knowledge, teachers will enhance their cognition and understanding of professional subjects from recognition and inquiry to understanding academic knowledge. They need to grasp the essence of "scientific inquiry" teaching and have the ability to ask questions, observe experiments, collect information, and form conjectures, as well as acquire technical design, communication, cooperation, and scientific thinking skills (He, 2006). The survey shows that teachers' understanding of the nature of science has more influence on teaching performance than other scientific knowledge and methods.

Guide Science Popularization Personnel, Especially Science and Technology Instructors, to Join Elementary Science Teachers

In 2015, there were 2.0538 million science popularization personnel nationwide. Science popularization personnel refer to organizers of science popularization activities and disseminators of science and technology. They have high scientific literacy, especially science and technology instructors, who are fully qualified to teach science in elementary schools. Schools can attract science popularization personnel who are willing to engage in education, especially science and technology instructors, to join the teaching profession. Professional science popularization personnel can guide elementary science teachers to take a step further into the professional field. Science popularization personnel can bring a wider range of related professional knowledge of natural science, social science, thinking science, philosophy, mathematics, and elementary school science. For example, elementary science teachers complete an open class with the theme "Waste Classification" and have science popularization of professional knowledge in the fields of social science and natural science: the knowledge related to social science is the value and cost of garbage recycling, and the knowledge related to natural science is the impact of waste landfills on the natural world. As is well known, garbage significantly affects the environment, but ordinary elementary school science teachers do not understand the deeper content, such as the following: "What kind of garbage buried underground will produce which kind of harmful gas or substance," "How long does it take to produce the harmful substance or gas," and "Which substances are decomposed and synthesized?" Landfilling alone can lead to a lot of specialized knowledge, including specialized knowledge in the fields of chemistry and mathematics. Therefore, it is difficult for students to complete the abovementioned expansion and inquiry learning that relies on elementary school science teachers. Thus, professional science popularization personnel play a major role and are able to provide professional knowledge and lead teachers and students to conduct relevant experiments.

Make Full Use of the Existing Science-Teaching Experimental Equipment and Out-of-School Science Education Resources
According to the previous analysis, in recent years, the basic facilities of elementary science education in China have improved each year, so we should make full use of the existing experimental equipment. Combining new curriculum standards and new textbooks, schools, and teachers can take advantage of the existing teaching experimental facilities for transformation and utilization. In recent years, China has increasing attention to the popularization of science, and an increasing number of science education resources are available outside school (Xu, 2018). Primary school science courses should make full and effective use of science education facilities and venues outside school, give full play to the role of social science resources, and ensure the full development of primary school science courses.

Make Good Use of New Media and Information Technology to Carry Out Teaching and Teacher Training
The new curriculum standard has been promulgated in the era of new media. We should make good use of new media technologies. Students can acquire scientific information and knowledge through various online public courses, teaching resources, education apps, or directly through online searches. Teachers can use new media technology to provide students with more vivid scientific teaching practices, as well as provide scientific learning experiences related to real life. This alleviates the lack of science teachers to a certain extent and meets the demand for teaching experimental equipment.

16.4.2 Carrying Out Regular Training in ESD Using Both Online and Offline Learning Methods and Compiling Training Materials for ESD

To carry out regular ESD training, we should use online learning platforms to enhance subject awareness and "follow" a series of WeChat official accounts to conduct microlearning and thinking. A series of WeChat public accounts focusing on science learning is a shortcut to improving the teaching and research ability of elementary science teachers. Practical public accounts are (1) public accounts that share science-related knowledge or skills (such as Love Science, Great Scientists, Good Teachers, Science Manual Class, Cool Science, Elementary School Science Teachers, Elementary School Science Teaching Resource Sharing, Elementary School Science Courses and Teaching, and Primary Schools Scientific high-quality teaching videos) and (2) public accounts of magazines that share excellent articles (such as Science Class, Experiment and Teaching, People's Education, and China Education News).

Subsequently, we should organize offline teacher training; learn the science, technology, engineering, arts, and mathematics (STEAM) curriculum concept; and change or improve teacher's pedagogy. Experimental schools carry out practical activities, train teachers to complete a public practice class, and organize teachers to attend and comment on each other's classes. Teachers identify shortcomings in their pedagogy through practice and attempt to improve or change it.

Furthermore, we should organize teachers to compile systematic teaching materials on ESD and to develop real-world lesson examples. In contrast to science courses that already have complete curriculum standards and teaching materials, the main methods of STEAM curriculum development are original creation, introduction and reference, expansion, and adaptation. In this process, teachers should focus on the development of lesson examples (Wang, 2018). After all, the introduced resources may not be suitable for the actual situation of their schools. Frontline teachers should lead students back to real life to find examples of lessons.

16.4.3 Carrying Out STEAM Education to Promote Teachers' Professional Level

STEAM education has gradually emerged in the field of Chinese subject education and has attracted increasing attention. In January 2017, the state promulgated a new compulsory education elementary school science curriculum standard (hereinafter referred to as the "curriculum standard"). One of the major changes is the introduction of the "technology and engineering field." The curriculum standard points out that it is necessary to promote STEAM education and project-based learning (Zhang, 2016). Two things need to be considered to achieve integration.

The Integration of Science Education and STEM Education

STEM stands for science, technology, engineering, and mathematics, but it should be called the integration of different disciplines. Therefore, we call STEM education a comprehensive education that integrates the four disciplines. It can effectively cultivate students' comprehensive literacy abilities and creative thinking and improve their ability to find and solve problems. Science education emphasizes the general rules of scientific inquiry, and scientific problems stem from certain phenomena. For example, how does the same substance of different colors absorb and dissipate heat? STEM project teaching stems from a problem that needs to be solved: for example, how to improve the heat absorption capacity of solar energy? Regarding the content of thermal knowledge, the basic practice of STEM education is to constantly ask questions to determine the problems that need to be solved and then determine a successful solution. In addition to scientific methods, STEM education uses a lot of technical knowledge to solve problems through engineering-related thinking to complete the STEM course.

The creative part of STEM is embodied in the fact that teachers can provide students with everyday materials, without much cost, or ask them to prepare their own materials for activities. For example, when teaching the fifth-grade course "The Making of Recycled Paper," the teacher asks the students to prepare the waste paper by themselves. The first step is to organize students to discuss how to recycle the paper in their hands. To respond to the question "What other materials are needed?" students conduct group research based on the characteristics of paper, dissolve the paper in water, and conclude that water is needed. Based on this, it is concluded that recycled paper requires water. Subsequently, to respond to "How to operate with water?" students proceed with group practice. They soak the paper, mash it, stir it into a pulp, spread it flat on the plastic board, and dry it. There is also another group practice process; that is, the paper is shredded and then soaked and stirred into a pulp. The subsequent processes are the same. However, other students might ask, "The above methods are all possible, but the paper produced is not delicate enough and takes time and effort. How can the efficiency and quality of recycled paper be improved?" After a few minutes of deliberation, the students concluded that to make recycled paper at home, one could use the shredder to help pulverize waste paper into a pulp; the pulp could be brushed on the tray and dried with a dryer, which would not only improve efficiency but also ensure the fineness of the paper.

The above teaching process embodied the core ideas of the STEM course. After teachers provide topics, students ask questions, think deeply, draw preliminary conclusions, find problems through practice, and solve and correct the problems. The expansion of students' thinking is like the butterfly effect; it entails broad expansion and in-depth exploration. The entire teaching process is student oriented, and teachers organize practical explorations from the side.

From the comparison of the integration of the abovementioned science courses and STEM courses, we find that in terms of cooperative learning and evaluation methods, the two have many points in common, but STEM education has its prominent advantages. This enables frontline teachers to reconstruct their teaching concepts. Based on the status of elementary science education, it is necessary and inevitable for elementary science teachers to carry out STEM education to improve teachers' professional quality and professionalism.

Integration of Goal Reconstruction and Project-Based Learning

The curriculum standard clearly states that "STEM is a project-oriented, problem-solving-oriented curriculum organization method that organically integrates science, technology, engineering, and mathematics. It is recommended that science teachers try to apply it to teaching practice." STEM education focuses on the integration of the content of science, technology, engineering, and mathematics, and on enabling students to integrate STEM subject knowledge and skills, use them to solve practical problems, explore the real world, and highlight the comprehensiveness of STEM. As an important method to implement STEM education, project-based activities have a certain activity cycle and meet the characteristics and requirements of learners in elementary school. Elementary school learners have just begun to acquire knowledge and skills and have strong learning needs for all aspects of knowledge.

STEM project learning focuses on allowing students to solve problems in real life and requires students to try their best to show their ideas with finished handmade products so as to help cultivate students' problem-solving abilities. Therefore, when teaching science, teachers can also try to adopt this diversified teaching method. In scientific experiments, teachers should continuously observe students' inquiry behaviors and guide them to choose high-quality and meaningful activities in a timely manner. In the process of "scientific inquiry," teachers need to conduct continuous observations of students through formal or informal channels to ascertain their understanding, interest, and needs of current inquiry activities to provide them with help and guidance in a timely manner. Observation is an important method for promoting student inquiry. Based on observations, teachers should apply appropriate questions to inspire and guide students' self-directed activities.

16.4.4 Towards Teachers' Local ESD and Its Research

The goal of the science curriculum has changed from simply emphasizing "dual bases," including basic knowledge and basic skills, to comprehensively improving students' scientific literacy, focusing on cultivating students' emotions, attitudes, scientific spirit, and other important content. It also includes strengthening students' experience of scientific learning processes and methods, emphasizing the relationship between science, society, technology, and life, especially the connection and integration between disciplines. The change in the goals of science courses has promoted changes in the method of science learning. This has enabled students to embrace an active learning method that actively explores and practices learning, facilitates cooperation and communication, and emphasizes practical problem-solving.

To better enable students to learn science, school teachers should tap into science education resources outside the school setting and, in addition to school education, allow students to "go out" to a setting outside the school to learn, experience, explore, and practice in the real world. Students can use the "school + environment," (i.e., "formal + informal") learning mode to acquire a more comprehensive and in-depth mastery of scientific knowledge and skills, improve their esthetic and cognitive levels, and enhance their scientific literacy. Therefore, how teachers can effectively connect and use science education resources in school science courses and settings outside the school has become an important issue that science educators need to pay attention to and study.

By combining real life, comprehensive practical courses were studied. In the comprehensive practice class, teachers need to focus on guiding students to optimize life problems, lead students to practice together to achieve teaching goals, and improve students' practical and innovative abilities (Zhang, 2019). For example, the "Environment and Us" unit in sixth grade allows students to understand the hazards of garbage based on their life experience, and it encourages students and their families to learn about garbage classification and exposes students to real-life situations.

Through experiments, students understand the hazards of garbage and clearly grasp their classification. Students are guided to think about how to solve problems, such as "how to dispose of the waste generated from daily necessities such as abandoned express boxes and shampoo boxes" and "how to turn it into treasure." Through the information review, students can make fruit peels into enzymes, which can be made into shampoos and dishwashing liquids, and can remove pesticide residues from fruits and vegetables. In this process, students apply interdisciplinary knowledge. For instance, the process of making enzymes requires proportional calculation and uses mathematical knowledge. The use of waste paper boxes to create clothes storage boxes requires mathematical and geometric knowledge and painting skills. This reflects the principle of discipline linkages. Through cross-disciplinary integration and the optimization of practical problems, students can effectively improve their innovative ability and comprehensive literacy. Therefore, teachers need to carry out practical activities in accordance with their actual lives so that classroom research can better meet the development needs of students and enhance classroom flexibility.

16.5 Conclusion

The basic principle of ESD in elementary school science teaching is to enable primary school students to have certain concepts and thoughts on sustainable development so that they can understand the true meaning of sustainable development and its importance to the development of human society (Zhang, 2016). To integrate ESD into elementary science education, teachers must first have a strong view of sustainable development, be able to accept their social responsibility, have strong basic qualities, protect the environment and resources, and innovate teaching methods so that they can treat the surrounding environment gently with a sustainable development mentality. Teachers must themselves have the basic qualities of sustainable development to influence and educate students so that students acquire the basic qualities of sustainable development.

References

He, S. (2006). Research on American science education teacher training and its enlightenment. *Comparative Education Research, 2*, 82–86. (in Chinese).

Ministry of Education of the People's Republic of China. (2018). *Standards on science curriculum in primary school of compulsory education in People's Republic of China*. Beijing Normal University Press. (in Chinese).

Shi, G. (2021). ESD: The innovation paradigm of the construction of quality education system. *Journal of Hongqi University, Xuanwu, Beijing, 1*, 9. (in Chinese).

Wang, Q. (2015). Ten years of education for sustainable development: Experiences of other countries and Chinese models. *Chinese Moral Education, 17*, 18–23. (in Chinese).

Wang, X. (2018). International education for sustainable development: Transformation and innovation in education and learning. *Shanghai Education, 32*, 26–31. (in Chinese).

Xu, X. (2018). The strategies on promoting education for sustainable development in science teaching. *Journal of Shanghai Educational Research, 8*, 82–85. (in Chinese).

Zhang, J. (2016). Education for sustainable development: China's effectiveness and global outlook. *Journal of Teacher Education, 3*(3), 7–16. (in Chinese).

Zhang, J. (2019). The path and practice exploration of ecological civilization education in primary and secondary schools. *Educational Vision, 5*, 44–46. (in Chinese).

Chapter 17
Developing Indonesian Preservice Science Teachers' Understanding of Education for Sustainable Development (ESD) Through Integrated Courses

Ari Widodo and Ida Kaniawati

17.1 ESD as the Heart of Education

The importance of education for sustainable development (ESD) in achieving the Sustainable Development Goals (SDGs) is understandably recognized. ESD plays a central role not only in improving quality education but also in achieving the other SDGs (UNESCO, 2017). The aims of ESD are more than just achieving the eight competencies, i.e., systems thinking, anticipatory, normative, strategic, collaboration, critical thinking, self-awareness, and integrated problem-solving (UNESCO, 2017), but more importantly, ESD is about changing participants' lifestyle toward a sustainable lifestyle. Indeed, a number of studies suggested that the implementation of ESD in schools promotes the quality of education (Laurie et al., 2016), but ESD should not end at this point. The changing lifestyle requires more than just developing knowledge, skills, and attitude. Rather, it requires changes in the belief system that is rooted deeper in the heart and mind. Sustainability should be a part of the way of life of every individual.

To be able to play such an important role, ESD should be treated as more than just an environmentally oriented education program, such as environmental education or eco-school program. As suggested by Hopkins (2012a, b), ESD should be treated as an effort to engage everyone involved in education as well as the public to address not only environmental issues but also economic and social. Without the integration of economic and social issues, ESD would not be able to contribute to the achievement of SDGs. Jucker (2011) suggested even further that a social

A. Widodo (✉) · I. Kaniawati
Universitas Pendidikan Indonesia, Bandung, Indonesia
e-mail: widodo@upi.edu

H. Fujii, S.-K. Lee (eds.), *Science Education for Sustainable Development in Asia*, Education Innovation Series, https://doi.org/10.1007/978-981-99-8711-5_17

movement at all levels is needed. ESD is not just a whole-school approach, but it should become a whole-society approach. It means that ESD generates participation and changes to the whole society.

Despite its necessities, reality shows that there are challenges to bringing ESD into practice in schools. Firstly, the current school curriculum is heavily loaded with subjects that an introduction of a new school subject would create an extra load to the existing curriculum. The discussion around the school curriculum is to simplify the curriculum structure by reducing the school subjects. Adding ESD into the existing school subjects may also create resistance from the schools. Therefore, introducing ESD as a new school subject is very unlikely.

Although introducing ESD would not be popular in Indonesia, since the last few years, however, the government introduced environmental education as an elective school subject. The government also introduced a green school program called Adiwiyata Green School of Indonesia, which is an optional environment promotion program at the school level. Environment education and green school program do not necessarily run together. These programs are more environmental education programs than ESD because ESD covers not only environmental issues but also social issues and economics (Hopkins, 2012b).

Secondly, school subjects are compartmentalized. School subjects are taught by different teachers, and every subject focuses on its own syllabus. Every subject is taught at a certain time slot. Finding a common program across subjects is not easy due to the difficulties to identify cross-cutting concepts among those subjects. As a result, initiatives to integrate ESD based on a school subject would be difficult. Attempts to integrate ESD would require the reorganization of the syllabus and a new managerial system that allows for different scheduling. For this purpose, commitment from leaders and other school regulators is needed.

Thirdly, most teachers are not well prepared to teach ESD. Reports of studies on the implementation of ESD in Indonesian schools indicated that Indonesian teachers need training and support to facilitate them in implementing ESD in the schools (Listiawati, 2011; Kelompok-Kerja-LSM-untuk-PPB-Indonesia, 2014). Training such a large number of teachers is indeed very demanding in terms of funding and time. Previous experience shows that training teachers for a new program requires relatively a long time.

17.2 Strategic Roles of Teacher

Teachers play a key role in the implementation of ESD (McKeown, 2014; Bürgener & Barth, 2017). Competence and committed teachers are key factors for the implementation of ESD in schools. Teachers are not only the agents of change for ESD, but more importantly, they bring multiple effects. Every year, a teacher educates a number of students, so cumulatively there will be many students educated with ESD by a single teacher. This multiplier effect is even bigger when ESD is successfully adopted in teacher education. Every year, a teacher education university produces teachers who can contribute to ESD. Therefore, teacher education institutions play even greater roles in bringing ESD into practice in schools (Bürgener & Barth, 2017).

As teachers are prepared at the universities based on the subjects they are going to teach, no teacher graduated as an ESD teacher. ESD teachers gain their ESD teaching competencies mostly through training. Indeed, there are initiatives to educate university students about ESD (Tabucanon, 2008; Harpe & Thomas, 2009; McMillin & Dyball, 2009; Holdsworth & Thomas, 2015; Otte, 2016); however, more initiatives are needed to support teacher education institutions. A review of the development of ESD in teacher education institutions (McKeown, 2012; Mckeown, 2014) indicated that some progress was made. In order to facilitate the implementation of ESD, teacher education institutions need support in the forms of finances, resources for teacher education, and other necessary supports to implement guidelines for reorienting teacher education. Indeed, efforts are made to support teacher education institutions with a guide in implementing ESD (UNESCO, 2018); however, no clear guidance was actually offered on how teacher education universities should prepare teachers to teach ESD. Among a few is the idea proposed by Bürgener and Barth (2017), who suggested an open-learning environment in which preservice teachers work together on real-world sustainability issues with experienced partners in the schools. The implementation of the idea, however, is not simple since the curriculum structure of teacher education institutions needs to be highly flexible, but unfortunately, it rarely happens. Another issue with teacher education is the lack of strategy to develop ESD teachers' competencies. Attempts to identify the competencies of ESD teachers initiated by Okayama University ESD Promotion Center (2020) managed to identify key ESD teacher competencies, but it does not address the strategy for preparing ESD teachers. Therefore, despite the pivotal roles of teachers, plenty needs to be done on the strategy to prepare ESD teachers.

17.3 Preparing Teachers for Teaching ESD

Teacher education is expected to play an important role in preparing teachers to teach ESD in schools (Kieu et al., 2016; Bürgener & Barth, 2017), but it is arguable that teacher education institutions are ready to play such a role. In the Indonesian school curriculum, ESD is not yet explicitly mentioned. Indeed, some schools offered "Environment Education" as an elective subject, but it is not listed in the national curriculum structure. Since at teacher education institutions there is no education program to prepare ESD teachers, there is no school teacher with ESD educational background. The absence of an ESD teacher has both negative and positive sides. On one hand, it is a "good" situation for schools to organize integrated and comprehensive ESD programs, but on the other hand, there is no one who assumes the responsibility to teach ESD.

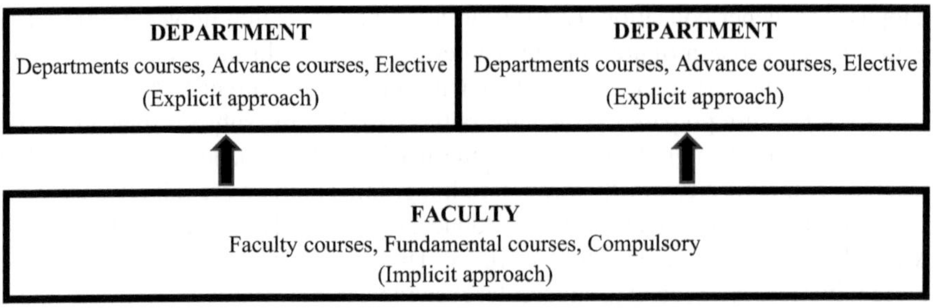

Fig. 17.1 Arrangement of ESD content in the curriculum structure

The Faculty of Mathematics and Science Education, Universitas Pendidikan Indonesia, realizes the situation and tries to respond by offering courses on ESD. Though ESD is not a part of science, ESD is closely related to science. In addition, a compulsory course on environmental education has been offered by the faculty for many years. Nonetheless, the transfer from environmental education to ESD was not easy as many faculty members and the curriculum developers are not yet familiar with ESD. With more than 3500 students enrolled in the faculty and around 70% of them are preservice teachers, educating them with ESD can make a significant contribution to the implementation of ESD at schools. Considering the situation and the effectiveness aspect, two implementation strategies were chosen, i.e., the implicit approach and explicit approach (Fig. 17.1).

17.3.1 The Implicit Approach

The implicit approach was implemented in two faculty-level courses, namely, "Mathematics, Science, Technology and Engineering" and "Application of Mathematics, Science, Technology and Engineering." As the two courses are faculty-level courses, the curriculum structures of all departments in the faculty have to list both courses as compulsory courses. Unlike department-level courses managed by each department, the content and the teaching strategies of faculty courses are developed by the faculty. The two courses are designed as consecutive courses offered in the first and second semesters. We believe that the ultimate goal of ESD is to create a sustainable culture, and preservice teachers should be enculturated as early as possible. The reasons to introduce ESD to first-year students are to build a foundation of sustainability at the earliest possible time and that first-year students are generally easier to shape.

Unlike the common courses that focus on content (mathematics, biology, chemistry, physics, and pedagogy), the two courses focus on two central competencies, i.e., scientific process skills and engineering design skills. Scientific process skills and engineering process skills are the backbone competencies of scientists and engineers that science teachers should develop to be able to teach science and ESD. It is expected that by developing those competencies, teachers can guide their students to do simple scientific investigations and develop a simple technology that takes into account sustainability aspects. Sustainability is expected to be a point of consideration in any scientific activities and technological solutions developed by students.

The two courses were designed to develop service teachers' basic understanding of the SDGs and ESD. In the first few meetings, preservice teachers explored the basic ideas of the SDGs and identified strategies to implement sustainability principles in their projects. The first cycle of the project focuses on scientific processes. Preservice teachers are required to identify problems related to sustainability and conduct research to promote a better understanding of the issue at hand. The groups follow steps of scientific processes, i.e., identify a problem, design an investigation, conduct an investigation, and draw a conclusion based on the data collected. Developing scientific process skills is considered a crucial step since throughout the study, they will have to do some investigations. In the second cycle (engineering processes), preservice teachers use their findings from the first cycle to develop technological solutions. The engineering processes include thinking of a technological solution to solve the problem, designing a technology, making a prototype technology, testing the prototype, and redesigning the technology. The technology they developed should adopt principles of sustainability. Therefore, sustainability is addressed both as an issue to be solved and as sustainability-oriented technology.

Attempts to enculturate students into sustainability culture are done not only by presenting ESD as the content of the courses but also by designing the courses that allow preservice teachers to develop ESD competencies (Fig. 17.2).

Firstly, to develop preservice teachers' normative competencies, collaboration competencies, critical thinking skills, and self-awareness skills, the courses are designed as team-based courses. Members of a group are coming from different departments. Throughout the course, preservice teachers have to work in groups. Unlike the conventional courses, those assessments are based on individual

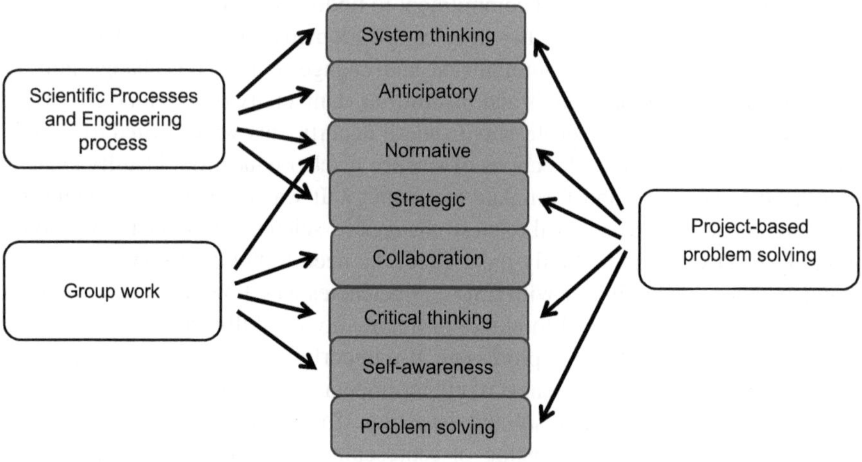

Fig. 17.2 The design of the courses and the contribution to ESD competencies

achievement; in these two courses, the assessment is fully based on group achievement. This approach of assessment provides a unique challenge for the preservice teachers since they have to build a team with individuals they never knew before. In the pandemic situation when students cannot meet the members personally, the challenge is even greater because they can only meet online. The assignments are deliberately designed as a team-based approach to provide students with real-life situations in that very often people have to collaborate with new people they never know before, and they may live far away that they cannot meet them in person.

Secondly, to develop preservice teachers' system thinking competencies, anticipatory competencies, normative competencies, and strategic competencies, the courses focus on scientific process skills and engineering skills. The courses are designed to develop preservice teachers' capacities in conducting scientific research (identify the problems, formulate research questions, design investigation, conduct investigation, analyze data, and draw a conclusion). Scientific process skills are essential to be able to understand phenomena scientifically. After developing an understanding of the problems and gaining knowledge related to the problem, the groups have to develop a technological solution. The process of developing the technology follows the engineering process, e.g. thinking of the technology, designing the technology, developing prototype technology, testing the prototype, and redesigning the technology.

Thirdly, to develop system thinking competencies, normative competencies, strategic competencies, critical thinking competencies, and integrated problem-solving competencies, the courses are designed as project-based, contextual courses. Each group is challenged to identify a contextual and real problem in daily life, e.g., energy crises, food shortages, and traffic jams; research the problem; and design a technology to solve the problem. Project-based assignments that require the groups to manage their work within the available time and resources make them learn how to run an efficient project. Since the problem is drawn from their daily life, the groups are very enthusiastic as they are part of the victims. The contextual problems raised students' ownership and commitment to finding solutions.

To ensure the success of the two courses, we adopted a four-step change management process, i.e., define, communicate and engage, embed the change into the system, and prepare individuals and institutions (University of Virginia, 2020). For the first step (define), key professors from all departments were invited to a workshop to discuss the future directions of science teacher education, identify changes that have to be done, and formulate the strategy. It was agreed to restructure the existing curriculum to adopt the latest changes in science and education. Though initially ESD was not explicitly mentioned, the need to build preservice teachers' competencies that are in line with ESD competencies was, however, acknowledged, such as creative and critical thinking, problem-solving skills, collaboration skills, and system thinking. The key professors also underlined the necessity for developing preservice teachers' awareness of environmental issues and the competence that enables them to contribute to solving the issues. To operationalize the ideas, a curriculum developer team consisting of representatives from all departments was assigned to formulate the technical and implementation strategy.

In the second step (communicate and engage), the ideas were presented to the management leaders of the faculty and departments. The developer team explained the technical details of the courses, which include the content, delivery, the consequences of scheduling, and academic administration. Since the new courses will affect scheduling and academic administration, the support of the management was very crucial. One of the consequences is the need for management at the faculty and the department to recalculate lecturers' workloads and to reallocate resources.

Upon receiving the initial agreement, we proceed to the third step, i.e., embed the change into the existing system. In this step, the developer team prepared the teaching materials, designed the implementation strategy, designed and developed monitoring and evaluation instruments, and prepared training materials. Since the courses are interdisciplinary, courses will be delivered by a team of lecturers with different academic backgrounds. A teaching team consisting of lecturers from different academics was considered the best option because each lecturer may contribute to different aspects. Such teaching team has never been done before a special arrangement for the academic and financial system is needed.

Finally, the last step of the change management process (preparing individuals and institutions) was conducted. A series of training for lecturers were conducted to prepare them for the new courses. Training for lecturers is very important not only because the courses are new but also because they will have to work in a team with lecturers from other departments. In total, 65 lecturers were participating in the training. The workshops were designed to provide lecturers with full experience as students and as a lecturer. In the first part of the workshops, the lecturers played the roles of students. They work in groups and do the tasks under the supervision of the developers. In the second part, they played the roles of lecturers who had to prepare lessons, conduct lessons, and assess students' learning. In addition to training for lecturers, training for supporting staff (administration and laboratory) was also conducted to familiarize them with the new course requirements.

Despite the implicit nature, the courses were designed to produce a significant impact on the promotion of ESD. To gain maximum benefits, a number of considerations were taken in developing the courses. The first consideration is the number of people who benefited from the courses. Every semester, there are approximately 650 participants who take the course. As a compulsory faculty-level course, it means that all students in the faculty have to take the courses. This ensures that all graduates of the faculty have developed ESD as the basic knowledge, skills, and behavior to live sustainably. The courses also involved a large number of lecturers. In total, there are at least 65 lecturers (nearly one third of the total faculty members) who are involved in delivering the courses. This strategy ensures that almost all persons in the faculty are exposed to ESD through these courses.

The second consideration is the duration of the courses. To give a longer intervention and to produce bigger impacts, the courses are set up as two consecutive courses to be taken in the first and second semesters. Therefore, preservice teachers are actually exposed to ESD for 1 year. Such an arrangement is new in the system as courses usually last for one semester only. Longer exposure to ESD should give more opportunities for preservice teachers to internalize the knowledge, skills, and attitudes they learned that, in turn, can change their lifestyle.

The third consideration is early intervention. Setting the courses as the fundamental course allows the courses to shape the fresh minds of the new students toward sustainably thinking and behaving. Lessons learned in the first year should be a foundation for the next semesters. In addition, new students are generally more open to new ideas, so it is expected to be easier to enculturate them with sustainability values.

At the end of the semester, two questionnaires were distributed to measure preservice students' sustainability consciousness and ESD teacher competencies. Sustainability consciousness was measured using an instrument developed by Gericke et al. (2018), while preservice teachers' perception of ESD teacher competencies was measured using an instrument based on a framework developed by Okayama University Promotion Center (2020). ESD sustainability consciousness consists of three aspects, namely, sustainability knowingness, sustainability attitudes, and sustainability behavior (Gericke et al., 2018). As Indonesians tend to pick options in the middle, for this purpose, the responses are modified to four scales instead of five.

As presented in Fig. 17.3, the preservice teachers show high sustainability consciousness (average score of 86). Since there is no previous data as a benchmark, we try to make sense of the data by comparing it to a finding of research conducted in Taiwan and Sweden (Berglund et al., 2019). A direct comparison is certainly inappropriate due to such a different condition. Nonetheless, in terms of average scores, our students produce slightly higher scores. Our result indicates that the integration of ESD principles into a course can raise preservice students' sustainability consciousness.

Similar to our findings, the study in Taiwan and Sweden also reported that the lowest score among the three sustainability consciousness indicators was sustainability behavior (Berglund et al., 2019). This indicates that developing sustainability behavior is probably the most difficult one. This finding aligns with the common understanding that knowledge and attitudes are not sufficient to change behavior. Unlike knowledge and attitudes that can change in a relatively short time, changing behavior seems to take a longer time and requires a suitable condition to occur, such as regulation and law enforcement.

Fig. 17.3 Preservice teachers' sustainability consciousness after courses with implicit approach

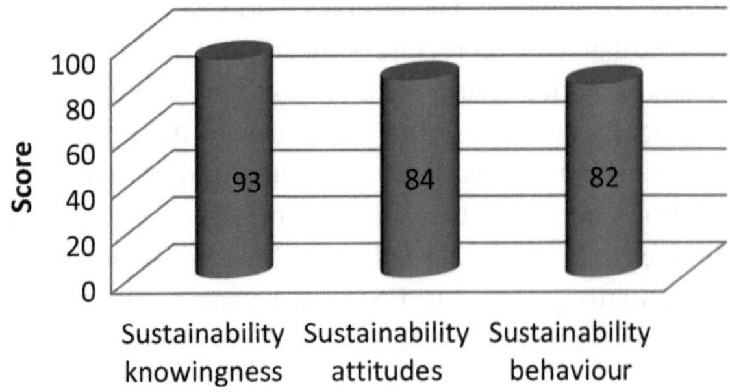

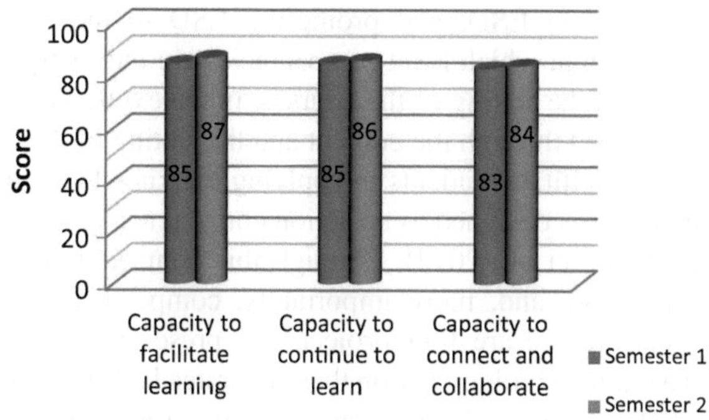

Fig. 17.4 Preservice teachers' scores on ESD teacher competencies after courses with the implicit approach

Preservice teachers' perceptions of ESD teacher competencies were measured at the end of the first semester and the second semester. The questionnaire was based on a framework formulated by Okayama University ESD Promotion Center (2020), namely, the capacity to facilitate learning, the capacity to continue to learn, and the capacity to connect, collaborate, and engage. The results of the questionnaire indicate that preservice teachers developed strong perceptions of ESD teaching competencies (Fig. 17.4). Figure 17.4 also shows that there were slight positive changes from the first semester to the second semester.

Analyses of the results of the implicit approach show that the approach produces a positive impact. There are advantages and disadvantages to offering ESD as an integrated course. On the negative side, only limited ESD content can be addressed in the course. Of the 17 SDGs, the courses can only address four of them. Some issues cannot be discussed, though they need urgent attention, e.g., health, pollution, and clean water shortage. On the positive side, the integrated course provides an exemplary model of teaching ESD. Since in the school curriculum ESD is not offered as a school subject, ESD is most possibly integrated with other school subjects. The experience of learning ESD in this course should develop preservice teachers' competencies in adopting strategies for teaching ESD. One of the unique features of the courses is collaboration. Working together is important for achieving a goal. The courses provide opportunities for the preservice teachers to develop their teamwork and collaboration skills. It is expected that later as teachers they can teach ESD by collaborating with fellow teachers from different subjects.

17.3.2 The Explicit Approach

The explicit approach was conducted by offering ESD as a course. In this part, we present the implementation of an ESD course, named "Biology Education for Sustainable Development," offered by the Biology Education Department. The course is a two-credit elective course offered in the fifth semester. The content of the course covers a general introduction to the SDGs and ESD, integrating ESD into a biology lesson, ESD as an extracurricular program, networking with organizations working on ESD, and promoting ESD to the community. Unlike the implicit approach in which ESD competencies are indirectly built through the implementation and the setting of the courses, in the explicit approach, ESD competencies are developed through the content and the setting of the course.

To facilitate students in applying the knowledge they learned in the course, the course was designed as an action competence course (Mogensen & Schnack, 2010; Hedefalk et al., 2014). Through this course, students develop knowledge, skills, awareness, and, more importantly, competence to perform real action on ESD issues. There are four projects the preservice teachers have to accomplish, i.e., designing a biology lesson that integrates ESD principles, designing a school-based ESD extracurricular program, building a network for initiating an ESD organization, and promoting ESD to the community using a variety of media. The assignments are designed as a sequential building block that facilitates preservice teachers' competencies to expand from an individual level, school level, and community level. Finally, preservice teachers have to demonstrate their competencies in real action.

Similar to the implicit approach, the course is also designed as a group-work project. In the first project, the groups have to identify biology content that fits with ESD principles. In the beginning, most groups perceived that only environment biology relates to ESD since they perceived ESD as environmental education. However, as they learned that ESD is not the same as environmental education and as they grasped the idea of ESD, they could easily integrate ESD into biology lessons. For example, they can link the lesson on food and the digestive system with SDG 1 (No Poverty), SDG 2 (Zero Hunger), SDG 3 (Good Health and Well-Being), SDG 13 (Climate Change), and SDG 14 (Life Below Water). Working on ESD in the curriculum not only promotes their sustainability awareness but also strengthens their understanding of biology content. As they tried to integrate ESD, they found connections between the content with other areas that they did not realize before. For example, a group of preservice teachers realized that meat and milk production could contribute to climate change due to methane produced by cow farming.

For the second task, each group is required to design a school-based extracurricular program. This task is designed to push preservice teachers to move beyond the biology content knowledge. In their design, they may integrate ESD into the existing extracurricular program or design a completely new program. Unlike curricular programs that can be independently designed by the teacher, extracurricular programs should involve school stakeholders, such as the school principal, teachers, parents, and the community. As the program involves school stakeholders, the design should take into account not only environmental aspects but also social and economic. The groups came up with some interesting ideas, such as hydroponic farming and selling the vegetables to the local market, mushroom farming and making mushroom snacks, and making soap from oil waste.

The third task of the course is networking with a variety of organizations or institutions that work on the SDGs or ESD promotion. Collaboration is essential to achieve our goals. The task is designed to develop preservice teachers' communication and collaboration skills. Each group is required to learn how the organizations are managed and to collaborate with them. Learning how an organization is managed should inspire the group on how to run an organization of their own. The task of initiating the SDGs/ESD organization is pushed further in the next project assignment. In general, the groups have no difficulty communicating with the SDGs or ESD organizations. Some groups move even further by contacting organizations from overseas. At the end of the assignment, some groups reported that they joined as a member of the organization, working as apprentices at the organization, and jointly organizing ESD events with the organization.

The final assignment of the course is designed as a demonstration of students' action competence. Students are required to set up an ESD promotion center for promoting ESD to the community. Although the assignment does not really meet the criteria of an "open learning environment," as proposed by Bürgener and Barth (2017), the task was designed to promote preservice teachers' active involvement in ESD and to multiply the effect of the course. It is expected that the benefits of the course are not only for students but also for the community. For this purpose, each group has to set up "an ESD promotion center" that educates the community to live sustainably. As an ESD promotion center, they have created promotional content and broadcast the content using available communication media, such as websites, Twitter, YouTube, Facebook, and TikTok. To motivate the participants, the final grade of the course is based on the number of people they can educate, as reflected in the number of followers on their media. The response from the community was very good, as measured by the number of followers. Three Instagram-based promotion centers (esdish.id, consumable.id, and meallion.id) that focus on raising ESD awareness on SDG 2 (No Hunger) received significant followers. Groups that based their promotion on TikTok (e.g., ESDforchange, Avirama.project, and ESDranger) also gained good attention from the community. As many Indonesians also use Facebook, some groups also base their promotion via Facebook (e.g., Ironmover and ESDforchange).

At the end of the semester, questionnaires on sustainability consciousness and ESD teachers' competence were distributed to the preservice teachers. As presented in Fig. 17.5, preservice teachers show strong sustainability consciousness. Sustainability knowingness scores the highest, followed by sustainability attitudes and sustainability behavior. There is a similar pattern with the scores obtained in

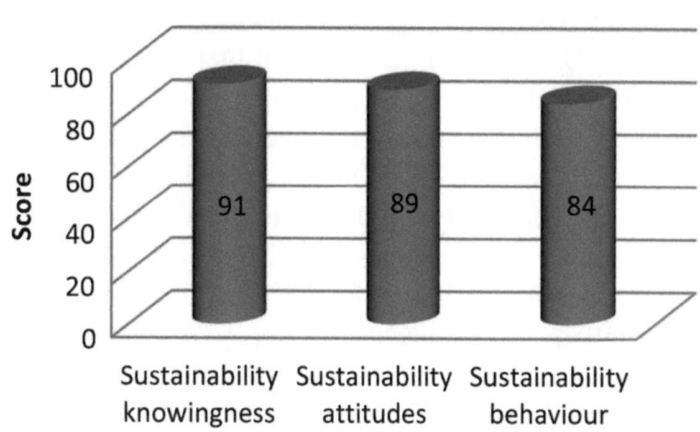

Fig. 17.5 Preservice teachers' sustainability consciousness after a course with explicit approach

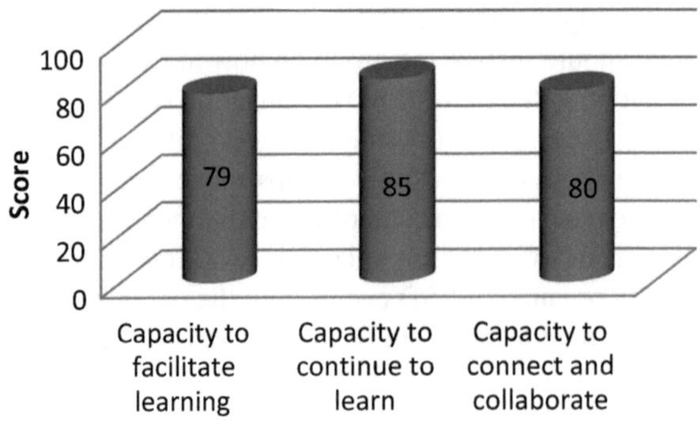

Fig. 17.6 Preservice teachers' scores on ESD teacher competencies after a course with explicit approach

implicit approach courses. Again, the findings indicate that changing sustainability behavior is the most difficult.

Preservice teachers' average score on ESD teacher competencies is relatively high (81), with the capacity to facilitate learning scores the lowest (Fig. 17.6). Compared to the scores of the implicit courses (86), the average score of explicit courses is slightly lower. The finding is, to some extent, surprising since participants of the explicit courses were given opportunities to integrate ESD into lessons. Additionally, they also have taken pedagogical courses. Indeed, comparing the scores is unfair given that the conditions of the two approaches are very different. Nonetheless, the scores indicate that ESD teacher competencies can be developed by either approach.

In general, there are four approaches to embedding ESD into teacher education curricula (Evans et al., 2017). Our attempts fall into the last two categories, i.e., infusing ESD as a component of a compulsory course and offering ESD as an elective course. As shown by the results measured at the end of the courses, both implicit approach and explicit approaches are effective for developing preservice teachers' sustainability consciousness and ESD teacher competencies. This indicates that depending on the condition, ESD can be integrated into the existing subjects, as an independent subject, or as a combination of both. Since no preservice students have taken both courses, it is interesting to see the impact of a combination of both courses. Developing preservice teacher ESD competencies with courses is not enough. A recurring education program starting from the first semester to the last semester is certainly preferable since ESD competence development is a continuous process (Brandt et al., 2019). The first year is a time to build the foundation, the second and third years to develop the competencies, and finally the fourth year to intensify and transfer the competencies into practice.

As elaborated in the document on the ESD competency framework published by Okayama University ESD Promotion Center (2020), there are three domains of ESD teacher competencies. In this study, ESD teacher competencies are measured using a perceptional instrument. Therefore, the scores reflect participants' perception of their competence (self-efficacy) instead of their actual competence. Although perceived self-efficacy plays an important role in the self-development, adaption, and change of an individual (Bandura, 2012), the lack of actual evidence on competency raises a question of whether teachers are really competent in teaching ESD. The current ESD competence framework is certainly a good foundation for measuring ESD teachers' competencies, but evidence of competencies should be measured in terms of self-efficacy as well as actual competency. The availability of instruments for measuring teachers' actions in teaching should be very helpful in setting up a standard for ESD teachers' competencies.

17.4 Lessons Learned and Ways Forward

Our attempts to prepare ESD teachers focused mainly on content-related courses without including pedagogy-related courses. Courses on pedagogy cover not only "pure" pedagogy but also other teaching and learning-related aspects, such as educational technology. The lack of ESD integration in pedagogy-related subjects may hinder the development of teachers' pedagogical content knowledge (PCK) or teachers' technological pedagogical content knowledge (TPACK). The development of technology and technological education demands teachers to employ technology in their teaching, including teaching ESD. Our studies on teachers' TPACK (Yulisman et al., 2019; Yulisman et al., 2020) suggest that to be able to integrate pedagogical knowledge, content knowledge, and technological knowledge into TPACK, teachers need to have a strong understanding of all that knowledge. With regard to technology, our study also shows that teachers' attitude toward technology plays a very important role in their ability to use technology in their lessons. As suggested by Buchanan et al. (2018), teachers' mastery of technology will contribute to their ability to deliver effective ESD lessons. They also suggest that the use of technology for environmental purposes should consider the scientific method, real-world issues, student learning, and their roles as agents of change.

Currently, every education program at teacher education institutions focuses only on preparing teachers to teach subjects in the school curriculum. Unless ESD is listed as a school subject, it is very unlikely that teacher education institutions will open a program for preparing ESD teachers. As a result, schools will never have special teachers on ESD. One of the feasible solutions to have teachers with ESD teaching competencies is offering ESD content as much as possible in the curriculum of teacher education, which can be done by creating ESD courses or infusing ESD into content-related and pedagogy-related courses. Integrating ESD into the existing will add content load to courses that university professors may be reluctant to do (Buchanan, 2013). Moreover, educating university professors with ESD can be the first and probably the toughest task to do. For this reason, a strong political will and the commitment of top management are needed to develop a comprehensive ESD program in teacher education.

The introduction of STEM (or STEAM) education is an alternative option to infuse ESD. In cases where having ESD as an independent subject is not possible, ESD principles can be introduced in every step of STEM teaching. Though such integration is not easy (Hopkinson & James, 2010), our experience suggests that infusing ESD values into a STEM-based course is possible. STEM and ESD are both interdisciplinary, and professors find the two ideas closely linked to each other and that a comprehensive solution to problems can be achieved by combining STEM and ESD. Our strategy to put scientific processes and engineering processes as the backbones of the courses seems to be a key success to getting lecturers' acceptance. Resistance from lecturers and professors is not very strong since the courses focus on common scientific practices, such as observation, experiment, critical thinking, and creative thinking. When ESD is to be integrated with STEM, we recommended that ESD should be introduced in the first meetings so that students can develop an understanding and awareness of ESD issues. Understanding and awareness of ESD will serve as foundation knowledge for students in developing technological solutions.

Training programs for preparing ESD teachers are needed. Since most existing teachers are not prepared to teach ESD, a special training program on ESD teaching is needed. As previously mentioned, the key to ESD is "action," so action competence should be the core of any teacher training. For this purpose, teacher training on ESD teaching should allow them to experience teaching and action on ESD issues. Since nongovernmental organizations Non-Governmental Organizations (NGOs) usually gain significant experience in ESD from their action program, the involvement of NGOs in teacher training should give a valuable contribution to participants' practical ESD knowledge and skills (Kieu et al., 2016). In addition to the main training program, post-training support is crucial for teachers to implement the result of the training. School administrators play a very important role in ensuring that teachers receive the necessary support in implementing the ESD program.

Another important issue related to ESD is its sustainability. ESD as a program should not be viewed as a temporary project; rather, it should be viewed as a common need in our lives. In the past, new ideas or initiatives in education were commonly introduced in the form of projects. As a result, many perceived the program as a temporary project rather than permanent and lasting. Using the framework proposed by Hart and Milstein (2003), to be sustainable, a program should be able to develop a sustainable value for all shareholders. The model consists of four quadrants made of a vertical axis that reflects the need to manage today's conditions while simultaneously creating technology and markets of the future. The horizontal line reflects the need to maintain growth but at the same time protect internal organizational capabilities. In short, to be sustainable, a project should keep on innovating and should be able to convince shareholders as a credible project. ESD activists should focus not only on promoting ESD but also on developing shareholders' sense of belonging. This means that to ensure the sustainability of ESD, the ESD program should cover both the content of ESD and its strategy.

Finally, ESD is not just about knowledge and skills. Presenting ESD as knowledge and skills would only result in individuals with understanding and skill on ESD issues but lacking heart involvement. We believe that only things that are coming from the heart may reach hearts. Therefore, anyone involved in ESD should try to involve the heart in designing and running the program.

References

Bandura, A. (2012). On the functional properties of perceived self-efficacy revisited. *Journal of Management, 38*(1), 9–44.

Berglund, T., Gericke, N., Boeve-de Pauw, J., Olsson, D., & Chang, T.-C. (2019). A cross-cultural comparative study of sustainability consciousness between students in Taiwan and Sweden. *Environment, Development and Sustainability, 22*(7), 6287–6313.

Brandt, J.-O., Bürgener, L., Barth, M., & Redman, A. (2019). Becoming a competent teacher in education for sustainable development. *International Journal of Sustainability in Higher Education, 20*(4), 630–653.

Buchanan, J. (2013). Sustainability education and teacher education: Finding a natural habitat? *Australian Journal of Environmental Education, 28*(2), 108–124.

Buchanan, J., Pressick-Kilborn, K., & Maher, D. (2018). Promoting environmental education for primary school-aged students using digital technologies. *EURASIA Journal of Mathematics, Science and Technology Education, 15*(2), em1661.

Bürgener, L., & Barth, M. (2017). Sustainability competencies in teacher education: Making teacher education count in everyday school practice. *Journal of Cleaner Production, 174*, 821–826.

Evans, N., Stevenson, R. B., Lasen, M., Ferreira, J.-A., & Davis, J. (2017). Approaches to embedding sustainability in teacher education: A synthesis of the literature. *Teaching and Teacher Education, 63*, 405–417.

Gericke, N., Boeve-de Pauw, J., Berglund, T., & Olsson, D. (2018). The sustainability consciousness questionnaire: The theoretical development and empirical validation of an evaluation instrument for stakeholders working with sustainable development. *Sustainable Development, 27*(1), 35–49.

Harpe, B. D. L., & Thomas, I. (2009). Curriculum change in universities: Conditions that facilitate education for sustainable development. *Journal of Education for Sustainable Development, 3*(1), 75–85.

Hart, S. L., & Milstein, M. B. (2003). Creating sustainable value. *Academy of Management Perspectives, 17*(2), 56–67.

Hedefalk, M., Almqvist, J., & Lidar, M. (2014). Teaching for action competence. *SAGE Open, 4*(3). https://doi.org/10.1177/2158244014543785

Holdsworth, S., & Thomas, I. (2015). Framework for introducing education for sustainable development into university curriculum. *Journal of Education for Sustainable Development, 9*(2), 137–159.

Hopkins, C. (2012a). Twenty years of education for sustainable development. *Journal of Education for Sustainable Development, 6*(1), 1–4.

Hopkins, C. (2012b). Reflections on 20+ years of ESD. *Journal of Education for Sustainable Development, 6*(1), 21–35.

Hopkinson, P., & James, P. (2010). Practical pedagogy for embedding ESD in science, technology, engineering and mathematics curricula. *International Journal of Sustainability in Higher Education, 11*(4), 365–379.

Jucker, R. (2011). ESD between systemic change and bureaucratic obfuscation: Some reflections on environmental education and education for sustainable development in Switzerland. *Journal of Education for Sustainable Development, 5*(1), 39–60.

Kelompok-Kerja-LSM-untuk-PPB-Indonesia. (2014). *The journey of ESD in Indonesia from the perspective of NGOs by the end ESD decade 2005–2024*. Kelompok-Kerja-LSM- untuk-PPB-Indonesia. (in Indonesian).

Kieu, T. K., Singer, J., & Gannon, T. J. (2016). Education for sustainable development in Vietnam: Lessons learned from teacher education. *International Journal of Sustainability in Higher Education, 17*(6), 853–874.

Laurie, R., Nonoyama-Tarumi, Y., McKeown, R., & Hopkins, C. (2016). Contributions of education for sustainable development (ESD) to quality education: A synthesis of research. *Journal of Education for Sustainable Development, 10*(2), 226–242.

Listiawati, N. (2011). The relevance of ESD values and teachers' readiness to implement them at schools. *Jurnal Pendidikan dan Kebudayaan, 17*(2), 135–152. (in Indonesian).

McKeown, R. (2012). Teacher education 1992 and 2012: Reflecting on 20 years. *Journal of Education for Sustainable Development, 6*(1), 37–41.

McKeown, R. (2014). The leading edge of teacher education and ESD. *Journal of Education for Sustainable Development, 8*(2), 127–131.

McMillin, J., & Dyball, R. (2009). Developing a whole-of-university approach to educating for sustainability: Linking curriculum, research and sustainable campus operations. *Journal of Education for Sustainable Development, 3*(1), 55–64.

Mogensen, F., & Schnack, K. (2010). The action competence approach and the 'new' discourses of education for sustainable development, competence and quality criteria. *Environmental Education Research, 16*(1), 59–74.

Okayama University ESD Promotion Center. (2020). *Guide for the effective dissemination of the Asia-Pacific ESD teacher competency framework*. Okayama University ESD Promotion Center. http://ceteesd.ed.okayama-u.ac.jp/pdf/200511.pdf

Otte, P. P. (2016). Integrating sustainable development in higher education through experience-based learning: Insights from experts in team (EiT) for developing a combined theoretical framework. *Journal of Education for Sustainable Development, 10*(1), 131–159.

Tabucanon, M. T. (2008). Asia-Pacific University Network formed to integrate ESD and SD into higher education. *Journal of Education for Sustainable Development, 2*(1), 73–75.

UNESCO. (2017). *Education for sustainable development goals: Learning objectives*. UNESCO.

UNESCO. (2018). *Integrating education for sustainable development (ESD) in teacher education in South-East Asia: A guide for teacher educators*. UNESCO.

University of Virginia. (2020). *UVA change management framework*. https://organizationalexcellence.virginia.edu/change-management

Yulisman, H., Widodo, A., Riandi, R., & Nurina, C. I. E. (2019). Moderated effect of teachers' attitudes to the contribution of technology competencies on TPACK. *Jurnal Pendidikan Biologi Indonesia, 5*(2), 185–196.

Yulisman, H., Widodo, A., Riandi, R., & Nurina, C. I. E. (2020). The contribution of content, pedagogy, and technology on the formation of science teachers' TPACK Ability. *Edusains, 11*(2), 173–185.

Chapter 18
Workshops to Enhance Teachers' Reflection as Reflective Practitioners and to Help Them Gain Professional Capacities for Education for Sustainable Development: A Case Study of Japanese Prospective Teachers

Hironori Sasaki

18.1 Introduction

In March 2020, the Second Asia-Pacific Regional Meeting on Teacher Education for Education for Sustainable Development (ESD) released the second report titled "Guide for the Effective Dissemination of the Asia-Pacific ESD Teacher Competency Framework" (Okayama University ESD Promotion Center, 2020). It presented the final version of the Asia-Pacific Education for Sustainable Development (ESD) Teacher Competency Framework, the objective of which is to contribute to the achievement of the United Nation's Sustainable Development Goals (SDG) targets 4.7 and 12.8 by 2030. It has provided a blueprint to support the training of both pre- and in-service teachers in the Asia-Pacific region and globally. Furthermore, it has outlined three key domains for the ESD: (1) capacity to facilitate learning; (2) capacity to continue to learn and create; and (3) capability to connect, collaborate, and engage.

This chapter has focused on the third domain because the ability to learn and create emphasizes the teachers' updating their knowledge to continue the ESD practice; additionally, it illustrates the circulation of "reflect," "innovate," and "transform" supported by thinking, attitudes, and motivation, as shown in Fig. 18.1. Reflection is positioned at the core of the cycle and plays the role of prompting the cycle to innovate and transform. Is "reflection" comprehended as a simple mindset such as remembering what you experienced? In this chapter, Sect. 18.1 clarifies the disposition of reflection in the context of ESD; further, Sects. 18.2–18.4 propose

H. Sasaki (✉)
Chugoku Gakuen University, Okayama, Japan

H. Fujii, S.-K. Lee (eds.), *Science Education for Sustainable Development in Asia*, Education Innovation Series, https://doi.org/10.1007/978-981-99-8711-5_18

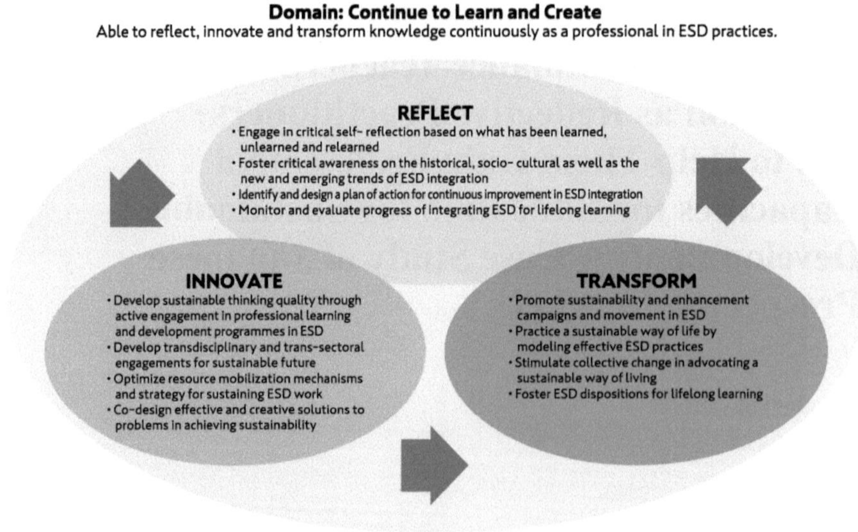

Fig. 18.1 Domain: continue to learn and create

three kinds of workshops to enhance teachers' reflection and help them acquire its methodology in the ESD practices.

18.2 Disposition of Reflection in ESD

Schön (1983) originally proposed the concept of "reflective practitioner" as a brand-new professional model. He insisted that a technical expert tends to apply rational rules to everyday situations, whereas a reflective practitioner is likely to use situational learning and adapt to various situations through reflection. Sato (1993) pointed out the future need for teacher education to be based on the reflective practitioner model, rather than the technical expert one that it has relied on, in order to pursue individualistic and creative education. He prescribed that the teachers' professional capacity entails the development of practical knowledge that is constructed by reframing their prior knowledge through reflection and deliberation. The latter is defined as a mode of thinking that teachers devise and utilize to modify the scientific theories flexibly, depending on the context, when faced with problems and required to make decisions regarding countermeasures. Reflection plays a crucial role since practical knowledge develops through it and deliberation. Accordingly, how is the third domain described according to development of practical knowledge? Innovation entails devising and modifying the prior knowledge and creating new one to solve the problem through reflection in the context of ESD. Transformation signifies reframing teachers' previous practical knowledge and changing the ways they perceive and think about the world through reflection in the ESD practices. Therefore, their updating of knowledge is regarded as changing the prior practical knowledge to the new one. Thus, the diagram of the "reflect," "innovate," and "transform" cycle vividly represents the teachers' professional development. As a sequel to inquiring its comprehension, reflection is not a simple mindset, and rather, a crucial element of a teacher's professional capacity as a reflective practitioner. Therefore, it is regarded as a specific methodology that teachers should acquire as a professional ability.

The SDG's learning objectives have been described by the United Nations Educational, Scientific and Cultural Organization (UNESCO) (2017), stating the following: "ESD can develop cross-cutting key competencies for sustainability that are relevant to all SDGs. ESD can also develop specific learning outcomes needed to work on achieving a particular SDG." There are eight key competencies for sustainability: systems thinking, anticipatory, normative, strategic, collaboration, critical thinking, self-awareness, and integrated problem-solving. Thus, reflection is not one of these eight main competencies. Regarding these abilities, the UNESCO has explained, "They are acquired during action, on the basis of experience and reflection," and "Key competencies represent cross-cutting competencies that are necessary for all learners of all ages worldwide (developed at different age-appropriate levels)."

It is necessary for learners of all ages worldwide to acquire the key competencies through reflection on action, which plays a crucial role in doing so. Therefore, they need to learn how to reflect on action and experience. Consequently, the methodology of reflection for teachers requires them to teach students how to reflect on action and experience in the ESD practices. Additionally, the UNESCO (2017) has proposed "Learning objectives for teachers to promote ESD." One of the eleven objectives is "Reflect on the concept of sustainable development, the challenges in achieving the SDGs." It is possible to say that reflection is disposed as an essential methodology for all teachers to acquire in order to facilitate their learners to achieve the SDGs. Above all, it has been indicated that reflection is a methodology that teachers must gain to develop practical knowledge as a professional capacity and to allow their students to achieve the SDGs. Accordingly, the following three workshops have been proposed for the teachers to acquire the method of reflection:

- Workshop for teachers to learn about the concept of reflection and to become aware of their reflection in teaching
- Workshop for teachers to learn how to reflect on teaching and to become informed of their practical knowledge through reflection
- Workshop for teachers to learn how to share their practical knowledge and change the prior practical knowledge through collaborative reflection

In each section below, firstly, the theoretical background has been presented; secondly, an example of a previous workshop has been introduced; and thirdly, an application workshop has been proposed in the ESD context.

18.3 Workshop for Teachers to Become Aware of Their Reflection in Teaching

18.3.1 Theoretical Background

For an extended time, the professional development of teachers has relied on the technical expert model that entailed changing it from theory to practice. This was the technical rationale that the teachers' professional capacity would be improved by formulating a lesson using the instructional method and materials they have learned. However, since Schön presented the reflective practitioner model as a new professional image in 1983, it has attracted the attention of researchers and teachers.

From the 1980s, the trends of constructivism and social constructivism appeared as a new paradigm to replace logical positivism, such as behaviorism and cognitive science. Within this trend, although Schön (1983, 1987) did not study the case of teachers directly, he assessed the professions urban engineering, architecture, psychoanalysis, and management consultancy. He proposed that the reflective practitioner model, based on reflection in the process of action, has replaced the traditional technical expert one for applying the scientific and technological expertise. Schön insisted that an excellent practitioner would reflect on a problem by fully using the accumulated tacit knowledge from experience, and expanding reflective thinking by interacting with the situation by building peer-to-peer relationships with customers, thus solving the complex problems occurring in such circumstances.

One of his concepts is "knowing-in-action," defined by the tacit knowledge embedded into daily action, that appears in action. The other includes both "reflection-in-action" and "reflection-on-action" that demonstrate the ways of thinking and the relationship of actions (Schön, 1987). The former is to think, make a decision, and act momentarily, thus interacting with the situation; the latter entails thinking about an action consciously after it. Schön indicated that the "frame" as a symbol of practice in action is important. Reframing would happen by comparing the new and prior experiences. Among the three aforementioned concepts, the most crucial concept is reflection-in-action. The teachers' professional capacity depends on how appropriately they can make a decision and act in the context of practice. However, it is difficult to assist in developing professional capacity by supporting reflection-in-action directly that works instantly in each teacher's mind. Nevertheless, it is not comparatively difficult to support reflection-on-action since it functions after the action. Therefore, it is crucial to help teachers develop professional capacity as well as reflection-in-action.

Thus, teachers need to know the concept of reflection that Schön advocated and acquire how to reflect on their experience in teaching. Therefore, a workshop with the same purpose should be proposed. In the subsection below, although the workshop was implemented for student teachers, it is also expected to help teachers develop professional aptitude.

18.3.2 Previous Example of the Workshop

Goal of the Workshop

The example of the workshop provided in this subsection was implemented on 29 student teachers on the subject "A Study for Teaching Practice in Elementary School" that this chapter's author was accountable for in 2015 (Sasaki, 2016). The subject consisted of 15 lessons and an internship in an elementary school. When the student teachers attended the internship, they were required to write the journals about their teaching practice daily. The journal was handed out to the student teachers and the concept of a reflective practitioner was taught in the subject's lesson before the internship. Subsequently, they attended the internship for 4 weeks, post which, they came back to the university and participated in the lesson again. The workshop was implemented in that lesson. The student teachers brought their journals of the internship, and reflection was assumed to have been concealed behind them. In the workshop, the student teachers extracted the episodes including reflection from their journals and classified it into three categories. Therefore, this workshop's goal was that the student teachers learn about the concept of reflection and become aware of the quality of their reflection through analyzing their journals.

Method

The workshop's procedure consisted of the following four phases. The student teachers:

1. Review the concept of reflection that Schön advocated
2. Extract the reflection episodes and fill in the reflection worksheet
3. Classify their own reflection into three categories
4. Share the results of classification

In the first phrase, the student teachers reviewed the concept of a reflective practitioner and reflection that Schön advocated. In the second phase, the blank worksheet regarding reflection episode shown in Fig. 18.2 was distributed to them. Subsequently, the facilitator explained how to extract the episode that includes reflection from the journal descriptions using an example, as displayed in Box 18.1. In the case of "episode 1," the student teachers described only what they observed in the lesson; it comprised only facts without any reflection. Therefore, the episode was abandoned. However, in the case of "episode 2," the first half included what the student teachers observed. In the second half, they reflected how their instruction was and considered how they should have paid attention to the children's behavior. Therefore, episode 2 was extracted because reflection was included. An episode that consists of reflection is called "a reflection episode," which does not include merely one sentence. If several sentences present one story, then it is regarded as an episode. Subsequently, the student teachers read their journals carefully and extracted many reflection episodes.

Worksheet of Reflection Episode

No. ( ) Name ( )

No.

	Episode	Category 1	Category 2	Category 3
		Time&Place	Method	Level of Reflection
#1	The teacher was not only teaching in the lesson, but also paying attention to the students, for example, whether he/she uses a celluloid sheet laid under writing paper and whether there are unnecesary things on his/her desk except a notebook and a textbook. (snip),therefore I also have to pay attention to such behavior.	Lessons	Observation	Technical
#2	Today in break itime, "Ms...Ms..I say, my friend ** is....".There were a lot of students to speak to me about their troubles. I tried to listen to them as much as possible, but I said "Wait a moment. Wait a minute." and I couldn't respond to all of them and left them as they were.(snip) I don't like to leave them behind. So I would like to keep in touch them with aware of speaking to them and to tell them something important.	Break time	Experience	Practical
#3	Today I tried to speak to the students that I didn't speak to positively. Because I was aware that I didn't find and I could'nt find the goodpoint of individual students even though the teaching practice has come close to the end. (snip) I desire to know about students more. Although only four days remain, I would like to speak to sutudents positively.	Others	Thought	Critical
#4				

Fig. 18.2 Worksheet of the reflection episode

Box 18.1: Examples of the Episodes

Episode 1

In the arithmetic lesson, the homeroom teacher taught the multiplication of a two-digit number. He prepared the multiplication table to stick on the blackboard and let the students practice it

Episode 2

The teacher was not only teaching the lesson but also paying attention to the students; for example, she checked whether the pupil was using a celluloid sheet laid under the writing paper and whether there were unnecessary things on his/her desk, except for a notebook and a textbook....therefore, I also thought that it is important to focus on their behavior hereafter

Table 18.1 Category 2

Method	Criteria
Thought	Description of a student teacher's reflection based on his/her thought (ideal and goal) without identifying the time and place
Observation	Description of a student teacher's reflection based on the observation of the teachers' responding to the incident with identifying the time and place
Experience	Description of a student's reflection based on an experience of his/her responding to the incident with identifying the time and place

In the third phase, the facilitator explained how to classify the reflection episodes. They were categorized into three different groups. Category 1 was the time and place (Box 18.2) that indicated when and where the incident related to the reflection episode happened. If it occurred in the lesson, the episode was classified into the group named "lesson."

Category 2 included a method of reflection (Table 18.1). The abstractive and conceptual reflection episode was classified under "Thought." During the first half of the internship, the student teachers observed the instruction of the homeroom teachers responsible for them or other teachers. The reflection episode related to observing the teachers' instruction was classified under "Observation." In the second half of the internship, the student teachers engaged in the teaching practice. The reflection episode based on one's own teaching practice was classified as "Experience."

Category 3 comprised a level of reflection (Table 18.2). Three different levels were introduced. The terms were quoted from those of Van Manen (1977), however, the meaning differed. Subsequently, the student teachers read their own reflection episodes carefully and repeatedly and grouped them into three categories.

Table 18.2 Category 3

Level	Criteria
Technical reflection	Reflection on the level of strategy responding toward the incident technically
Practical reflection	Reflection on the level of verifying the validity of his/her practical strategy and revising it
Critical reflection	Reflection on the level of grasping a strategy from different viewpoints and reconstructing his/her frame of thinking

Box 18.2: Category 1

-lesson -meeting -break time -lunch time -sweeping time -out of school -after school
-sports festival -others

In the fourth phase, after finishing the classification, the facilitator requested a female student teacher to go to the front and projected her reflection episode description and classification on the screen. She explained how to classify the reflection episodes from the three categories using an example of description shown in row #1 of Fig. 18.2. Firstly, in Category 1, it was classified into "lessons" since this episode occurred in the lesson. In Category 2, it was classified into "observation" since this episode was based on his/her observation. In Category 3, it was categorized into "technical reflection" as this episode was only technical level reflection. Similarly, two other episodes were classified as shown in rows #2 and #3 of Fig. 18.2.

Finally, the facilitator collected the reflection worksheets and the 90-minute workshop ended. The facilitator totaled up each number of classification of the three categories into the table. On the following day, the students discussed about the tendency and features of their reflection from the table in the other lesson.

Through this workshop, the student teachers learnt about the concept of reflection and became aware of its quality by analyzing their journals. Consequently, it was suggested that the workshop's goal was achieved.

18.3.3 Proposal of the Workshop for ESD

Goal of the Workshop for ESD

When teachers tackle the challenge of an ESD lesson, the journals of ESD teaching are required in the workshop. After engaging in the ESD lesson, the teachers participate in the workshop and learn about the concept of reflection. In the workshop, the teachers analyze the descriptions of their journals, behind which reflection is assumed to have been concealed. As an activity of the workshop, the teachers extract the reflection episodes from the journals and classify them into specific categories. Subsequently, they recognize the tendency and features of their reflection. Therefore, the workshop's goal is that teachers learn about the concept of reflection and become aware of the quality of their reflection by analyzing their journals regarding ESD teaching.

Method of the Workshop for ESD

The workshop's procedure consisted of the four phases as well as a previous training. The first and second phases continued similar to the previous workshop. In the third phase, several categories were needed to classify the extracted reflection episodes. In the previous workshop, three categories were organized. Category 1 was time and place that indicated when and where the incident related to the reflection episode happened. In the context of the teaching practice in the internship, various situations were arranged. However, regarding teaching ESD, the situation of time and place was limited, such as to "lesson" or "field," since the ESD lessons are usually implemented in a certain external lesson or field. Therefore, this workshop's category of time and place was eliminated.

Category 2 was a method of reflection that consisted of "thought," "observation," and "experience." It was possible to set up a similar category as Category 2. Some words of the criteria were modified; the new Category 1 has been shown in Table 18.3. An abstractive and conceptual episode is usually described without identifying a specific situation, such as the time and place. However, a particular episode is usually described with the identification of the time and place. Therefore, the words "identifying time and place" of the criteria were deleted. The ESD lessons were ordinarily implemented in the school project, which indicated that multiple teachers attended the project. Thus, the teachers engaging in the ESD could observe the other teachers' instruction of it.

Category 2 was set up as the level of reflection as well as the previous Category 3 since the reflection episodes in the ESD context could also be classified into "technical," "practical," and "critical reflections."

In addition to the two categories, a new category was established to evaluate the quality of the ESD reflection episodes. In the booklet of "Learning objectives for teachers to promote ESD," the UNESCO (2017) proposed the description of the specific learning objectives for all SDGs described in the cognitive, socio-emotional, and behavioral domains. Therefore, Category 3 was the learning objective domain, as shown in Table 18.4. The reflection episodes related to the students' knowledge and thinking skills; social skills, self-reflection skills, values, attitudes, and motivation; and action competencies were classified in the "cognitive," "socio-emotional,"

Table 18.3 Category 1

Method	Criteria
Thought	An abstractive and conceptual description of a teacher's reflection based on his/her thought (ideal and goal)
Observation	Specific and concrete description of a teacher's reflection based on the teacher's observation of responding to the incident
Experience	A vivid and particular description of a student's reflection based on the experience of his/her responding to the incident

Table 18.4 Category 3

Domain	Criteria
Cognitive	Description of a teacher's reflection concerning the student's knowledge and thinking skills necessary to better understand the SDG and the challenges in achieving it
Socio-emotional	Description of a teacher's reflection concerning the student's social skills, self-reflection skills, values, attitudes, and motivation
Behavioral	Description of a teacher's reflection concerning action competencies

and "behavioral domains," since the teacher reflected on his/her teaching of ESD to enhance the former's understanding of the SDG, to motivate their challenging, and to conduct their improved behavior, respectively. Subsequently, the teachers read their own reflection episodes carefully and repeatedly and classified the episodes into three categories.

In the fourth phase, after finishing the classification, some teachers went to the front and projected their descriptions of the reflection episode and classifications on the screen. Finally, the facilitator collected the reflection worksheets and the 90-minute workshop ended. On the other day, the teachers discussed about the tendency and feature of their reflections. This workshop's goal was believed to be achieved.

18.4 Workshop for the Teachers to Become Aware of Their Practical Knowledge Through Reflection

18.4.1 Theoretical Background

The student teachers became aware of the quality of their reflection by analyzing their journals of the elementary school internships, as presented in the previous section. Sato (1996) prescribed that the teachers' professional capacity entails the development of practical knowledge. However, it was unclear whether the student teachers themselves became aware of their gains in their practical knowledge. Therefore, a development of a procedure to help them become aware of their practical knowledge gains is needed.

The source of the concept of reflection originated from Dewey (1933), because of whom reflection was considered as a form of thinking different from the other logical thinking. When a person confronted a problem situation, embarrassment and hesitation occurred at first. During the subsequent stage, "inquiry" was implemented in order to resolve the problem. Finally, the thinking was accomplished by executing decision-making and the practice of problem-solving. He stated that it was reflection that was thought to suspend the improvisational action. The process of reflection was that of inquiry, which was to clarify the problem, analyze the goal, identify whether the means were appropriate to form a hypothesis, and verify practically. The authentic knowledge is acquired finally through the inquiry process.

Dewey's inquiry process is implemented in five logically distinct steps (Dewey, 1938): (1) an experienced difficulty, (2) its location and definition, (3) suggestion of a possible solution, (4) development by reasoning of the suggestions' bearing, and (5) further observation and experiment leading to its acceptance or rejection, that is, the conclusion of belief or disbelief.

Dimova and Kamarska (2015) reconstructed the inquiry process into three situations and four steps to adapt to the peculiarities of science education. It was as follows: (1) indeterminate situation-formulating questions; (2) problematic situation – (a) formulating problems, (b) observation of facts (discussing solutions), (c) choice of hypothesis, and (d) observation and experimenting; and (3) determinate situation-solution.

Accordingly, teachers need to know the inquiry process that Dewey advocated and acquire how to gain the necessary practical knowledge as an authentic knowledge. It was assumed that the practical knowledge is gained through the inquiry process. Therefore, the workshop for teachers to learn about the inquiry process and become aware of their practical knowledge through it has been proposed. In the subsection below, although the workshop was implemented for the student teachers, it is expected to help teachers to develop professional capacity as well.

18.4.2 Previous Example of the Workshop

Goal of the Workshop

The example of the workshop provided in this subsection was implemented for 20 student teachers in 2017 under the same condition of the internship as the workshop of Sect. 18.2 (Sasaki, 2018a). Practical knowledge was assumed to have been obtained through arranging the reflection episodes into the worksheet of the inquiry process. In the workshop, the student teachers extracted the episodes including reflections from their journals and arranged the episode according to the inquiry process. Therefore, this workshop's goal was that the student teachers learn about the inquiry process and become aware of their practical knowledge by organizing their reflection episodes of teaching practice.

Method

The procedure of the workshop consisted of the following four phases. The student teachers:

1. Review the concept of reflection that Schön advocated
2. Extract the reflection episodes
3. Learn about the inquiry process and arrange their reflection episodes into the worksheet of the inquiry process
4. Share the inquiry process and practical knowledge

In the first phase, the student teachers reviewed the concept of a reflective practitioner and reflection that Schön advocated. In the second phase, the facilitator explained how to extract the episode that includes reflection from their journal descriptions as well as the workshop of Sect. 18.2. Subsequently, the student teachers read their journals carefully and extracted their reflection episodes from their journals. In the third phase, the blank worksheet of the inquiry process (Fig. 18.3)

Worksheet of Inquiry Process

No.( ) Name( )

No.

Formulating questions	Japanese language/polite expression. How to encourage the students to present their ideas?
Formulating problems	In the drill, although many students got the correct answers, they didn't raise their hands to present their ideas
Observation of facts, Discussing solution ideas	The students looked unconfident in their answers.
Choice of hypothesis	If the teacher encourages the students, they present their ideas.
Observation and Experimenting	I talked to them and admired their writing. Then some of them began to raise their hands.
Solution	The students could present their ideas when the teacher encouraged them.
Practical Knowledge	It is important that the teacher has to support the students since presenting their ideas and responding to them results in their understanding of the learning contents and motivations to learn.

Fig. 18.3 Worksheet of the inquiry process

Box 18.3: Example of an Episode

...In the second period, I observed the Japanese language lesson about polite expressions. Its content include only repeating the teacher's explanation and writing on the blackboard. Therefore, the students did not seem to enjoy it. Since the teacher told me that a teacher occasionally has to teach such lessons, I now understand that it is impossible to make every lesson enjoyable. In the second half of the lesson, they tried performing the drills about polite expressions. When she instructed them to present their answers, only the students who often raised their hands did so enthusiastically. While they were answering, I walked around the class and checked how they were performing. Subsequently, I found that there were many students who had written the correct answers. However, they looked unconfident. Thus, I talked to them and admired their writing, after which, some of them raised their hands. I felt that the students could present their ideas when the teacher encouraged them. I thought it is important for a teacher to support the students since presenting their ideas and responding to them results in their understanding and learning of the content and motivation to learn.

was handed out to help the student teachers arrange the reflection episodes according the inquiry process. The facilitator explained about the inquiry process and the method to arrange the reflection episodes into the worksheet. The student teachers were required to write short statements regarding the first step of "formulating questions" through to the sixth and final step of "solution" in the right order. Finally, what the student teachers considered to gain through the inquiry process was written as the practical knowledge at the bottom. Subsequently, they began to arrange the reflection episodes according the inquiry process.

In the fourth phase, the facilitator requested one female student teacher to come to the front and share the inquiry process and her own practical knowledge to the other student teachers. Thenceforward, she showed the reflection episode (Box 18.3) that she extracted and began to explain how to arrange the reflection episode into the inquiry process (Fig. 18.3).

Finally, the facilitator collected the worksheets of the inquiry process and the 90-minute workshop ended. The facilitator totaled up all the practical knowledge into a table. On the other day, the student teachers discussed the practical knowledge they gained from the table in the other lesson.

Through this workshop, the student teachers learnt about the inquiry process and became aware of their practical knowledge by arranging their reflection episodes of teaching practice. Consequently, it was suggested that this workshop's objective was achieved.

18.4.3 Proposal of the Workshop for ESD

Goal of the Workshop for ESD

The journals of teaching in the ESD lesson were required in this workshop as well. After engaging in the ESD lesson, the teachers participated in the workshop. The practical knowledge of ESD was assumed to have been obtained through arranging the reflection episodes into the worksheet of the inquiry process as well. In the workshop, the teachers extracted the reflection episodes from the journals and arranged the episode according to the inquiry process. Therefore, the workshop's goal was that the teachers learn about the inquiry process and become aware of their practical knowledge by organizing their reflection episodes of teaching in the ESD lesson.

Method of the Workshop for ESD

The workshop procedure consisted of the four phases as well as a previous workshop. The first and second phases continued similar to the previous workshop. In the third phase, the worksheet of the inquiry process was handed out to the teachers. Subsequently, the facilitator showed an example of a reflection episode that a male homeroom teacher wrote (Box 18.4) and explained how to arrange the reflection episode into the inquiry process (Fig. 18.4).

Box 18.4: Example of an Episode

Recently, my school has been engaging in environmental learning where the students observe the environment around the local river. Thus far, the homeroom students of grade four have studied about organisms such as plants, insects, and birds that live around the river. This year they have started learning about the flow of the river and the geographical feature around it. Today, they had gone on a field trip and observed the river from the riverbank. Simultaneously, they were aware of abundant garbage at the riverside. The garbage included plastic bottle, styrene foam, plastic bags, wood pieces, and more. The students were surprised by the amount of plastic bottles and trash in and around the river, since they had learned the 3Rs that stand for recycling, reusing, and reducing plastic bottles at the school. A female student mumbled "Who trashes this garbage in the river? We are tackling the 3Rs and classification waste dumping, though." After a while, she said, "No matter how much effort we put in, I feel like nothing would change." When I heard that, I felt she experienced a sense of attachment to the river since they had engaged in the inquiry study about the river. I thought, "She might be very sorry to witness the garbage. It cannot be helped; she felt helpless. However, I want to motivate her to preserve the river. I have to talk to her later." After finishing the homeroom meeting, I told her, "I understand your sorrow about the garbage, nevertheless, I want you to maintain the affection to the river without saying 'nothing would change.'" She smiled and answered, "Yes," in a soft voice. I was happy to hear that.*

* Three Rs means Reduce, Reuse, and Recycle.

Worksheet of Inquiry Process

No.( )Name( )

No.

Formulating questions	Field Trip in the environmental study around the local river. How do I allow the students to observe the environment?
Formulating problems	There is a student who had a question about their effort for preserving the environment.
Observation of facts, Discussing solution ideas	The student feels like helpless.
Choice of hypothesis	If a student had a wrong idea, a teacher should talk to the student and correct the idea.
Observation and Experimenting	As I talked to her, the student said yes.
Solution	The student changed her idea since I talked to her.
Practical Knowledge	It is important that the teacher talks to the student whenever the student had a wrong idea as soon as possible.

Fig. 18.4 Example of the inquiry process worksheet

Thenceforward, the teachers began to arrange their reflection episodes into the worksheet according to the inquiry process. In the fourth phase, after finishing the arrangement, some teachers went to the front and shared their practical knowledge. Finally, the facilitator collected the reflection worksheets and the 90-minute workshop ended. This workshop's goal was expected to be achieved.

18.5 Workshop for the Teachers to Share and Change Their Practical Knowledge Through a Collaborative Reflection

18.5.1 Theoretical Background

The student teachers became aware of practical knowledge through arranging their reflection episodes into the worksheet, as introduced in Sect. 18.3. However, the workshop required considerably more time than expected to complete the inquiry process worksheet. Therefore, the worksheet was simplified based on Schön's reflection process (Sasaki, 2018b). It was trialed in the workshop of a pre-nursery class; it was concluded that the worksheet was useful and efficient in facilitating the student teachers' gains in practical knowledge. However, the student teachers' practical knowledge tended to pertain to their underlying values and assumptions, and to be stable, because the worksheet was written by each student teacher individually and their values and assumptions were based on their belief. This raised the question of how the practical knowledge should be shared among the student teachers and be expanded beyond their own individual underlying values and assumptions in the workshop. Argyris and Schön conducted studies on the relationships of learning between individuals and organizations. Thus, in this section of the workshop, the student teachers were allowed to share and change their prior individual practical knowledge, referring to Argyris and Schön's work.

Theory of Action

Argyris and Schön (1974, 1978) suggested that there is a theory consistent with what people say and one with what they do. According to Argyris et al. (1985), the distinction is not between "theory and action but between two different "theories of action." Those concepts are "espoused theory" and "theory-in-use." The former is regarding the values and assumptions people believe their behavior is based on; the latter is about the values and assumptions implied by their behavior, or the mental maps they use to take action. Argyris and Schön developed models that seek to explain the processes that create and maintain people's theory in-use, the model of which has been shown in Fig. 18.5.

The governing variables are values and assumptions that people attempt to keep within an acceptable range. Individuals have many kinds of governing variables. Action strategies are those employed by people to keep their governing variables within an adequate range. The consequences are the results of an action. These can include both intended and unintended results. Even if it is the latter, people rarely change their governing variable. Instead, they usually modify their action strategy in order to achieve a satisfactory consequence.

Single- and Double-Loop Learning

It has been suggested that the first response to this mismatch between intention and outcome is to search for another strategy that will satisfy the governing variables (Argyris et al., 1985). The new action strategy is used to fulfil the existing governing variable. The change occurs in the action only, and not in the governing variable itself. Such a process is called single-loop learning (Fig. 18.6).

An alternative response would be to examine the governing variables themselves. The associated action strategy might be to discuss the issue openly. When a person cannot find another action strategy, regardless how he/she attempts to search for it,

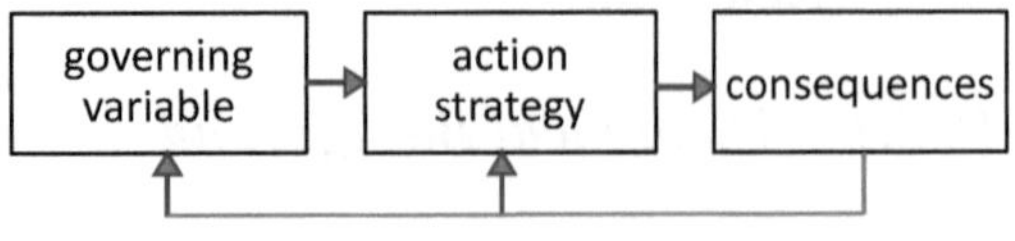

Fig. 18.5 Model explaining the process of developing the theory-in-use

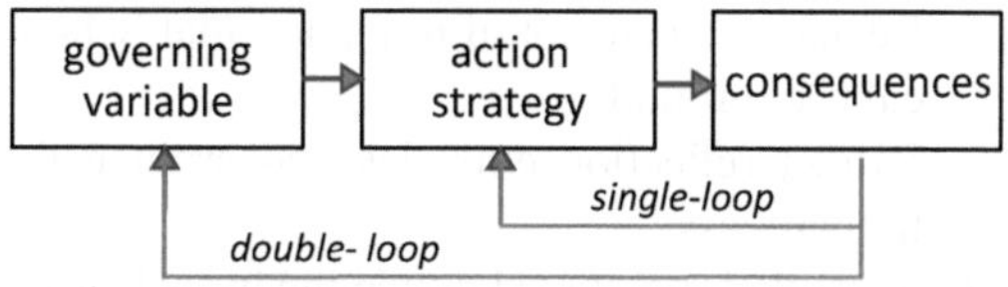

Fig. 18.6 Single- and double-loop learning

this may lead to discarding the prior governing variable and substituting it for a new one. Under the new governing value, an action strategy would be set up. Therefore, both the governing variable and the action strategy have changed. This would constitute double-loop learning (Fig. 18.6).

The student teachers' practical knowledge was based on their theory-in-use. In single-loop learning, people modify their action strategies according to the difference between the expected and reached outcomes. Conversely, in double-loop learning, people alter not only their action strategies but also their governing variables, which are the frameworks for their operating values and assumptions. Their assumptions underlying the current views are questioned, and hypotheses about their behavior are examined openly with others during the workshop.

In summary, teachers need to know the single- and double-loop learning processes that Argyris and Schön advocated; additionally, they need to acquire the method to share and change their practical knowledge. It was assumed that both single- and double-loop learning change their prior practical knowledge. Therefore, the workshop for teachers to learn about them and to change their previous practical knowledge through the collaborative work in groups has been proposed. In the subsection below, even though the workshop was implemented for the student teachers, it is predicted to help teachers develop professional capacity as well.

18.5.2 Previous Example of the Workshop

Goal of the Workshop

An example of the workshop provided in this subsection was implemented for 60 student teachers in the subject "Theory of Nursery Teacher" that this chapter's author was responsible for in 2019 under the same condition of the internship as the workshops of Sect. 18.3 (Sasaki, 2020). Practical knowledge was assumed to have been changed beyond their own individual underlying values and assumptions, through sharing their knowledge in collaborative group work. In the workshop, the student teachers extracted the episodes including reflection from their journals and shared their practical knowledge in each group. Therefore, this workshop's goal was that the student teachers learn about single- and double-loop learning and share and change their prior individual practical knowledge through collaborative group work.

Method

This workshop's procedure consisted of the following four phases. The student teachers:

1. Review the concept of reflection that Schön advocated and learn about theoretical background
2. Extract reflection episodes and arrange them in the worksheet of the reflection process
3. Share their practical knowledge in each group and fill in the rethinking worksheet
4. Reconsider their practical knowledge

In the first phrase, the student teachers reviewed the concept of reflection that Schön advocated. The facilitator explained the theoretical background by contrasting the workshop's procedure and theory of action. When the student teachers encounter a problematic situation, they produce a hypothesis and challenge to solve it based on the espoused theory. However, the action of response might be practically based on the theory-in-use. Therefore, provided that they reflect their practice, their practical knowledge is merely a product of the theory-in-use. It is uncertain whether the student teachers are aware of their theory-in-use. Therefore, the opportunity to share and explain their practical knowledge to other student teachers is needed. In addition, the opportunity for other student teachers to offer comments and advice on how to respond to the issue is required. After listening to a student teacher's theory-in-use and comments from other student teachers, if he/she changes their strategy, this would constitute learning. In the case that he/she retains their governing variable within the acceptable range and modifies only their action strategy, this would signify single-loop learning. Conversely, if they alter their governing variable and action strategy, this would constitute double-loop learning. Changing the governing variable concerning the reframing of their concept would be regarded as professional development as a reflective practitioner.

In the second phase, the student teachers read their journals carefully again and extracted their reflection episodes. Subsequently, in order to help the student teachers arrange the episodes they extracted, the blank reflection process worksheet (Fig. 18.2) was prepared. The format of the worksheet was simplified than that of Sect. 18.3. The facilitator explained the reflection process and the method to arrange the reflection episodes into the worksheet. The student teachers were required to write short statements in the empty row of "Problematic Situation" through "Reflection-on-Action" in the right order. Finally, what the student teachers considered to have gained through the reflection process was written as practical knowledge at the bottom. Thenceforward, the student teachers began to arrange the reflection episodes according the reflection process.

In the third phase, the student teachers were divided into small groups of four or five; further, the rethinking worksheet (Fig. 18.7) and the sticky notes were handed out. In a group of five student teachers, the numbers 1 to 5 were designated to them. Firstly, student #1 shared his/her reflection process and practical knowledge, and simultaneously, the other student teachers wrote their understandings of student #1's practical knowledge and provided comments or advice to student #1 on a sticky note. Secondly, the other student teachers turn-wise stated their understandings of student #1's practical knowledge. What the other student teachers mentioned was regarded as the theory-in-use. Thirdly, the other students gave comments or

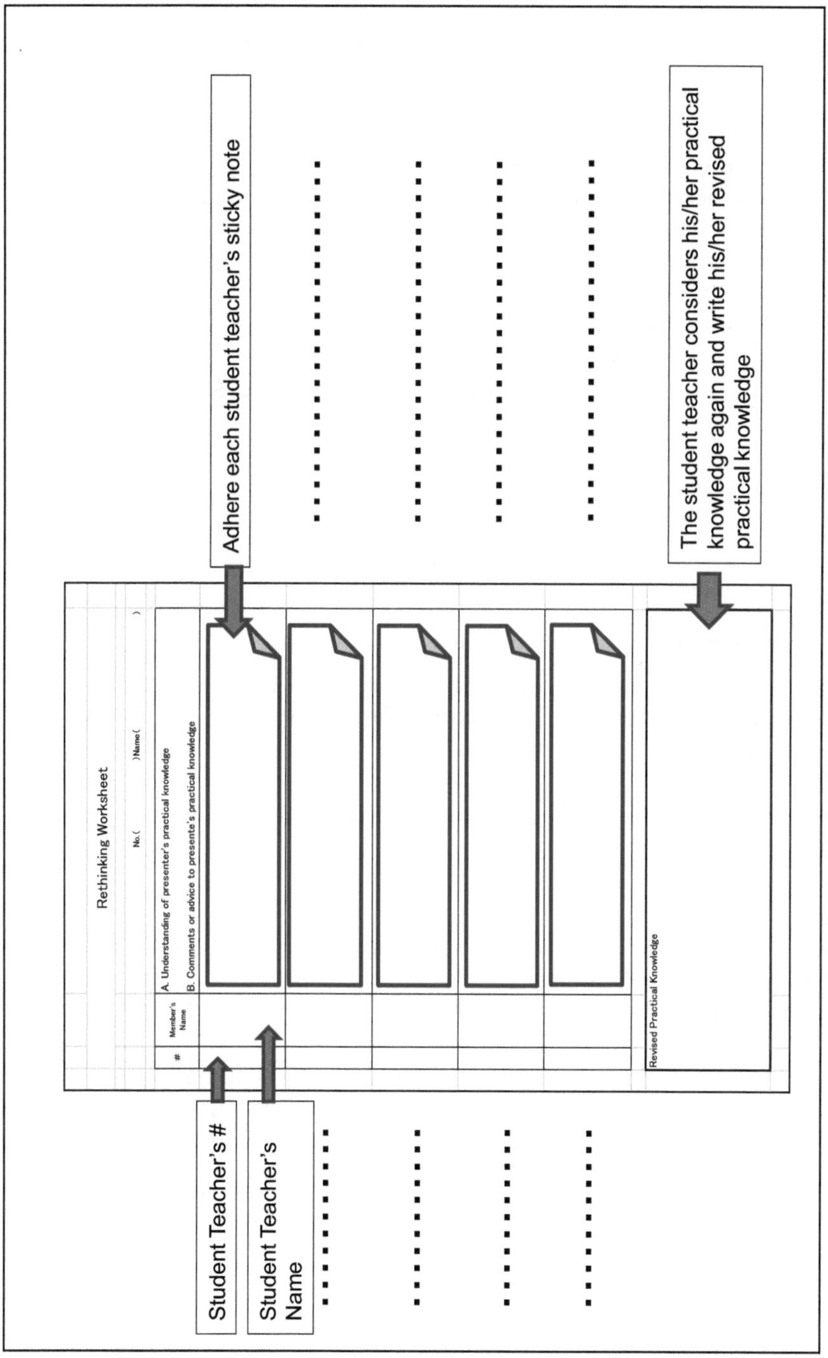

Fig. 18.7 Rethinking worksheet

recommendations to student #1, who adhered the sticky notes on his/her rethinking worksheet. This three-step procedure was repeated for each student in the group.

In the fourth phase, the student teacher read the descriptions of the sticky notes in his/her rethinking worksheet and considered his/her practical knowledge again. If he/she changed it, he/she wrote their revised practical knowledge at the bottom of the rethinking worksheet. After finishing writing the revised practical knowledge, the facilitator requested one female student teacher to come to the front and explain how to revise her practical knowledge. Subsequently, she showed the reflection episode (Box 18.5) and began to explain how to arrange it on the reflection process worksheet (Fig. 18.8).

Worksheet of Reflection Process

No.(　　　　)Name(　　　　)

No.

Date, Period	20-Jul-18
Age, Domain	Three years old, Human Relationship
Problematic Situation	Some children scrambled for the toys and one child who got angry threw a toy and hit a classmate when they were playing at housekeeping.
Reflection-in-Action	Those two children began to quarrel. I thought I had to mediate them soon.
Reflection-on-Action	It is difficult for some children to express their feelings and solve the problem. Therefore the teacher have to act the role of messenger that listens to what they want to say and speak for the other child.
Practical Knowledge	Nursery teachers should listen carefully to what the children want to say and speak for those who cannot express themselves.

Fig. 18.8 Worksheet of the reflection process

Box 18.5: Reflection Episode

I am in-charge of 3-year old children. Today I encountered the problem that some of them scrambled for the toys and one child who got angry threw a toy and hit a classmate when they were playing with the housekeeping. Consequently, those two children began to quarrel. I thought I had to intervene soon. Subsequently, I came close to them and stopped the quarrel. After that, I reflected on the problem that it is difficult for some children to express their feelings and solve the issue. Therefore, the nursery teachers have to play the role of a messenger. We, nursery teachers, should listen to what the children want to say and speak for the other child.

Subsequently, she explained how to change the practical knowledge through the collaborative work. After writing her reflection process worksheet, the student teachers were divided into groups. During the collaborative work, she shared her reflection process and the other student teachers wrote about their understandings of her practical knowledge and gave comments or advice to her on a sticky note. One female student teacher read her sticky note, "Nursery teachers should take the role of a messenger, since children are poor at expressing their feelings," and gave her the following advice: "It is extremely hard to listen what the children want to say, however, I believe that such behavior of the nursery teachers might build a good relationship with the children." The other student teachers also discussed their understandings of her practical knowledge turn-wise. After listening to the other student teachers' recommendations, she gathered the sticky notes to adhere to her rethinking worksheet. She reviewed the description of the sticky notes, and consequently, revised her practical knowledge. She mentioned, "Although it is difficult to understand the children's feelings, it would be possible if teachers and children form an interpersonal relationship. Building associations helps with everyday communication and expressing what they poor at vocalizing, and it gives them a sense of comfort." Some other student teacher shared their reflection process and practical knowledge as well.

Finally, the reflection process and rethinking worksheets were collected by the facilitator, and the 90-minute workshop ended. The researchers who were charged with nursery education evaluated the student teachers' descriptions in the worksheets. Consequently, 18 and 2 student teachers who revised their practical knowledge were examined as having achieved single- and double-loop learning. The facilitator totaled up each number evaluation into a table. On the other day, the students discussed their reflection process and revised their practical knowledge from the other lesson's table.

Through this workshop, the student teachers learnt about single- and double-loop learning, and shared and changed their prior individual practical knowledge through the collaborative group work. Consequently, it was suggested that the workshop's goal was achieved.

18.5.3 Proposal of the Workshop for ESD

Goal of the Workshop for ESD

The journals of teaching the ESD lesson were required in this workshop as well. After engaging in the ESD lesson, the teachers participated in the workshop. Their practical knowledge was assumed to have been changed beyond their own individual underlying values and assumptions through sharing their knowledge in collaborative group work as well. In the workshop, the teachers extracted the reflection episodes from the journals and shared their practical knowledge in each group. Therefore, this workshop's goal was that the teachers learn about single- and double-loop learning and share and modify their individual practical knowledge about the ESD through the collaborative group work.

Method of the Workshop for ESD

The procedure of the workshop consisted of the four phases as well as a previous workshop. The first and second phases continued similar to the prior workshop. In the third phase, the teachers were divided into small groups of four or five; further, the rethinking worksheet and sticky notes were distributed. The facilitator showed the example of practical knowledge in the ESD and explained how to share in groups. As shown in Box 18.5 of the previous workshop of the ESD, the male homeroom teacher wrote, "It is important that the teacher talks to the student at the earliest whenever the latter has a wrong idea," as stated in the practical knowledge row shown in Fig. 18.4. In the collaborative work, the numbers 1–5 were designated to each teacher. The homeroom teacher was assigned as #1. He shared his reflection process and practical knowledge; simultaneously, the other teachers wrote their understanding of his practical knowledge on the sticky note. Teacher #2's description A stated the following: "It is important that the teacher talks to the student whenever the latter has an incorrect idea regarding the subject study, of course." His description B stated the following: "In the case of ESE, the teacher needs to wait until the student is aware of his/her own action to preserve the river." Teacher #3's description A stated, "In the arithmetic lesson, the teacher has to correct the student's wrong idea." Her description B explained, "What is the goal of ESD in your school? Let us go back to the fundamentals and read the goal of ESD again." The homeroom teacher gathered the sticky notes and adhered to his rethinking worksheet, as shown in Fig. 18.9. Subsequently, he read the understanding and comments from the other teacher and considered his/her practical knowledge again, and finally, he revised his practical knowledge that "It is more important that the homeroom teacher expects the growth of the students and waits for their independent actions to preserve the local river."

In the fourth phase, after revising his/her prior practical knowledge, some teachers went to the front and shared their revised practical knowledge. Finally, the facilitator collected the worksheets and the 90-min workshop ended. The workshop's goal was expected to have been achieved.

Rethinking Worksheet

No. (   ) Name (   )

#	Member's Name	A. Understanding of presenter's practical knowledge B. Comments or advice to presente's practical knowledge
2	H.Sasaki	A: It is important that the teacher talk to the student whenever the student had a wrong idea in the subject study, of course. B: In the case of ESE, the teacher needs to wait until the student is aware of his/her own action to preserve the river.
3	H.Fujii	A: In the arithmetic lesson, the teacher has to correct the student wrong idea. B: What is the goal of ESD in your school? Let's go back the fundamental and read the goal of ESD again.
4	K.Sato	A: …… …… B:…… ……
5	T.Sugita	A: …… …… B:…… ……
6	M.Tanabe	A: …… …… B:…… ……

Revised Practical Knowledge
It is more important that the homeroom teacher expects the growth of the students and wait for their independent actions to preserve the local river.

Fig. 18.9 Rethinking worksheet

18.6 Conclusion

This chapter focused on the third domain, "to continue to learn, reflect, innovate, and transform," of the Asia-Pacific ESD Teacher Competency Framework. Reflection was grasped as a specific methodology that teachers should acquire as a professional capacity. Section 18.1 clarified the disposition of reflection in the context of ESD, and Sects. 18.2–18.4 introduced workshops. This chapter's author predicted that the workshops would be implemented to support the training of both pre- and in-service teachers in the Asia-Pacific region and globally in the ESD practices. Furthermore, the workshops would help all teachers acquire the methodology of reflection.

References

Argyris, C., & Schön, D. (1974). *Theory in practice: Increasing professional effectiveness*. Jossey Bass.

Argyris, C., & Schön, D. (1978). *Organizational learning: A theory of action perspective*. Addison Wesley.

Argyris, C., Putnam, R., & Smith, D. M. (1985). *Action science: Concepts, methods, and skills for research and intervention*. Jossey-Bass.

Dewey, J. (1933). *How we think: A restatement of the relations of reflective thinking to the educative process*. D.C. Heath & Co.

Dewey, J. (1938). *Logic – The theory of inquiry*. Henry Holt and Company.

Dimova, Y., & Kamarska, K. (2015). Rediscovering John Dewey's model of learning through reflective inquiry. *Problems of Education in the 21st Century, 63*, 29–39.

Okayama University ESD Promotion Center. (2020). *Guide for the effective dissemination of the Asia-Pacific ESD teacher competency framework*. Okayama University ESD Promotion Center. http://ceteesd.ed.okayama-u.ac.jp/pdf/200511.pdf

Sasaki, H. (2016). A student teacher training program renewal and an analysis of practical instructional competence of the student teachers. *Chugokugakuen Journal, 15*, 1–8.

Sasaki, H. (2018a). Developing a procedure to assist student teachers to recognize gains in their practical knowledge during teaching practice. *Chugokugakuen Journal, 17*, 11–16.

Sasaki, H. (2018b). An analysis method of journals to find out practical knowledge which student teachers gained. In *Proceedings of the 21th conference of the Japan society for studies on educational practice* (pp. 130–131). (in Japanese).

Sasaki, H. (2020). Trial of a workshop adapting single/double loop learning for nursery student teachers. *Chugokugakuen Journal, 19*, 1–9.

Sato, M. (1993). Teachers' reflection and wisdom=foundation of professional competence. *Japanese Society for Teacher Education Annual Report, 2*, 20–35. (in Japanese).

Sato, M. (1996). *Study of educational methods*. Iwanami Shoten, 137. (in Japanese).

Schön, D. (1983). *The reflective practitioner. How professionals think in action*. Basic Books.

Schön, D. (1987). *Educating the reflective practitioner*. Jossey-Bass.

UNESCO. (2017). *Education for Sustainable Development Goals: Learning objectives*. UNESCO.

Van Manen, M. (1977). Linking ways of knowing with ways of being practical. *Curriculum Inquiry, 6*, 205–228.

Chapter 19
Issues and Perspectives on Teacher Education for Education for Sustainable Development in Kazakhstan

Zharasbek Baishemirov

19.1 Introduction

> It is widely agreed that education is the most effective means that society possesses for confronting the challenges of the future. Indeed, education will shape the world of tomorrow. Progress increasingly depends upon the products of educated minds: upon research, invention, innovation and adaptation. (UNESCO, 1997)

Education for Sustainable Development (ESD) is a fairly new and unfamiliar topic, although it is closely related to traditional approaches in education and its foundation is laid in traditions, cultures, and ways of life of our ancestors. An official campaign to integrate sustainable development into education and training was launched by the United Nations Educational, Scientific and Cultural Organization (UNESCO) in the early 1990s. Subsequently, educators in many sectors of the world began to think about how to reflect "The Millennium Goals" in the educational process, raise people's awareness of environmental and developmental issues and the linkage between these issues, and equip people knowledge and skills need to solve the problems facing humanity. Obviously, in order to effectively educate people with insights into their interests in sustainable development, education should cover the physico-biological and socio-economic components of life and identify and study their inextricable linkage and interdependence. It is important to integrate the ideas of sustainability into all areas of education and develop people's perceptions of economic, environmental, and social factors leading to stable harmonious development or, on the contrary, to a crisis. The goal of ESD is to foster educated people with active and civic positions and make them enable to think critically, assess the situation, and anticipate the consequences of their activities from the perspective of impact on social development and the environment.

Z. Baishemirov (✉)
Abai Kazakh National Pedagogical University, Almaty, Kazakhstan

H. Fujii, S.-K. Lee (eds.), *Science Education for Sustainable Development in Asia*, Education Innovation Series, https://doi.org/10.1007/978-981-99-8711-5_19

19.2 Teacher Education in the Context of Sustainable Development

The concept of ESD has been evolved through UNESCO's responses to the following questions: (1) Why and how the world around has changed? (2) What awaits humanity tomorrow? (3) Is it possible to reconcile the demands of nature, economy, and society? and (4) How can we change the world better? In today's world, the importance and role of science and education go beyond the possibility of acquiring new knowledge and applying it in practice. Education and enlightenment have always been an effective tool for changing the stereotypes of peoples' thinking and behavior and modernizing national society and economy generally. Furthermore, education today has played a central role to creating the changes needed to achieve sustainable development (Fig. 19.1).

Notedly, ESD is a dynamic concept that includes all aspects of education and training like learning, upbringing, and acquiring knowledge and skills necessary for people of any age and any social affiliation to achieve sustainable development. ESD can and should use all the positive achievements of traditional education while enriching it with social, economic, and cultural context and taking it into account local characteristics. In ESD, students discuss and understand environmental protection, sustainable lifestyle, cultural diversity, social justice, etc., but most importantly, they learn to anticipate the future. In order to create a sustainable world, students need to be able to envision a sustainable future and realistic ways to reach it. Therefore, teachers should more often discuss issues on implementing certain plans and programs of ESD in country, city, and local community levels.

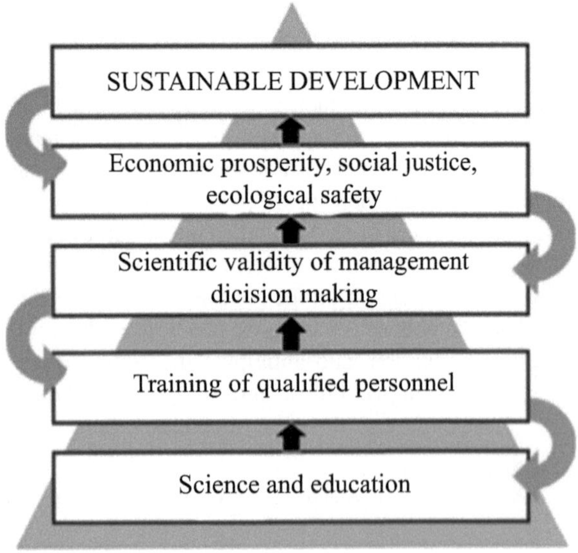

Fig. 19.1 Science and education for sustainable development

Teacher education should systematically fit sustainable development issues in various educational programs. At the same time, it is necessary to incorporate the issues into special courses and activities of each discipline, such as biology, chemistry, geography, economy, culture, and so on. ESD is not a new discipline or program, but a call for reorienting education. One of the main goals of ESD is to help teachers integrate the ideas and objectives of sustainable development into their curriculum, pedagogy, and assessment. Of course, this is a difficult task because ESD requires teachers to be motivated, interested, develop educational practices thoughtfully, and then transform themselves to create a sustainable society. Education is a vital means to solve the problems facing society. It is necessary to teach students to better understand what happens in our environment, economy, and society and to find optimal solutions to problems with a long-term perspective in light of the interests of all members of today's society and nature (Webster et al., 2005).

ESD is a process of continuous learning. Teachers should be always prepared to review and rethink previously their own established views in response to changes that are taking place and will occur in the world. In addition, teachers mainly have the following tasks: (1) to help students to understand why sustainable development is important for everyone; (2) to actively involve students in the discussion of sustainable development issues; (3) to deal with issues from different perspectives; and (4) to encourage students to reflect on the issues not only in the classroom but also outside of the formal education system (Azizov & Akinshina, 2009).

The main problem of ESD is that there is still no clear idea of what exactly an individual should know in order for him/her to consciously, competently participate in creating a more ecological and economical sustainable society. Therefore, it is recommended for teachers to involve students in self-education process and encourage their independence in learning. Before teachers decide to integrate the elements of sustainable development into their educational practices, they need to understand this complex subject and answer some questions below (Azizov & Akinshina, 2009).

1. Is the concept of sustainable development really relevant to us?
   Considering the three pillars of sustainable development, environment, economy, and society, in conjunction with cultural and personal aspects, will better represent the relevance of topics related to sustainable development. Analyzing the causes of problems and ways to solve them in the classroom, rather than describing the symptoms, will contribute to develop holistic, systemic thinking.
2. Is reorienting education in the interests of sustainable development a temporary action or a long-term, permanent process?
   An objective and realistic analysis of the situation, especially based on local context, will stimulate a long-term thinking about sustainable development, and then enable to understand that ESD is not just another campaign but a need for the future and there is no other way.

3. How can sustainable development issues be better represented in the curriculum of specific school subjects?
   Considering the resource potential of our planet, region, or country inevitably leads to a conclusion that it is crucial to reduce resource and energy consumption, manage waste, introduce environmental friendly technologies, convert to a green economy, respect social justice, and take into account cultural and local traditions. All these topics, to one degree or another, can be revealed in the lessons of biology, chemistry, geography, economics, etc.
4. How accessible and reasonable ways to solve specific sustainable development issues are provided?
   Only open discussion of issues related to lifestyle, social justice, cultural values, traditions and behavior, laws, accessibility of information, etc., with the support of advanced technologies, provides an opportunity for students to participate in analyzing the prospects for the future and finding ways to solve the issues.
5. What is the key to effective transformation in education?
   Pursuing effective means of changing students' attitudes toward sustainable development involves finding effective mechanisms and methods of the influence at the levels of individual, community, business, media, and government. The speed with which the concept of sustainable development is disseminated to the general public will depend on how quickly and justifiably a way is found to change students' attitudes toward sustainable development, particularly among teachers.
6. Will ESD contribute to the development of democracy, citizenship, and patriotism?
   Discussion about the rights of current and future generations stimulates young people to develop a sense of environmental and civic responsibility to act and commit to the value of democracy. It is necessary to explain and show how education can help students better understand their personal and public interests in a sustainable future that provides more opportunities for decent living conditions.
7. Are there available educational literatures and information resources on ESD?
   In Kazakhstan, there is a limited number of books available on sustainable development and ESD in English, Russian, and Kazakh languages. However, there are a significant number of Internet sites devoted to ESD issues, which are constantly updated with new information.
8. Will ESD become an integral part of the retraining and professional development of teachers, and if so, when?
   Reorienting the existing education system in the interests of sustainable development enables that short-term training seminars in which teachers stuff ready-made recipes are debugged and more educational approaches are discussed to integrate ideas and methods of ESD into educational processes. Upgrading should result teachers redefining what they is doing for ESD in their specialty (Azizov et al., 2007).

One of the prerequisites to effectively introduce ESD into learning processes is a student-centered approach. In this approach, the role of teachers is to create educational environments necessary for personal development, namely, (1) using situations where students will succeed in the classroom, (2) providing supports and assistances to students, (3) adapting democratic and respectful communication styles, (3) stimulating students' motivation for acquiring knowledge and skills, and (4) involving each student in the learning process of ESD taking into account their abilities and characteristics.

Of course, not all teachers will immediately approve and support the reorientation of education in the interests of sustainable development and not everyone will agree that learning is, first of all, an interactive and reflective activity and not the transfer of knowledge from teacher to student. In addition, it will not be easy to convince teachers that environmental education, which provides information about how nature functions and how it can be conserved, is not the equivalent of ESD, although they are closely related fields (Environment for Future Generations, 2004).

19.3 Challenges and Perspectives of Teacher Education in the Context of Sustainable Development

19.3.1 Changes in Teacher Education in Kazakhstan

The effective operation of the educational system mainly links to train teachers, on whom the development of all other levels of the educational system depends. It must be recognized that the quality of personnel training is the most important. At present, in light of the dramatic changes in the requirements for Kazakh teachers, there is a need for a profound reflection on new approaches to the training and professional development of individual teachers. In connection with the expansion and qualitative change in the field of teachers' professional activities, it is necessary to clarify the principles and approaches to form and organize the content of teacher training. At the same time, the focus of the training is on targeting the specific needs of educational institutions of all types and the subjects of the educational services market while preserving the individual's freedom to choose the educational trajectory. In addition to traditional functions of teachers, for example, teaching, upbringing, development assistance, educational evaluation, and working with parents, new professional functions such as forecasting, designing, and organizing the content and procedural aspects of education and the sociocultural environment are realistically required in modern conditions (Sadykov, 2005). In addition, communicative functions like social dialogue skills and social partnership in relation to social, cultural, economic, and environmental activities are in demand. With the transition to information society characterized by a constant increase in the amount of knowledge, teachers' ability to conduct independent research work and their professional motivation become increasingly important.

The importance of university education for prospective teachers consists in the fact that the fundamental bases of education have become more scientific, functional, and professional orientation deeply saturated with humanistic meaning. The scientific nature of education does not in any way mean the abolition of functional content, but fundamentality does not mean only technological training. In turn, professional orientation does not prevent the education of teachers from their development.

The modern paradigm describing teachers' personality and activities also includes their characteristics such as variability, creativity, criticality, and manufacturability, but does not, however, divert teacher training from the aspiration to create a holistic model as a specialist through the gradual formation of the personality. The construction of upcoming professional activities on a new conceptual basis requires a change from the established stereotypes and the formation of a new approach to educational problems. This is impossible without the development of a meaningful approach to the scientific explanation of the essence of modern educational processes at all levels of educational systems, and without the creation of teacher training programs, textbooks, teaching aids, etc. At the same time, we should draw attention to teacher education institutions for theoretical and methodical training of future teachers.

Today in Kazakhstan there is a very acute question of how, in the current conditions of university education, it is optimal to combine the special scientific training of physicists, mathematicians, philologists, etc. with the pedagogical, psychological, and methodological training. It is obvious that the reduction of special and scientific training is justified. There is a second way that enriches scientific education with a pedagogical orientation as seen in Western education system. At the same time, fundamental science should be studied not only as a complex of knowledge but also as a type of activity that enriches learning in solving heuristic problems, questions, and research tasks and develops students' abilities necessary for a teacher. The educational process itself becomes for students a model of training in their chosen specialty. Thus, we are pursuing the goal of ensuring the integration of fundamental theoretical knowledge that students receive at the university, with the positive experience of psychological, pedagogical, and methodological training accumulated in former pedagogical institutes.

The accumulated experience of university education indicates that a highly professional specialist is formed not only on the basis of classical knowledge, which is provided by curricula and mandatory for the student, but also nontraditional, the so-called additional knowledge for one or another specialty. Among the set of skills (organizational, communicative, analytical, and constructive) required for a university graduate, in the first place, it is legitimate to put the ability to analyze pedagogical phenomena, situations, facts, to establish cause-and-effect relationships and dependencies between them. On this basis, it is necessary to achieve flexibility in the pedagogical thinking of students, teach them to correlate the general things with the specific, independently deviate from the learned pattern. So, curricula and programs of a university education should focus on the formation of such qualities (CAREC, 2003).

The multistage system of pedagogical personnel offers a wide range of possibilities for rationalizing educational processes, combining fundamental and applied training. There is a need for higher education institutions in Kazakhstan to introduce bachelor-master degree courses with the obligatory preservation of the traditional training of teachers. To this end, the concept of continuous higher pedagogical education in the bachelor-master system should be developed and adopted in accordance with the current and future needs of the education system, state requirements for the minimum content, and level of training of bachelors in the areas of higher pedagogical education. State educational standards for teacher education should fully implement the principle of continuity of various levels and stages of education. Thus, the principle of continuity is one of the necessary conditions for ensuring lifelong pedagogical education, which presupposes general conceptual approaches to the implementation of content at all stages and levels of lifelong education.

At the present stage, one problem is about the absence of a legal framework and economic mechanisms for implementing the concept of lifelong pedagogical education, as well as the absence of scientifically based approaches to the creation of consistent educational standards and programs of teacher education. This will be facilitated by the creation of a unified educational and methodological association of higher and secondary vocational educational institutions for pedagogical education. The basis on the foregoing, in the near future, it is necessary to update the state educational standards of higher and secondary vocational pedagogical education in order to ensure the continuity of the levels and stages of education and to strengthen its practical orientation (Sadykov, 2005).

The targeted development of teacher education in Kazakhstan as initially based on a regional principle. In each regional center, pedagogical schools to training primary school teachers and pedagogical institutes to train middle school teachers are opened. For a long time, pedagogical schools/institutes and universities had developed as two separate educational systems. Despite in both systems specialists of the same profession were trained similarly and the same courses were prepared, there was no unity of the educational concept and the continuity of educational programs. Pedagogical schools/institutes in secondary level built their training based on sound practical training according to a principle "Do as I do," while universities in tertiary level gave future teachers solid theoretical training. In both systems, curricula and programs for studying disciplines were uniform throughout the state, which significantly limited the possibilities for mutual convergence of educational levels at different levels and at the regional level.

The differences in the organization and implementation of teacher education, in particular different approaches to the correlation of theoretical and practical training, was quite legitimate in itself. However, the experience of recent years shows that the possibility of ensuring the continuity and consistency of educational system institutions had been excluded. In accordance with the concept of lifelong education, proclaimed in Kazakhstan in the early 1990s, the integration of various stages of professional-pedagogical training began to be carried out. The "college-university complex" unified pedagogical schools/institutions and universities formed on a contractual basis (Sadykov, 2005). At the moment there are 85 institutions that have been granted the license for teacher training in Kazakhstan.

The principle, nature of the activity, and the scale of the college-university complex vary from region to region. The newly organized pedagogical colleges prepares the educational program for 3 years and connects to the third grade of universities. These college-university complex have created several advantages: for example, the level of pedagogical works in the college is enhanced, and the staffing, logistics, and library capacity are used more effectively. Furthermore, the complex has made education an economical benefit for students and their families. In fact, the number of students have enrolled in pedagogical colleges for a shorter-term study is increasing compared to that in universities. The creation of these complexes will optimally ensure the interaction of pedagogical theory and innovative practice in the professional training of teachers and the restructuring of the pedagogical process at school.

Taking into account the fact that teacher education in Kazakhstan is built on a regional basis, it seems optimal to form a regional system of continuous teacher education by creating the complex that unite all the stages from preschool education to the training of scientific and pedagogical personnel. This system will enable (1) to effectively carry out pre-university training to attract talented youth to pedagogical specialties; (2) to ensure the relationship and continuity of the goals of professional training and retraining of teachers at various stages of their professional growth; (3) to correct and develop the content of education at school, college, and university by combining the efforts of scientists and teachers, creating a cooperative atmosphere; and (4) to diversify in the variability of forms and methods of teaching at school, college, and university by stimulating the pedagogical creativity of teachers, scientists, and students – future teachers. Unfortunately, an educational system of pedagogical colleges has not been studied in Kazakhstan from scientific standpoints. Theoretical and methodological analyses to consider whether the system is more effective in teacher training are necessary.

19.3.2 Training Teachers as a Researcher and Pedagogical Education

Modern teachers are first and foremost a researcher since it is not only necessary to have a professional knowledge of their work but also to have high professional mobility, to be able to independently navigate themselves to vast scientific, technological, pedagogical, psychological, and sociopolitical information and to know modern scientific concepts. All of these cannot be achieved without active participation in research work at the stage of professional training. Therefore, one of the most important tasks of teacher trainers is to develop teachers' cognitive and creative abilities that are manifested in the desire to acquire new knowledge.

This challenge can only be solved if, during the period of university education, students as prospective teachers is involved in active scientific research. In order to be ready for the creative implementation of the knowledge, skills, and abilities obtained at the university, students must have formed scientific thinking and master the methodology of scientific activities. It is particularly relevant for prospective teachers as educational profession to have clear ideas about the internal mechanisms, laws, and methods of science since they have to form the foundations of creative thinking in school children. It is extremely difficult to acquire a research culture in the courses of educational and practical activities. Prospective teachers should prepare for scientific research in a targeted manner. This should be offered as organization of Educational Research Work of Students (ERWS) and Scientific Research Work of Students (SRWS).

The main task of ERWS is to teach students the skills of independent theoretical and experimental works with its methodology and research methods. The specificity of ERWS that distinguishes it from traditional types of training is that while doing this type of work, students do not act as a passive learner under pedagogical influence, but as an active learner with cognitive process. ERWS should be carried out at all stages of the preparation of teachers as future specialists in the field of education. It should be, therefore, carried out by students within the framework of the courses provided by curricula of the teaching profession. SRWS is carried out beyond and outside the curricula. It should be implemented to a system of complex tasks, guide a solution that leads to continuous enrichment of the research experience, and encourage personal and professional self-determination of students. The main task of SRWS in the process of training pedagogical personnel is to create the prerequisites for the upbringing, formation, and self-realization of the personal creative abilities of students; to ensure the most effective professional selection of capable, gifted, and talented youth for further education in masters and doctoral studies; and to replenish scientific and pedagogical personnel.

Developing the content of lifelong pedagogical education should be based on the principles of fundamentality and integration. The learning process of the education should include disciplines and courses that will be able to reflect the fundamental elements of the dual process of integration and differentiation in science and use the achievements of knowledge at the intersection of sciences. Furthermore, basic and interdisciplinary courses that contain fundamental knowledge aimed at creating a holistic view of the world based on the complementarity of content and unity of purpose and requirements should play a primary role (Shalgimbayev, 2005).

Scientific and methodological support of pedagogical education requires its further development of the system of information technology and methodological and scientific supports, which involves the development and implementation of educational program materials focused on the individual choice of the learning paths by students. Based on these supports, it is necessary to develop new pedagogical education programs in bachelor and master levels focused on prospective teachers.

19.3.3 Human Development and Quality Improvement of Teachers

The strategic objectives of countries in the field of social and economic growth are closely related to the growth of the Human Development Index. The accumulation of intellectual resources to ensure human development is evidenced by the growing proportion of teachers with higher education. One of the educational tasks in Kazakhstan is to achieve high parameters of human development established in the world, and for this, it is important to determine the best way to pursue a balance between the demand and supply of teachers. This balance should be achieved based on a clear definition of the needs of teachers, which will allow the rational use of public funds. In practices in the world, it is customary that calculations to substantiate staffing needs are based on demographic forecasts, which make it possible to determine in advance the number of people involved in the educational process in the coming period. In other words, the strategic task is to ensure the unity of predicting the age structure of the population of a given region and the methodology for determining its general, additional, and relative need for teachers, because this need depends, in the first place, on the contingent of students in primary and secondary schools and vocational schools such as lyceums and colleges.

Demographic calculations should cover areas such as the natural and mechanical movement of populations and their reproduction patterns. The number of school-age children, especially those in primary and middle schools, is increasing, which has a certain impact on the needs for teachers with high qualification, designed to teach this category of the population. For instance, according to the population census, for the period from 2010 to 2020, the number of school-age children increased by 785.6 thousand people (Bureau of National Statistics, n.d.)

Regional features of the formation of the students' contingent are important, and its presence may be due to both natural and mechanical movement of the population. The solution to the problem largely depends on government measures for the rational distribution of teachers throughout Kazakhstan, especially the optimal placement of educational institutions in the relevant specialties, retraining, and advanced training of teachers.

The dynamics of the need for teachers depends on both external and internal regional mechanical movement of the population. However, the parameters of this movement are taken into account most often in the case of the development of new rich deposits of minerals, agricultural land, and territories. The strategic challenge here is that the possible increase of the population in a particular region is predicted in advance to identify problems of the social order in general and the proper organization of the educational process, in particular. The solution to this problem is focused mainly on the mechanism of state distribution of young teachers of high qualifications, whose training is financed from the state budget. In our view, teachers in rural and small schools, as well as educational organizations in depressed areas, can be provided with state educational credit through the granting of credit repayment benefits.

One of the most important strategic tasks is closely related to the notification to Kazakhstan of the Lisbon Convention in 1997, which makes it possible to carry out legalized state measures to enter the European Educational Space. This means, first of all, that at the level of secondary general education there appears the feasibility of a gradual and consistent transition to 12-year education in schools and it is intended to gradually switch to 12-year education from the year 2023. Consequently, it is impossible to achieve this objective without first taking into account the increasing demand for teachers in the relevant disciplines. The process of meeting this demand is long-term, and therefore, the complex of these issues should be reflected in the strategic plans for the development of secondary and higher pedagogical education in the country and individual regions, both financially and educationally. Currently, there is a need to create a special monitoring service for higher education based on marketing research, predict the needs of the labor market, and, accordingly, develop recommendations on the structure and scope of teacher training (Sadykov, 2005).

In order to retain young teachers in schools, it is necessary to change the current approaches to their admission to universities. The state order for the training of pedagogical personnel should be formed based on regional needs and, in our opinion, should be approved mainly for regional universities, since, as experiences show, admission to pedagogical specialties is carried out by 90% of the residents of the region. In this regard, we consider it necessary to amend the standard rules for admission to higher education institutions, in terms of the choice of the holders of state educational grants and university loans for training, setting restrictions on the volume of state orders.

This approach to the development of secondary education was the main reason for the emergence of negative trends in the educational sphere, in particular, the deterioration of indicators reflecting the needs of society in the number of schools and teachers. This is evidenced by the data in Table 19.1 (Bureau of National Statistics, n.d.). It is clearly seen that from 2015 to 2020, the number of students increased significantly, while the number of teachers increased slightly. Furthermore, the number of general education schools decreased. This situation objectively leads to a decrease in the quality of school education due to overcrowding and multiple shifts in schools.

Qualified human capacity is one of the most important factors ensuring the quality of education. Regarding the qualification of teachers in Kazakhstan, six out of ten teachers (57.3%) work in schools in rural areas, 89.5% of them have higher education qualification including 52.1% belonged to the highest and first category, compared to the indicators of teachers in urban areas are 92.4% and 56.8%, respectively. The shortage of highly qualified teachers persists in some regions. This is due to low social status of the teaching profession and the low salary level. In 2018, the need for teachers was 5267 people.

During the period of implementation of the State Program of Education and Science between 2016 and 2019, the number of teaching staff exceeded half a million people. At the same time, despite the constant mass graduation of specialists in the field of education, the need for teaching staff continues. There are no effective mechanisms for attracting and retaining qualified teachers in schools and

Table 19.1 Dynamics of numbers of students, teachers, and schools in Kazakhstan

Year	2015/2016	2016/2017	2017/2018	2018/2019	2019/2020
Number of general education schools	7511	7450	7414	7393	7398
Number of students	2,799,585	2,930,583	3,050,770	3,186,234	3,337,783
Number of teachers	325,184	319,167	334,205	338,755	347,052

universities. Mainly applicants with low scores of the unified national testing enter pedagogical specialties. The education system is not conducive to retaining talented teachers. Up to 35% of teachers leave for other areas. There is no alternative entrance to the teaching profession for motivated applicants.

Thanks to the introduction of a new system of teacher career growth, the salary has been increased from 30% to 50% of the official salary (teacher, teacher moderator, teacher-expert, teacher-researcher, teacher-master). In addition, teachers who teach subjects in English have an additional payment in the amount of 200% of the base official salary. However, wages in education remain one of the lowest in the economy (102,875 tenge with an average salary in the country of 163,000 tenge), which is a negative factor that constrains the inflow of qualified personnel. According to the results of the OECD TALIS 2018 survey, 96% of Kazakhstan teachers believe that an increase in their salaries fees should become the highest priority when increasing the education budget. Thus, despite the positive changes in the system of training and development of teaching staff, a number of problems remain that negatively affect the attraction and retention of qualified teaching staff. In this regard, the "State Program for the Development of Education and Science of the Republic of Kazakhstan for 2020–2025" provides for measures to address problems in the development of the country's pedagogical potential (Institute of Legislation and Legal Information, 2019).

19.3.4 Competency-Based Education Reform and Advancement of Teacher Education

Modern education makes new demands on human resources such as training teachers for work within the framework of a competency-based approach to education, providing teachers with health-saving technologies and developing teachers' willingness to work in conditions of significantly increased individualization of the educational process. Therefore, teachers need to be prepared to accept the methodology and content of the new "State Education Standard for General Secondary Education," so that they can change the educational program and methodology and the objectives and methods of educational activities.

The updated content of general secondary education should be based on competency-based approach. The development of theory of key competences and the inclusion of competence-based approach in educational practice have been becoming more and more relevant to educational science and practices around the world. In the developed Western countries, the objectives of education are to form students' competencies, manifested in their ability to integrate, transfer, and use knowledge in different life situations. Everything related to competences is related to the experience and activity of the subject, and outside the situation and activity of the competences do not manifest themselves.

Under the initiative of the Ministry of Education and Science, the system of advanced training of teachers has been modernized. The "Center for Pedagogical Excellence" and the "National Center for Advanced Studies-Orleu" have been established, which unites 17 branches in all regions in Kazakhstan. Orleu carries out systematic advanced training of teachers, ensuring the professional growth of teachers and high quality of education in accordance with the states' program for the development of education from 2011 to 2020 and the development strategy of the national centers. However, it is difficult to overestimate the scale and importance of modernizing the country's education system. First of all, the reason is that the new system of advanced training of teachers covers all levels of education: preschool, school, higher, vocational and technical, and university teachers in accordance with world standards.

The main purpose of the advanced training system is to develop the professional competencies of teachers who is able not only to apply new pedagogical knowledge and technologies but also to develop them and to make reflexive changes in their own professional activities in the context of ongoing reforms and current trends. The solution lies primarily in updating the content, based on a new concept of development. Therefore, commissioned by the Ministry of Education and Science, Orleu is carrying out large-scale work on the development of multi-vector modular programs, teaching aids, and digital educational resources for advanced training, in accordance with world trends and the challenges of reforming the system.

All educational programs of Orleu and Center for Pedagogical Excellence are aimed at the development of conceptual knowledge of school and university teachers, helping them to form the ability to build their work. Future teachers learn on the basis of the integration of action learning, theory and practice, and active independent work of students so that their attention is always focused on the formation of students' functional literacy (Zaminova, 2015).

The analytics and monitoring departments in all branches of Orleu and the Center for Pedagogical Excellence have monitored the effectiveness of advanced training toward improving the quality of education. For the first time in the history of the advanced training system, diagnostic tools and technologies for conducting monitoring studies were developed, and objective information was obtained on the state of the advanced training system, analyzing the changes in professional activities of teachers in the post-course period, the satisfaction of teachers with the quality of course training, achievements, and problems that need to be solved. As the results of the research show, the introduction of conceptual ideas of the level programs has been generally well received by teachers throughout Kazakhstan.

Institutions in the advanced training system are designed to improve the professional competence and skill of teachers, where the concept of continuous multilevel professional development of teachers is implemented in order to achieve a high level of professional and methodological competence. The modern system of professional development allows for continuous professional development of the teacher (Institute for Economic Research, 2019).

19.4 Further Steps in Implementing ESD Goals

Further steps in implementing ESD goals in Kazakhstan are as follows:

- Continuing to work to ensure accessibility and inclusiveness of education
- Expanding public-private partnerships and attracting private investment in education
- Digitalization of education and the creation of digital educational content, the dissemination of methods of interactive and distance learning
- Revision of approaches to attracting and training future teachers, improving the qualifications of today's teaching staff, and the prestige of the teaching profession
- Creation of an educational hub and development of the export of educational services and products

Kazakhstan will strive to become the educational hub of Central Asia (Institute for Economic Research, 2019).

References

Azizov, A., & Akinshina, N. (2009). *Education for sustainable development* (Teaching aid). UNDP. (in Kazakh).

Azizov, A., Akinshina, G., Nishonov, B., & Salikhova, G. (2007). Education for sustainable development in Central Asia. *Document of international scientific and practical conference in central Asian countries and Kazakhstan "Education for sustainable development: Innovative approaches in the system of advanced training of pedagogical personnel"*.

Bureau of National Statistics. (n.d.). *Agency for strategic planning and reforms of the Republic of Kazakhstan*. Bureau of National Statistics. (in Kazakh).

Central Asia Regional Economic Cooperation Program (CAREC). (2003). Overview of educational progress for sustainable development in Central Asia: Achievements, good practices, and proposals for the future. *Document for the Belgrade Ministerial Conference "Environment for Europe"*, prepared by CAREC in cooperation with Central Asian countries with the support of the European Commission.

Environment for Future Generations. (2004). *Environmental education textbook for high school*. CAREC. (in Kazakh).

Institute for Economic Research. (2019). *Voluntary National Review 2019. Implementing the 2030 agenda for sustainable development*. Institute for Economic Research. (in Kazakh).

Institute of Legislation and Legal Information. (2019). *State program for the development of education and science of the Republic of Kazakhstan for 2020–2025*. Institute of Legislation and Legal Information. (in Kazakh).

Sadykov, T. S. (2005). *Problems of sustainable development of higher education in the conditions of globalization*. Sciences. (in Kazakh).

Shalgimbayev, S. T. (2005). *Conceptions of stable development in educational programs*. Vestnik RUDN University. (in Russian).

UNESCO. (1997). *Educating for a sustainable future: A transdisciplinary vision for concerted action*. UNESCO.

Webster, M. A., Zhelvakova, P. N., & Kirilov, N. I. (2005). *Koryakina. From environmental education to education for sustainable development*. Saint Petersburg. (in Russian).

Zaminova, I. (2015). *Features of the level professional development system*. Vestnik journal, "Orleu" –kst. (in Kazakh).

Chapter 20
Development of Frameworks for Building Teachers' Competencies in Education for Sustainable Development at Bankeun Teacher College, Lao PDR

Sompong Siboualipha

20.1 Introduction: Principle and Reasons

20.1.1 Key Concepts on Education for Sustainable Development

Education for Sustainable Development (ESD) is concerned with transforming today's society to achieve sustainable societies in the future (UNESCO, 2012). For this, ESD reorients existing educational programs toward sustainability. This involves rethinking what is taught and how it is taught. ESD involves the formation of values, skills, attitudes, and behaviors that prepare learners to anticipate, critically think, tackle, and resolve problems relating to sustainability.

For example, the enforcement plan announced by the Japanese liaison committee among ministries and agencies on the United Nations Decade of Education for Sustainable Development (UNDESD) includes the following goals: "(1) everyone can enjoy the benefits of high-quality education, (2) the principles, values, and actions needed for sustainable development can be taken into all fields of education and learning, and (3) education and learning can bring the changing of people's actions for realizing a sustainable future in the aspects of environment, economics, and society" (Fujii & Ogawa, 2016). These concepts relate to the natural, social, and economic environments surrounding human beings, regions, societies, nations, etc. The objectives of ESD are to equip learners with the necessary skills, knowledge, and compassion to act responsibly for a sustainable future. It enables learners to understand the challenges and transgressions faced in social, environmental, and economic spheres (Pillai & Achilles, 2015). In this study, we explained the meaning

S. Siboualipha (✉)
Bankeun Teacher College, Bankeun, Lao PDR

H. Fujii, S.-K. Lee (eds.), *Science Education for Sustainable Development in Asia*, Education Innovation Series, https://doi.org/10.1007/978-981-99-8711-5_20

of sustainable development and described its most obvious features to the participants (pre-service teachers). Participants were expected to integrate ideas on ESD into the curriculum and lessons to solve problems in education that they encounter.

20.1.2 Bankeun Teacher College and the Performance of Training System

Bankeun Teacher College (BTC) was established in 1968 in agreement with the Ministry of Education and Sports, at Lao PDR. Initially a local teacher training school, it developed into a teacher college in 1990. It is located at road No. 10, Keun Neua village, Thoulakhom District, Vientiane Province, with approximately 167,006 square meters and is surrounded by villages and fields. With its role and mission of developing kindergartens, primary teachers, and secondary school teachers, BTC has developed its faculty by building knowledge, skills, and ethics of teaching-learning arrangements. The college aims to build a creative thinking teacher education system and achieve a high rank through excellence in biology and the Lao language by promoting multifaceted cooperation in the establishment of regional and international linkages.

Recently, BTC has developed its teacher personnel, courses, teaching methods, and the management and evaluation of its curricula. Moreover, it has established seminars for pre- and primary school in-service teachers, focusing on the understanding of and attitude toward career-oriented development and teachers in societies. Therefore, an understanding of and positive attitude toward ESD has played an important role in improving curriculum and teaching-learning experiences since 2012.

20.1.3 Applying the Context, Input, Process, and Product (CIPP) Evaluation Module

To successfully identify program providers' learning needs and beneficiaries, the context evaluation components were implemented using the CIPP evaluation model, as seen in Stufflebeam (2003) and Zhang et al. (2011). According to Zhang et al. (2011), the input evaluation component can help prescribe a responsive program that can best address the identified needs, while the process evaluation component monitors the program process and potential procedural barriers and identifies program adjustment needs. Finally, the product evaluation component measures, interprets, and judges program outcomes and interprets their merit, worth, significance, and probity. Holland (2011) suggested a more comprehensive evaluation model for assessing service-learning, based on a goal-variable-indicator-method design, best characterized as an objectives-based evaluation approach. So, our context evaluation model was adapted from this CIPP module.

20.2 Research Objectives and Methods

This chapter outlines the team's recommendations for building a comprehensive new system of professional education anchored in a practical, cross-disciplinary approach toward continuous learning throughout the life cycle of sustainable development. This report aims to describe the following: (1) How did science teachers at the BTC enhance the knowledge of and attitude toward ESD through the training program? (2) How did the teachers use curriculum mapping and curriculum analysis to integrate ESD solutions into the curriculum?

The framework for developing the 2016 training program was used in the CIPP evaluation model. To develop a model lesson, teachers were briefly explained the lesson study and integration of ESD. The program comprised issue analysis, lesson analysis, new lesson development with ESD integration, teaching-learning materials' analysis, and conducting and observing new lessons (research lessons).

From September to November 2017, an existing curricular was mapped, identified, connected, and integrated with ESD concepts, as indicated in Fig. 20.1. Similarly, in 2018, the biology curricular was mapped, identified, connected, and integrated with ideas on ESD for curriculum completion.

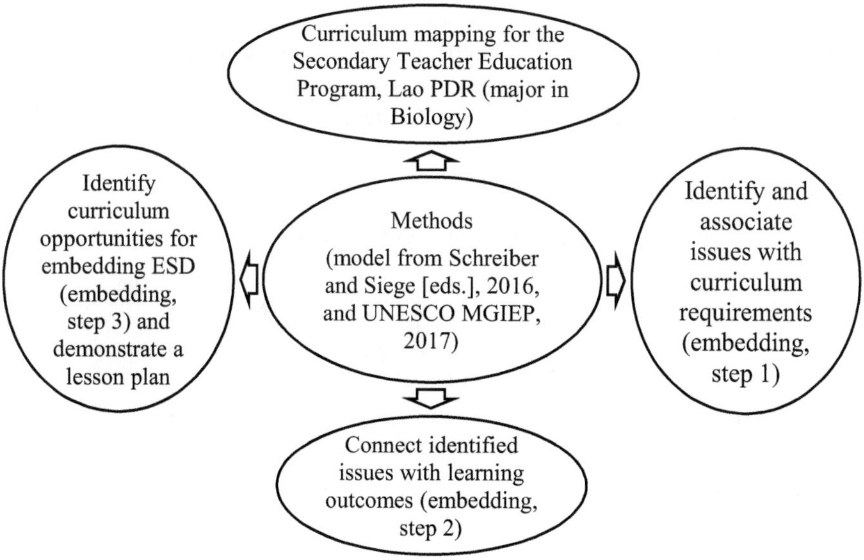

Fig. 20.1 Mapping, identifying, connecting, and integrating curriculum

20.3 Results and Discussion

20.3.1 Development of Science Teachers' Competences in ESD

Key findings from analyzing the questionnaires evaluating the four components were as follows. The data were collected via a survey from nine science teacher trainers (BTC) on 8 February 2016. A total of nine science teacher trainers (including two women) were assessed to probe their perception, attitude, and practice in teaching.

Context Analysis: Analysis of Pre-Test
A pre-test was used to assess participants' training needs and understanding of ESD. Figure 20.2 presents the questions assessing their understanding of ESD, indicated by the percentages of correct and wrong answers.

The institute is currently encountering these problems (the most pressing environmental, societal, and economic issues): low scientific background of students, low income of staff, imbalance in the recognition of the importance of education, incorrect usage of environmental resources, lack of environmental conservation and protection, and other factors such as global warming, climate change, natural disasters, air pollution, hunger, and species extinction.

Competencies or skills that participants expected from this training program (range 0–3) are indicated in Table 20.1.

Input Evaluation
For the input evaluation component, the program content was prescribed corresponding to the identified training needs, such as critical thinking and problem-solving. Acquiring these skills would allow the trainers to develop innovative learning opportunities within science lessons, etc. The input content began with an introduction to ESD and related lessons. Content and activities for this part of the program are described below:

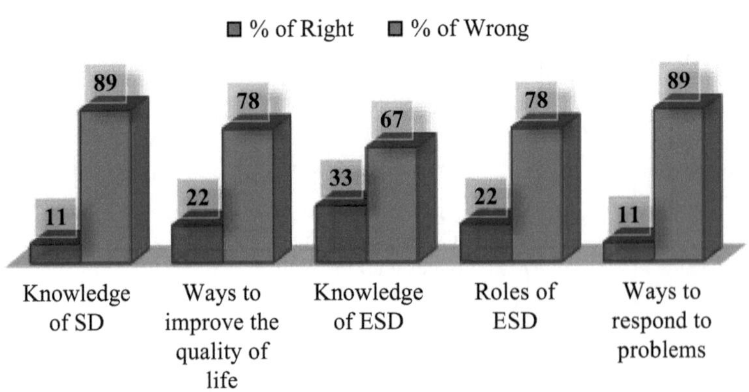

Fig. 20.2 Percentages of right and wrong responses to the pre-test questions

Table 20.1 Competencies/skills that respondents expect from the training program (Ranking from the least important to the most important, 0–3)

Range of expectation	Competencies
2.3	Context rational thinking
2.7	Critical (systems) thinking and problem-solving
2.7	Adopting research methods in response to the project goals
2.3	Cooperate with different people
2.7	The know-how to develop innovative learning processes
2.7	Integrating sustainable development into education
2.6	Issue analysis
2.7	Using ESD to support qualitative learning-teaching
2.4	Reorienting the curriculum
2.6	Lesson study and analysis based on ESD
2.4	Teaching and learning materials' analysis

An Introduction to ESD ESD supports five fundamental types of learning to provide quality education and foster SD, such as (1) learning to know, (2) learning to be, (3) learning to live together, (4) learning to do, and (5) learning to transform. Participants discussed and presented the components of ESD.

Competencies for ESD Teachers There are three overall competencies: (1) teaching, (2) reflecting or envisioning, and (3) networking. The five domains of the competencies were explained to the participants: knowledge, systems thinking, emotions, ethics and value, and action. Participants demonstrated knowledge building for the following ESD competencies: holistic approach, envisioning change, and achieving transformation. For knowledge building, participants were asked to play a game by placing cards that belong to the provided competency statement.

Integrating Sustainable Development into Education The participants were introduced to the relationships between BTC's vision and ESD's principles and asked to integrate the context of the ESD model and fill the statements in Table 20.2.

Process Evaluation

The Y Project (lecture and Seminar) Teachers reorient current curriculum units to address sustainability. The Y project was used (toolkit adapted from Regina Rizzi model (UNSECO, 2006). Three lessons were identified: (1) waste disposal, (2) the heat produced by a black box, and (3) soil and its characteristics (Fig. 20.3).

Issue Analysis There were three groups of participants (biology, chemistry, and physics). The analysis was conducted under three categories: issues, causes, and findings. The findings of each group are mentioned below.

Table 20.2 Basic components of ESD (a case study in BTC)

	Environment	Society	Economy
Issues	Air pollution due to the reckless garbage disposal	Depression Natural disasters and plague Filth, infections, and diseases	Cannot run any business, such as restaurants, guest houses, etc. Low productivity Waste funds Slow economic growth
Knowledge	Find out the causes for pollution Find ways to manage and reduce pollution Get to know the characteristics of good and bad environments	Everybody should know and understand environmental problems Know how to confront and respond to natural disasters Educate communities on garbage disposal	Running environmental friendly businesses Running a mixed business Revenue from the sale of garbage
Skills	Building knowledge and improving garbage disposal Ways to segregate and recycle Ways to manage waste	Open discussions on ways to improve Capacity to adapt and resolve problems Create rules and ways to improve garbage disposal by sharing ideas with communities	Capacity to reduce, reuse, and recycle Capacity to profit from waste management
Perspective	Good environment, should not recklessly dispose garbage Environment should be kept clean Garbage control and management	Gathering information from communities on environmental problems Good society, good health, and good waste management! Sustainable social consumption	Campaign and/or support social activities Create natural tourist sites with a campaign for waste management Sustainable consumption
Values	Waste segregation, reduce waste, and reduce pollution! Reduce waste with 3R (reduce, reuse, recycle)	For good health, decrease air pollution Good hygiene and good life Consciousness in maintaining the environment Reduce waste and create a healthy society Pollution control for the community with full participation	Good economics and good environment Promoting environmentally friendly products

Group 1: Waste disposal (biology group)

Issues	Causes	Findings
(1) Students could not classify and manage waste	(1) Students did not understand the types of waste and attitudes toward waste management	(1) Educate students (2) Fieldwork study at the disposal sites (survey the sources, causes, and effects) (3) Encourage students to give ideas on waste management in groups and then summarize them

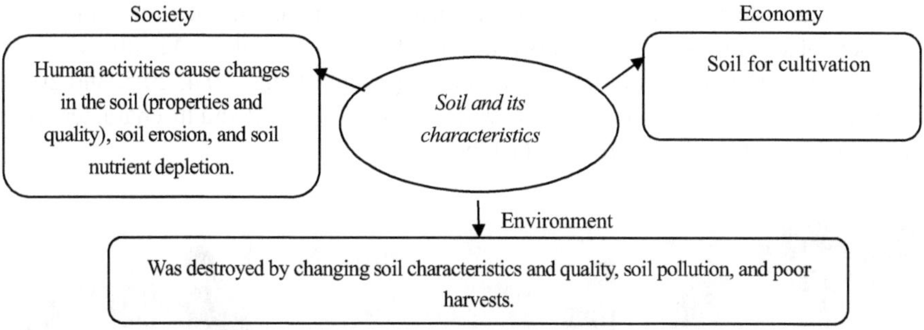

Fig. 20.3 Three lessons identified

Group 2: The heat produced by a black box

Issues	Causes	Findings
(1) Students cannot explain and understand the experiment results (2) Students do not know how to apply the knowledge about the subject in their daily lives	(1) Many students do not read the lesson before class (2) They are not interested in learning (3) They do not participate in any activities (4) Too many members in a group work (5) Too many content/topics for learning in a session	(1) Ask students to read lessons before class as their homework (set the points to read) (2) Stimulate them by assigning scores (3) Try to give an opportunity to all of them to increase their attention during activities (4) Reduce group size and divide into sessions (5) Classify and sequence the more important content/topics that should be learned in a class or at home as homework or individual work, etc. (6) Build students' attitudes toward heat and energy

Group 3: Soil and its characteristics

Issues	Causes	Findings
(1) Students cannot explain the characteristics of soil		(1) Economy: educate/raise students' awareness to choose the right soil to increase production
		(2) Society: raise students' awareness and attitude toward soil protection
		(3) Environment: the knowledge on soil and causes of pollution, types of soil, and the correct usage of each soil type for growing the right plants

Lesson study In this lecture session, Siboualipha started to motivate participants by using the following questions for discussion: (1) before conducting a lesson in your class, what did you prepare? (2) How did you prepare? (3) How did you assess your lesson after the class? After the discussion, we summarized the steps to conduct a lesson: plan (design a better lesson), do (teacher conducts the lesson while other teachers observe), and see (teachers reflect on the lesson together). Members in each group analyzed issues and topics that would be taught (Figs. 20.4, 20.5, and 20.6). The chosen topics or lessons were (1) waste disposal; (2) soil and its characteristics; and (3) the heat produced from a black box.

Fig. 20.4 Teachers analyzed issues and topics, which will be taught for lesson study (plan)

Fig. 20.5 (**a**) Teaching and learning ESD lesson (biology: waste disposal) (do). (**b**) Teaching and learning ESD lesson (chemistry: soil and its characteristics) (do). (**c**) Teaching and learning ESD lesson (physics: heat produced from a black box) (do)

Fig. 20.6 Feedback session for each lesson plan after conducting classes

Product Evaluation

Evaluation of Lesson Implementation The feedback of each lesson was implemented by all participants. Each of them shared and exchanged ideas on the strengths and areas of improvement related to the three lessons as indicated in Table 20.3.

Analysis of Post-test A post-test was used as an assessment for product evaluation based on the participants' training needs and understanding of ESD. According to a pre-test (seven questions) analysis, one more question was added to a posttest (from this experience, what are the differences between your past teaching plan and current lesson plan based on ESD?). Analysis of the percentages of right and wrong answers are indicated in Fig. 20.7. Problems faced by their community are excessive drug usage, inadequate township development, the expansion of agricultural land, deforestation, odor pollution, increased waste, poverty, poor teaching/learning material, unstructured lessons, and the lack of innovative learning experiences. The competencies/skills that participants acquire the most from this training program are specified in Table 20.4.

Assessing Participants' Perspectives Through the Integration of the ESD Training Program At the end of the program, participants were asked to complete a checklist (Yes or No) regarding their participation during the program. The checklist comprised 13 questions specified below. From this assessment, we found that all of the participants responded "YES" to these questions. It means that all of them have an understanding of and positive attitude toward ESD. In conclusion, participants did the following:

- Contributed to the identification of problems or issues
- Investigated the issues using a range of different knowledge resources
- Proposed realistic strategies for what needs to be done
- Helped with decision-making on what to do next
- Cooperated with colleagues
- Identified social, economic, environmental, and cultural aspects of the issue
- Identified why and how these different aspects contribute to the issue
- Learned the importance of ESD
- Realized that ESD builds their commitment and capabilities to implement and develop their lesson plans

Table 20.3 Strengths and areas of improvement for the three lessons

Group 1: waste disposal (lesson plan)
Strengths
Met all the objectives of teaching and learning
Both teacher and students had active interaction, and students were interested in each activity as they all participated and shared ideas in the discussion
Almost all of them understood the contents of the lesson as well as waste disposal and its basic management.
Areas for improvement
Time management in each activity
Few students didn't understand lessons such as waste types (hazardous chemicals, waste recycling, and/or transformation)
Activities in the lessons are basic. We should apply more scientific skills for students to practice and think critically
Very simple teaching-learning material
The activity to make the leaners conscious and have a positive attitude and perspective on sustainable development was just basic
Teachers should organize more activities to support learning and campaigns for students and the community
The heat produced from a black box (lesson plan)
Strengths
Met all the objectives of teaching and learning
Both teacher and students had healthy interaction, and students were interested in each activity as they all participated and shared ideas in the discussion
Almost all students understood the lesson, especially the terms related to heat absorption and dissipation, and heat produced from a black box
Teaching-learning materials were apt for students' learning
Areas for improvement
Time management in each activity
Should add more activities to get students to think critically
The activity to make the leaners conscious and have a positive attitude and perspective on sustainable development was just basic
Teachers should organize more activities to support learning and campaigns for students and the community to save energy or use renewable energy
Soil and its characteristics (lesson plan)
Strengths
Met all objectives of teaching and learning
Both teacher and students had healthy interaction, and students were interested in each activity as they all participated and shared ideas in the discussion
Almost all students understood the presented content. They can now identify types of soils and describe their properties
Students have a positive attitude toward agricultural land usage and choosing suitable plants to produce high-quality products and save the land
Areas for improvement
Time management in each activity

(continued)

Table 20.3 (continued)

Knowledge area is very basic because they have used only one property (ability to absorb water) to identify the types of soils
Teachers should provide more activities to get students to think critically
Scientific skills should be added to support transformative learning
The activity to make the leaners conscious and have a positive attitude and perspective on sustainable development was just basic
Very simple teaching-learning materials
Teachers should organize more activities to support learning and campaigns for students and the community to protect agricultural land
Should provide fieldwork activities to support students' learning, especially on some local agricultural lands

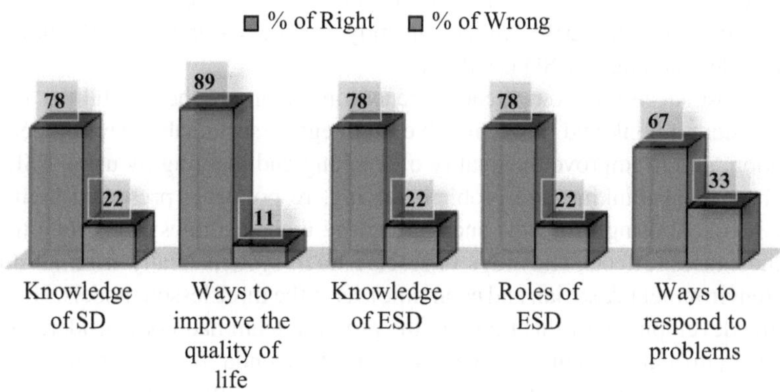

Fig. 20.7 Percentages of responses to the posttest questions

Table 20.4 Competencies/skills that participants acquire the most from the training program (Ranking from the least important to the most important, 0–3)

Range of benefit	Competencies
2.4	The ability to assess a certain topic from different viewpoints ("context rational thinking")
3.0	The capacity for critical (systems) thinking and problem-solving
2.8	The capability to adopt research methods in response to project goals
2.6	The ability to cooperate with people
2.6	The know-how to develop innovative learning processes
3.0	Integrating sustainable development into education
2.4	Issue analysis
3.0	Improve the quality of teaching and learning by using ESD support
2.3	Reorienting curriculum
2.4	Lesson study and analysis based on ESD
2.4	Teaching and learning materials' analysis

- Developed an understanding of the contribution of ESD to quality education
- Developed an understanding of how to implement ESD as a cross-curricular theme
- Developed an understanding of the relevance of all subjects to ESD and the process for integrating ESD into subject teaching
- Developed the skills to use a variety of teaching and learning approaches to achieve the wide range of ESD objectives

Based on this checklist, we realized that respondents understood how to improve everyone's quality of life the most, including future generations, followed by an understanding of SD and ESD and how they can make it happen and the roles of ESD in societies, environment, economics, and culture. Similarly, after investigating ESD, respondents could identify problems that their institute encounter. Furthermore, participants' understanding of ESD ranged from the majority focusing on the environment, followed by social and economic conceptions. This corresponds with Summers (2013). Since our study was the first in Lao PDR, our participants' understanding of ESD is still basic.

The most successful competencies benefitting them are the capabilities to critically (systems) think and problem-solve, to integrate sustainable development into education, and to improve the quality of teaching and learning by using ESD support. In critical thinking and problem-solving, respondents presented high issue analysis by choosing problems encountered by their institutes; then, they tried to find the causes, effects, and ways to resolve them systematically from all dimensions (environmental, social, and economic), as in the three lesson models. However, until the development of this training, the program team had not considered developing the curriculum to include the sustainability agenda. It was just aimed to provide participants with an understanding of and positive attitude toward sustainability and ESD. Further programs should be added to assess their understanding and practice, such as developing the curriculum to include the sustainability agenda and demonstrating good practices at the institute and/or national level. Colleagues or other institutions should participate and explore such programs.

20.3.2 Developing Lower Secondary School Students' Learning Science Programs

Subject: Plant Classification – Monocotyledon and Dicotyledon

- Problem: Low learning outcome of students, even after lesson demonstrations and repetitions after the teacher; misunderstanding the criteria when classifying plants; low understanding of the role of plants in daily life.
- Causes: Low scientific skills in students; they don't know how to apply lessons in their daily lives; some of them do not pay attention to the activities conducted, and insufficient number of activities.
- Findings (by integrating ESD): Three main activities were provided. They are as follows: (1) comparing the characteristics of seeds (monocotyledon and dicoty-

ledon germinating seeds), by observing how cotyledons grow (have students observe how seeds germinate); (2) comparing leaves and classifying into groups (encourage them to the use and learn the importance of each leaf); and (3) comparing the flowers of each group and identifying them with the appropriate groups (have them understand how flowers become fruits, the distribution of seeds such as the grass, etc.).

To implement this lesson plan, a teacher familiarized herself with the teaching materials, ways of teaching (including activities), and methods for evaluation before starting the lesson. Then, she conducted a lesson, while others observed the teaching, the learning outcomes, and the effects of the teaching materials. Finally, she was provided feedback for the lesson.

Students' performance was analyzed by comparing two classes on day 1 and day 2, as specified in Figs. 20.8 and 20.9. The figures show the high percentage of students' learning outcomes.

Benefits of Using ESD Integration for Model Lessons

The advantages of developing a lesson plan during classes are summarized in the following list:

- Met all learning objectives.
- Good interaction between the teacher and students and among students as well.

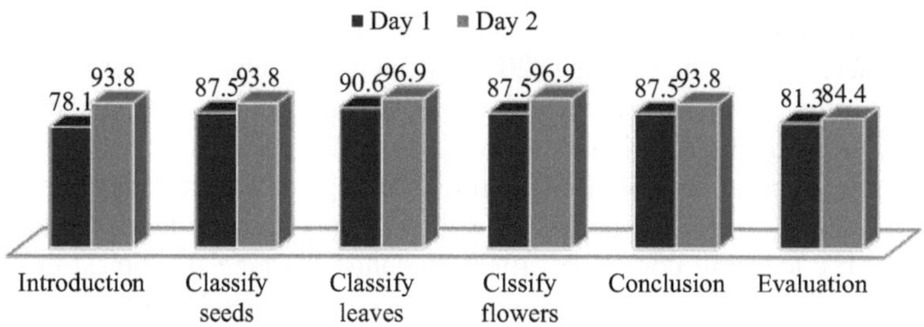

Fig. 20.8 Evaluation of learning outcomes (in percentages)

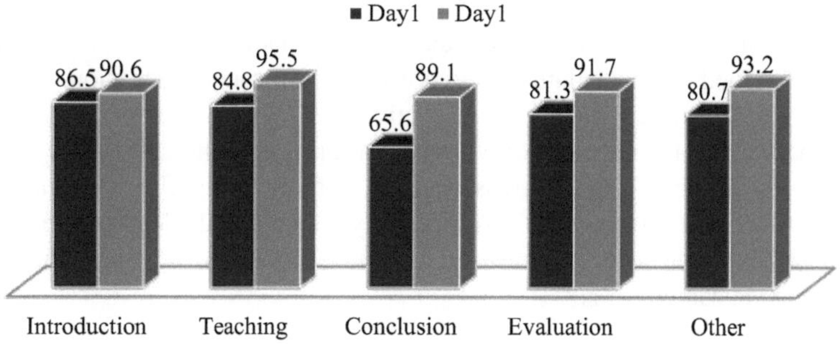

Fig. 20.9 Evaluation of the teacher's implementation of the lesson (in percentages)

- During activities, students learned from real situations and material related to their daily life, thereby helping them understand the content.
- Students demonstrated and exchanged ideas with each other on both the classification (parts of plants) and the importance of plants (including usage and conservation).
- Activity 1: students identified types of plants using cotyledon structures and determined that seeds germinate by water or humidity and that some of these seeds serve as food for humans.
- Activity 2: they identified groups of plants using leaves' structures, and classified those that could be food, herbs, grass, etc.
- Activity 3: they examined a flower's structure and explored knowledge on pollination by winds, insects, birds, etc.

Areas for improvement are as follows:

- Time management and controlling the classroom.
- The teacher's demonstration should precede the students' demonstration (especially leaves' structures).
- In-depth explanation of the structure of flowers and the criteria for classification.
- Assessments were conducted only in groups.
- For assessments, various assessment methods should be applied, such as playing a game with vocabulary cards, interviews, etc.

The process of using the lesson study cycle to developing this research lesson was conducted according to Baba (2007); it includes stating problems in a classroom or during teaching-learning, and then developing, conducting, and evaluating a lesson, respectively. The issues to be discussed included teacher competence, classroom interaction, school infrastructure, ability to demonstrate lessons, and the ability to identify living organisms as well as plants using criteria. According to Vegas (2005), teaching conditions affect students' learning outcomes. Our study explained the effects of those conditions. It also corresponds with Isoda (2010), which consisted of a study cycle, an implementation model (teacher model and observers), some differences between the title and objectives, no fixed lesson plans, teacher's commitment, and study results.

In activity 1 (87.5–93.8%, from days 1 and 2, respectively), students applied knowledge on monocotyledons and dicotyledons using seeds for germination, selection for food (beans and nuts), and preservation through discussion. In activity 2 (90.6–96.9%, from days 1 and 2, respectively), they identified parts of different leaves and then grouped them into types by demonstration; herewith, they explored knowledge of leaves, from the grass, garden, or farm plants. Finally, in activity 3 (87.5–96.9%, from days 1 and 2, respectively), they could identify structures of flowers and explored seeding. However, activities should utilize games, interviews, studies outside campus, etc.

20.3.3 *Integrating Curriculum to Address Sustainability: 2017 Practice*

Subject-Based Curriculum Mapping

A curriculum for the secondary teacher education program, with a major in Biology at BTC, was mapped according to the college's calendar. The model used for mapping included essential questions, contents, skills, and assessments as a part of the curriculum (Table 20.5).

Integration to Address Sustainability

In this example, we use the potential of the Science, Technology, Society, and Environment (STSE) approach to embed ESD into our curriculum, including the three following steps: (1) issue identification and association with curriculum requirements (Table 20.6), (2) connect issues with learning outcomes (Fig. 20.10), and (3) identify curriculum opportunities to embed ESD (Fig. 20.11). The issue was liver flukes (flatworm, phylum *Platyhelminthes*), common parasites found in central and southern Laos, constituting a major public health problem in the country.

Implementing a Lesson Plan

Topic Animal classification, phylum *Platyhelminthes* (flatworms).

Table 20.5 Curriculum mapping for secondary teacher education program

	Week 5
Essential questions	What are the structures and the functional characteristics of Phylum Platyhelminthes? What are representative examples and their classification level? What is the significance of *Platyhelminthes* to the environment and humans?
Content	Phylum Platyhelminthes
Skills	Reading fact sheet about *Platyhelminthes* (body form, feeding, reproduction) Classifying flatworms Analyzing and evaluating the diagram of life cycles, ecology, infection, and prevention Determining the most important types of worms
Assessment	Worksheet questions and demonstrations on the selected worms' ecology and life cycles

Table 20.6 Issue identification and association with curriculum requirements

Learning outcomes (original)	Suggested ESD learning outcomes
Learners should be able to: Use the principles established in the classification of animals	Learners should be able to: Use the classification system to group animals and identify relationships among them (role in ecology, importance, or impact on the environment, as well as respond to the issue in health consumption)

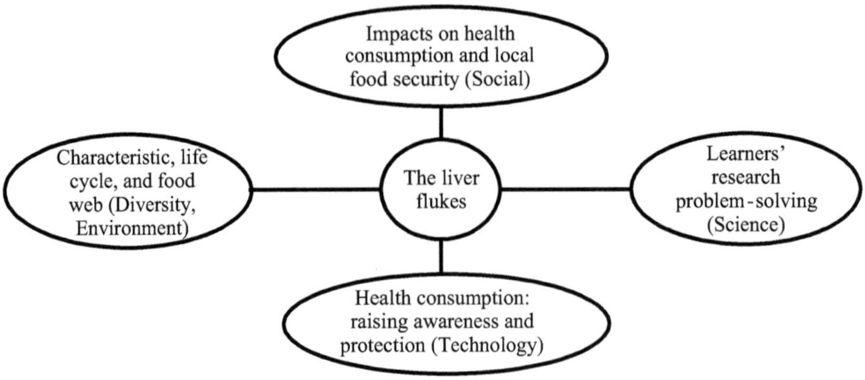

Fig. 20.10 Connecting issues with learning outcomes

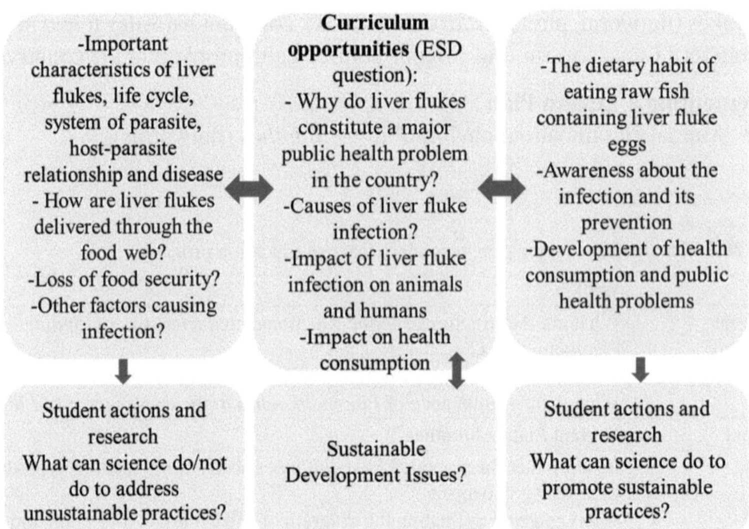

Fig. 20.11 Identifying curriculum opportunities to embed ESD

Deepen Students were asked to discuss and explain their body parts, habitats, movements, and eating habits after filling out the worksheet (and focus on the free-living flatworm). Then, the teacher presented a brief lecture on flatworm characteristics and its classification. After which, students were asked to classify figures of other flatworm species and group them into parasites or non-parasites (free-living) (Fig. 20.12).

Elaborate Students were asked to investigate the effect of parasitic flatworms on humans, especially liver flukes, which constitute a major public health problem in the country. Students worked in groups to understand the life cycle of flatworms,

Fig. 20.12 Teaching-learning animal classification – flatworms

their role or effect on humans, and how they cause diseases. They presented their work on large sheets and designed them to answer the following ESD questions from the teacher:

- Important characteristics (life cycle) and relationships between host-parasites and delivery
- Causes and effects (to animal/human, society, economy, and environment)
- Ways to resolve, spread awareness, and conserve healthy consumption patterns

Classroom resources include science books, videos, and science websites.

Based on this study, a curriculum for the secondary teacher education program of BTC was mapped into the college's calendar, comprising the following: (1) essential questions, (2) content, (3) skill, and (4) assessment. The most important skill to integrate is analyzing and evaluating the diagram of life cycles, ecology, infection, and prevention and determining the most important ones (for animal classification, subject-based). Three steps were included for integration: (1) issue identification and association with curriculum requirements, (2) connect issues with learning outcomes, and (3) identify curriculum opportunities for embedding ESD. To connect issues and develop a lesson plan, an STSE model was used. Problem-based learning was implemented as the teaching-learning method. From the implementation, we found that students were engaged in understanding healthy consumption patterns and displayed positive attitudes toward it. The students demonstrated a classification of animal groups, evaluated their role and importance, and were aware of their life cycles when teaching future students.

20.3.4 *Integrating Curriculum to Address Sustainability: 2018 Practice*

Subject-Based Curriculum Mapping
A curriculum for the secondary teacher education program, with a major in Biology at BTC, was mapped according to the college's calendar. The model used for mapping included essential questions, contents, skills, and assessments (Table 20.7).

Analyze/Identify Issues and Connect with the Curriculum Requirements
Topic Arthropods: characteristics, ecology, importance, and conservation.

Issue Arthropods are a large phylum of animals. Some students could not identify the specific characteristics of each body's systems, their habitats, and their importance.

Table 20.7 Curriculum for secondary teacher education program

	Week 13
Essential questions	What are the general characteristics of Arthropods? How many groups of Arthropods exist and their examples. Explain the exoskeleton, muscle, and circulatory and respiratory systems of some Arthropods? Explain the ecosystems of some Arthropods and their roles? Analyze the importance, uses, and effects of some Arthropods in their local environments What are some ways to conserve important insects or arthropods, and how to prevent or avoid the negative effects of some insects?
Contents	Unit 9: Arthropods (exoskeleton, muscle, circulatory and respiratory systems)
Skills	Learners should be able to: Tell the general characteristics of some Arthropods Explain the exoskeleton, muscle, and circulatory and respiratory systems of some arthropods Identify and explain the characteristics of their ecosystems and their roles in them Analyze the importance, uses, and effects of some Arthropods in their own local environments Find ways to conserve some important insects or arthropods and know how to prevent or avoid the negative effects of some insects
Assessment	Understanding general characteristics and processes of body systems from worksheets and tests Understanding of and developing a positive attitude toward animals' ecosystems and their roles from worksheets and tests Analyzing or classifying their importance, usages, and the effect of some insects and other arthropods on their local environments Understanding of and awareness about the conservation and protection of useful insects

Develop a New Lesson Plan After Connecting or Integrating Issues to the Curriculum Requirements

The topic was "Arthropods: Characteristics, Ecology, Importance, and Conservation." Activity 1 was to study the physical characteristics, exoskeleton, muscle, and circulatory and respiratory systems of Arthropods. Students were asked to fall into their designated groups to study animals' structures, body parts, exoskeletons, muscles, and circulatory and respiratory systems from slideshows, videos, and live animals. After filling out the worksheets, they introduced themselves to the class and exchanged ideas with other groups. Each group received different species of arthropods, such as honey bee, freshwater prawn, grasshopper, butterfly, and spider. From here, we found that students could explain about animals' body systems to the class and also exchange feedback with each other.

Activity 2 was to study arthropods, their ecology, and their importance. The students in the individual groups were asked to read and understand short stories and illustrations of these animals, such as habitats, the status of increment, reduction or extinction, importance, and usage and roles in the food webs. Then, they were encouraged to analyze these stories and diagrams by filling their own statements into blank spaces below the diagrams. Finally, they presented their findings to the class to share and exchange ideas with other groups.

Activity 3 was to make posters on Arthropods. The students in the group were encouraged to design their own posters from the worksheets and diagrams by arranging them on a board. On the board, the name of the animal, body systems or structures, habitats, life cycle, ecology, uses, and methods to conserve or protect them were included. After rearranging, students were encouraged to practice and finally present their posters. The assessment found that this poster making activity is the primary way to generate a positive attitude toward and awareness about living organisms.

From the assessment post the lessons, students presented an understanding of what makes insects a unique part of the animal kingdom and how they enter our bodies. They received an introduction to the type of arthropod groups by studying the main organs in systems used to classify insects or other arthropods. They gained an understanding of how insects adapt to survive and thrive in various habitats. They were introduced to the many roles that insects play in natural communities. Furthermore, students were able to define endangered or extinct, identify relationships between insects and their habitats, identify human activities that harm insects, and suggest ways to protect insects by working in groups and creating posters to spread awareness.

However, this lesson was still basic. So, future investigations should be more specific, such as having them discuss what they could do in their own homes to help protect insects that live around them, guiding them to evaluate each other's works, and displaying posters in a designated area of the school.

20.3.5 Perception and Recommendations of Respondents Through the Program

The most useful competencies inculcated include the capacity for critical (systems) thinking and problem-solving, integrating sustainable development into education, and improving the quality of teaching and learning by using ESD support. At the curriculum level, a curriculum for the secondary teacher education program at BTC was mapped according to the college calendar and provided essential questions, contents, skills, and assessments. The most important skill to integrate ESD is analyzing and evaluating diagrams of life cycles, ecology, infection, and prevention and determining the most important ones (for animal classification, subject-based). For its integration, three steps were included: (1) issue identification and association with curriculum requirements, (2) connect issues with learning outcomes, and (3) identify curriculum opportunities for embedding ESD. To connect issues and develop a lesson plan, a model of the STSE approach was used based on a problem-based learning method.

As for implementation, we found that students displayed engagement in understanding good health consumption. They demonstrated a classification of animals' groups and evaluated their role and importance, and were aware of their life cycles, when teaching future students. Otherwise, the implementation was basic for integrating ESD into the curriculum. Teachers agreed and suggested that integrating ESD into the curriculum at the unit levels should consider how to engage students to discuss what they could do in their own homes to help protect the environment, culture, animals, plants, and livelihoods around them. In the learning process, the teacher should provide guidance and opportunities for them to evaluate each other's works and display posters or diagrams in a designated area of the school and communities.

This study developed a primary sketch of the teachers' professional life experiences and daily practices. Teachers' reform involved efforts to promote an in-depth student learning reflection, collaboration, and effective working relationships among teachers at the institute and demonstrated schools. In the classrooms, teachers engaged their students and provided opportunities for interactive learning, addressing challenges that inhibit students' learning. Teachers improved their professional skills by engaging in many different professional development activities, including observation, reflection, interaction, collaboration, and training programs for their professional growth and development.

Acknowledgments This program was fully supported by Bankeun Teacher Training College, Ministry of Education and Sports, Lao PDR, and the Japan Society for the Promotion of Science, Core-to-Core Program, B. Asia-Africa Science Platform.

References

Baba, T. (2007). How is lesson study implemented? In M. Isoda, M. Stephens, Y. Ohara, & T. Miyakawa (Eds.), *Japanese lesson study in mathematics: Its impact, diversity and potential for educational improvement* (pp. 2–7). World Scientific Publishing.

Fujii, H., & Ogawa, H. (2016). Innovative lesson plans in chemistry education for brooding sustainable society. In M. Chiu (Ed.), *Science education research and practice in Asia* (pp. 385–407). Springer.

Holland, B. (2011). A comprehensive model for assessing service-learning and community-university partnerships. *New Directions for Higher Education, 114*, 51–60.

Isoda, M. (2010). Lesson study: Problem solving approaches in mathematics education as a Japanese experience. *Procedia Social and Behavioral Sciences, 8*, 17–27.

Pillai, J., & Achilles, V. (2015). *Learning with intangible heritage for sustainable future: Guidelines for educators in the Asia-Pacific region*. UNESCO/UNESCO Bangkok Office.

Stufflebeam, D. L. (2003). The CIPP model for evaluation. In T. Kellaghan, D. L. Stufflebeam, & L. A. Wingate (Eds.), *International handbook of educational evaluation. Part one: Perspectives* (pp. 31–62). Kluwer Academic Publishers.

Summers, D. (2013). Education for sustainable development in initial teacher education: From compliance to commitment – Sowing the seeds of change. *Journal of Education for Sustainable Development, 7*(2), 205–222.

UNESCO. (2006). *Education for sustainable development toolkit: Learning and training tool*. UNESCO.

UNESCO. (2012). *Education for sustainable development: Source book*. UNESCO.

Vegas, E. (Ed.). (2005). *Incentives to improve teaching: Lessons from Latin America*. The World Bank.

Zhang, G., Zeller, N., Griffith, R., Metcalf, D., Williams, J., Shea, C., & Misulis, K. (2011). Using the context, input, process, and product evaluation model (CIPP) as a comprehensive framework to guide the planning, implementation, and assessment of service-learning programs. *Journal of High Education Outreach and Engagement, 15*(4), 57–84.

Chapter 21
Introducing Education for Sustainable Development to In-Service Science Teachers in Myanmar

San Aye, Yin Mar Win, and San San Maw

21.1 Introduction

The concept of sustainable development was described by the Brundtland Commission Report in 1987 (UNESCO, 2012) as development that meets the needs of the present generation without compromising the ability of the future generations to meet their own needs. In 2005, the United Nations (UN) Decade of Education for Sustainable Development (ESD) was launched to enhance the role of education in promoting sustainable development. At the UN Conference on Sustainable Development in 2012, the international community agreed to "promote education for sustainable development and to integrate sustainable development more actively into education beyond the UN Decade of Education for Sustainable Development"(UNESCO, 2014). ESD is a type of education that encourages students to become active citizens who have a role in transforming how our societies operate. Through ESD, children will be encouraged to view themselves as people who have an active role in shaping a better future. This goal of ESD is reflected in the subject matter and in the way that it is taught (Henderson, 2009).

With regard to the ESD concept, there is a difference between education on sustainable development and education for sustainable development. The former is an awareness of the lessons or theoretical discussions, and the latter is the use of education as a tool to achieve sustainability (Mckeown, 2006). Development is a holistic approach in which people systematically enhance their capacity to solve their own problems while promoting their cultural, social, and economic well-being. Sustainable development occurs when this process is conducted in a manner that can be sustained over a long period of time (Wiltshire, 2008). Young people are the future, and teachers play a key role in the appropriate socialization of the youth for

S. Aye · Y. M. Win (✉) · S. S. Maw
Sagaing University of Education, Sagaing, Myanmar

H. Fujii, S.-K. Lee (eds.), *Science Education for Sustainable Development in Asia*, Education Innovation Series, https://doi.org/10.1007/978-981-99-8711-5_21

sustainable development. To do this effectively, teachers need to understand what ESD is and how it can be implemented. Henderson (2009) described that it is impossible to attempt to understand ESD by providing a simple definition because it is a complex and evolving concept. Different cultures and societies perceive sustainable development in several different ways. A more meaningful method of understanding ESD is to examine its rules, principles and value, and the methods used in its delivery. However Wals (2009) claimed that ESD is already a part of formal education. Depending on each country's historical background and educational perspectives, various approaches or strategies exist around the world to integrate ESD into formal education. Furthermore, he suggested that the approaches vary from the conventional methods, for example, integrating the ESD concept and sustainable development issues in existing school subjects, to more innovative ones, for instance, adopting a whole school approach to ESD and sustainable development issues. According to the baseline survey (Aye et al., 2019), although the Myanmar in-service teachers have positive attitudes toward the ESD concept, they have inadequate awareness of ESD and on the need to improve the teaching techniques that foster ESD awareness in students. Therefore, depending on the Myanmar context – cultural, social, and economic situations – this study was conducted to embed the ESD concept into the existing middle school science curriculum.

We initiated the ESD introduction project with the following objectives:

- Promote the in-service science teachers' understanding of the ESD concept in teaching and learning.
- Foster encouragement in the teachers to develop the science lesson plans focused on the ESD concept.
- Practice, analyze, and revise the ESD-focused lesson plans.

On completion of the introduction of the ESD project, the participants will be able to successfully complete the below tasks:

- Understand the theme of ESD in teaching and learning.
- Develop ESD lesson plans.
- Analyze and revise ESD lessons.
- Practice sustainable activities with their students.

21.2 Education for Sustainable Development

The concept of sustainability leads to a paradigm shift for envisioning a future in which the environmental, social, and economic considerations are balanced in the pursuit of development and an improved quality of life. Sustainability is the goal of sustainable development, and it is an unending quest to improve the quality of people's lives and surroundings and to prosper without destroying the life-supporting systems on which the current and future generations of human beings depend. To live sustainably means to find ways of improving everyone's quality of life without damaging the environment and without creating problems for the future generations or transferring them to people in the other parts of the world. It requires an understanding that lack of action on the existing problems has its consequences, and we must find ways to innovate and change at all levels within society (UNESCO, 2010). Education is a key instrument to bring about the changes in people's values and attitudes, skills, behaviors, and lifestyles consistent with sustainable development (UNESCO, 2009). Thus, ESD is a tool for addressing sustainability issues. The question that arises is "how can we implement ESD?" Therefore, the existing curriculum needs to be reoriented to address sustainability and include the important concepts for sustainable development (Mckeown, 2006).

21.2.1 Components of ESD

ESD is more than a mere knowledge base related to the environment, economy, and society. It also addresses the learning skills, perspectives, and values that guide and motivate people to seek sustainable livelihoods, participate in a democratic society, and live in a sustainable manner. Moreover, it involves studying the local and, when appropriate, global issues.

Mckeown (2006) explained the essential components to be addressed in a formal curriculum to reorient toward sustainability.

Knowledge First, students can gain knowledge on ESD from the natural sciences, social sciences, and humanities to understand the principles of sustainable development, how they can be implemented, and the complications of their implementation. Essentially, people need to have an insight on the importance of a balance between the rates of use of the renewable resources and the rate of their regeneration as well as a balance between the rates of pollution emission and the assimilative capacity of the environment. Then, they need to minimize the use of nonrenewable resources, prevent the erosion of renewable resources, and employ other methods to live in a sustainable manner.

Issues The second component concerns the global problems that people need to be aware of. The Earth Summit held in Rio de Janeiro in 1992 identified the issues in Agenda 21 including the following: population and sustainability, protecting and promoting human health, changing the consumption patterns, combating deforestation, protecting the atmosphere, conservation of biological diversity, and strengthening the role of farmers.

Skills With the increasingly faced concerns, people possess practical skills that will enable them to continue learning after leaving school, have a sustainable livelihood, and live sustainable lives. Therefore, teachers need to help their students to develop the ability to communicate effectively, think about systems, critically reflect on value issues, move from awareness to knowledge to action, and work in cooperation with others.

Perspectives It is important to understand the international and local issues in a global context. Observing the roots of an issue and forecasting the possible futures based on different scenarios are part of ESD. Students should have the different perspectives associated with ESD, namely, that social and environmental problems change over time and have a history and a future and that the contemporary global environmental issues are linked and interrelated between and among them. Therefore, considering differing views before reaching a decision is necessary.

Values The fifth component is values, which is an integral part of ESD. Understanding one's own worldview and other people's viewpoints is essential. Understanding one's own values, the values of the society one lives in, and the values of others around the world is a central part of education for a sustainable future. For example, it is necessary to develop respect and care for the community, maintain ecological integrity, and promote social and economic justice.

Countries across the world have implemented and continue to implement ESD. Teachers are the main agents in providing ESD, and thus, it is essential to integrate ESD in teacher education.

21.2.2 Integrating ESD in Teacher Education

ESD will empower learners to take informed decisions and responsible actions for environmental integrity, economic viability, and a just society for the present and future generations while respecting cultural diversity. It is about lifelong learning, and it is an integral part of quality education. ESD is a holistic and transformational education, which addresses the learning content and outcomes, pedagogy, and learning environment. It achieves its purpose by transforming society. To integrate ESD in teacher education, learning contents, such as climate change, biodiversity, disaster risk reduction (DRR), and sustainable consumption and production (SCP), should be integrated into the curriculum. As learning outcomes, students need to develop core competencies, such as critical and systemic thinking, collaborative decision-making, and taking responsibility for the present and forthcoming generations. To achieve these outcomes, teachers should develop and use pedagogy, such as designing teaching and learning in an interactive, learner-centered way that enables exploratory, action oriented, and transformative learning and rethinking of the learning environments – physical and virtual and online – to inspire the learners to act in sustainable ways. Another target of integrating ESD in teacher education is societal transformation. Empowering learners of any age and in any education setting to transform themselves and the society they live in implies equipping them with the skills for green jobs and motivating people to adopt sustainable lifestyles. In addition, it means empowering individuals to be global citizens who engage and assume active roles, both locally and internationally, to face and resolve the global challenges, and ultimately become proactive contributors in creating a more just, peaceful, tolerant, inclusive, secure, and sustainable world. There have been earlier attempts at integrating ESD in schools and educational instructions at all levels (UNESCO, 2018). Examples of the current curriculum mainstreaming practices are as follows:

- Creating a standalone ESD-specific subject
- Embedding ESD in the existing subjects or across the curriculum
- Adopting a thematic, issue, or problem-based approach, such as climate change, air pollution, or deforestation, based on the local realities
- Incorporating the use of teaching and learning methodologies consistent with ESD principles of learner-centered and participatory approaches, such as field trips
- Conducting ESD-based cocurricular or extracurricular activities, such as the use of student clubs and associations and activities
- Engaging with the local community, often through project-based learning opportunities

21.3 General Background of Basic Education in Myanmar

Myanmar has a long coastline, abundant fertile lands, rich natural resources, and one of the lowest population densities in the region. It is in the ongoing process of political, economic, and social transformation. After the country's government system changed from a military regime to a democracy, the new administration has made education reform a national priority. Education is widely recognized as a critical building block for nation building, national unity, and sustainable development. Education and poverty reduction have been identified as the two key drivers to support the democratic and peace-building process and to achieve the national goal of Myanmar (Ministry of Education, 2016). There is broad consensus that major shifts are required in the coming years to transform the national education system and ensure that all students progress through the education cycle, achieve quality learning standards, and fulfill their career and lifelong learning goals and aspirations. In Myanmar, the Ministry of Education (MOE) is the main provider of education and is responsible for the provision of the basic education subsector. Therefore, the MOE has undertaken a 3-and-a-half-year Comprehensive Education Sector Review (CESR) involving the following three phases: phase 1: rapid assessment; phase 2: in-depth research and analysis of the critical subsector challenges; and phase 3: drafting and building ownership for an evidence-based and costed National Education Strategic Plan (NESP) for the period 2016–2021.

21.3.1 Challenges in Basic Education

Myanmar is attempting to make education reform a national priority. However, its education system is currently facing multidimensional challenges, and most of them deal with the key concepts in education reform, such as access, quality, and equity. The major challenges in certain areas are described in the following section.

Preschool and Kindergarten Education Children living in the rural and remote areas have limited access to quality preschool and kindergarten education, and this negatively affects school readiness, student learning achievement, community well-being, and national economic development. In addition, preschool service providers need to improve the quality of their services, including supporting keen parental involvement in the child's educational development. The quality of the preschool teacher training needs improvement, and a national preschool database is needed for the effective management and planning of preschool services. A major priority is to design, develop, and implement a kindergarten curriculum that is developmentally appropriate, culturally responsive, and educationally relevant to ensure that all children can access appropriate and quality kindergarten learning.

Basic Education Children's access to schools and their retention and completion of basic education need to be significantly improved. Moreover, the dropout rates during the transition from primary to middle school and from middle to high school are still high. A school quality improvement framework is required to focus attention on measuring and addressing teaching, school management, and standards of the school facilities. Furthermore, a quality assurance system needs to be designed and implemented to improve school performance and ensure greater accountability from the head teachers, teachers, and education managers. With poverty being a key factor affecting access to basic education, greater efforts are required to enable the students living in the remote and rural areas to enroll in primary and middle schools.

Basic Education Curriculum The major challenges in the curriculum are redesigning the basic education curriculum in line with the new KG + 12 structure with a focus on twenty-first century skills and upgrading the professional capacity of curriculum development teams as well as the capacity of head teachers and teachers to successfully implement the new curriculum.

21.3.2 Basic Education Subsector Reform

The Constitution of 2008 is the fundamental law of Myanmar and also the foundation for the nation's transition toward democracy and the reform of its education sector. Article 28 of the Constitution stipulates that the Union shall earnestly strive to improve the education and health of the people; enact the necessary laws to enable citizens to participate in matters of their education and health; implement free, compulsory primary education system; and implement a modern education system that will promote all-round correct thinking and good moral character contributing toward the building of the nation. In addition, Article 366 states that every citizen, in accordance with the educational policy laid down by the Union, has the right to education; shall be given basic education, which the union prescribes as compulsory by law; and has the right to conduct scientific research, explore science, work creatively, and develop the arts and conduct research freely in the other branches of culture (Union of Myanmar, 2008).

To meet the challenges mentioned earlier, the government set the following goal: "Improved teaching and learning, vocational education, and training, research and innovation leading to measurable improvements in student achievement in all schools and educational institutions."

According to the National Education Strategic Plan (NESP), the MOE commits to achieving this goal statement by the end of the 2020–2021 fiscal years. Therefore, it has identified nine transformation shifts to collectively contribute to the achievement of the NESP goal. Importantly, these shifts will enable the MOE to achieve the Sustainable Development Goal for Education, that is, the SDG 4, which is to ensure inclusive and equitable quality education and promote lifelong learning opportunities for all. Thus, the MOE has categorized the main basic education subsector reform areas as access, quality, and inclusion; curriculum; student assessment and examinations; and teacher education and management.

Access, Quality, and Inclusion The National Education Law (NEL) recognizes the right of all citizens to education, and in particular free, compulsory primary education. Furthermore, the law mandates the establishment of an education quality assurance system; extends the basic education system to 13 years (including a kindergarten year); allows for the learning of the minority nationalities' languages and culture and the use of their languages as a classroom language; and provides a definition of, as well as a commitment to, inclusive education and commits to a decentralized education system. A number of other types of nongovernment basic education schools are also recognized by the NEL. However, the basic education system continues to face major challenges related to school access, retention, inclusion, equity, and quality assurance of education standards.

Basic Education Curriculum MOE has undertaken the basic curriculum reform at the primary, middle, and high school levels that is more relevant to all students, thereby motivating them to stay in school and complete their basic education. It redesigned the curriculum emphasizing the twenty-first century skills using the basic education curriculum reform program. Subsequently, the students are able to appreciate and respond to a more relevant basic education curriculum that emphasizes these contemporary skills. Teachers and students are actively and successfully using the new curriculum materials, including the textbooks and teachers' guides.

Teacher Education Major changes are needed to improve the quality of teaching, learning, infrastructure, and management in teacher education institutions (TEIs) – Education Colleges (ECs), Universities of Education (UOEs), and the University for the Development of National Races (UDNR). The NEL in 2014 and the NEL Amendment in 2015 provide a clear legal framework for progressive, integrated, and comprehensive teacher education and management reforms. The government has undertaken teacher quality assurance and management programs, and after attending the teacher training courses, the teachers achieve a certain level of competency according to the framework's standards.

21.4 Exploring ESD Practices and Introducing ESD Concepts

Although the teachers' perception toward ESD is generally positive, their ability to use teaching techniques to integrate the ESD concept into the teaching subjects is weak and needs improvement (Aye et al., 2019). This suggests that an examination of the current ESD-related practices of science teachers in schools is required. Therefore, we conducted a 3-day workshop with 20 middle school science teachers from four basic education schools in Sagaing Township, upper Myanmar.

On the first day, we interviewed the participants to explore their current ESD-related practices in their schools. The interview focused on the ways of garbage collection, systematic use of water, and sanitation. All participants gave the same response that the simple emptying method with containers is used. However, they do not teach systematic ways of waste management to the students. For instance, they do not educate their students on how to divide the waste materials into organic waste, such as food leftovers, paper, wood, and plant remains, which are degradable, and inorganic waste, such as glass, plastic, and electronic materials, which are not easily decomposed. All schools have resources for clean water and efficient drainage systems for sanitation. One of the factors for this lack of knowledge on waste segregation is that the schools in this study are rural ones which have large school campuses. Second, dry runs of the general content analysis of science textbooks (grades six and eight) were done to explore which concepts are directly related to the ESD themes. They identify such topics as human beings and their environment, environmental problems, causes of damages, environmental degradation, consequences of the problems, and environmental conservation of the earth (Curriculum, Syllabus and Textbook Committee, Ministry of Education, 2016). With regard to the teaching techniques for these topics, the teachers discussed how to use techniques which focus on the conceptual understanding of the students rather than the mere transmission of factual knowledge. For example, one of the questions to the students was "list the reasons for environmental damage." Instead of this, the question "how will you protect the environment from damage?" would prompt the students to present their ideas on what they need to do and how they would achieve their goals.

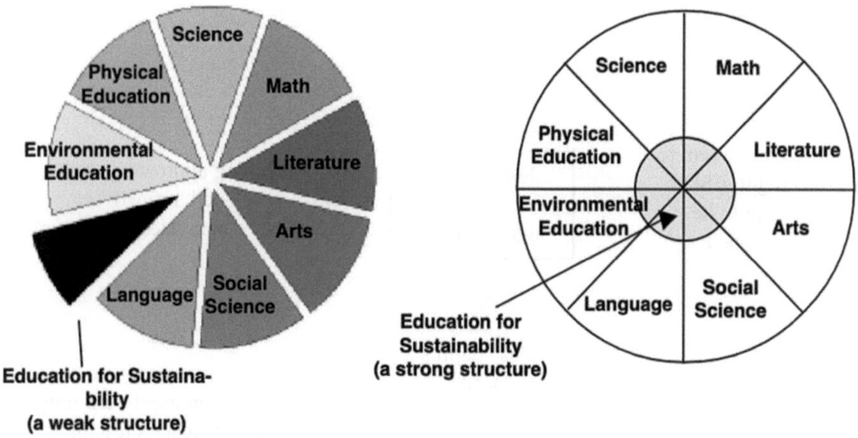

Fig. 21.1 Education for sustainability as an individual discipline and as a multidisciplinary subject. (Reproduced from Mckeown (2006))

With the help of worksheets, we discussed a few principles of sustainability and the components of ESD. Then (Fig. 21.1), the teachers were given a figure sheet to discuss and select what type of ESD should/could be taught to the students. They discussed and suggested that education for sustainability could be taught as a multidisciplinary subject. This means that the teachers realize that ESD can be taught as an integrated concept while teaching any other subject in their classrooms. Finally, the workshop encouraged the teachers to initiate the implementation of the integration of ESD with science teaching.

21.5 Lesson Study

Lesson study is a highly specialized form of classroom action research focusing on the development of teacher practice knowledge. It involves groups of teachers, collaborative planning, teaching/observing, and analyzing the teaching and learning in research lessons. Furthermore, it blends all the features of professional learning that improve learning and teaching: the professional learning takes place over a period of time; it happens in real classrooms with real groups; it involves an element of collaborative enquiry or experiment between teachers who are trying to solve a problem or improve an approach (Dudley, 2011). A lesson study consists of a cycle of at least three research lessons (RL) with the initial meeting of the lesson study group to determine the areas of improvement: joint planning of RL 1, teach/observe RL 1, interview pupils and post RL 1 discussion, and initial plan for RL 2 (cycle 1); joint planning of RL 2, teach/observe RL 2, interview pupils, post RL 2 discussion, and initial plan for RL 3 (cycle 2); joint planning of RL 3, teach/observe RL 3, interview pupils and post RL 3 discussion, and agree on the overall findings (cycle

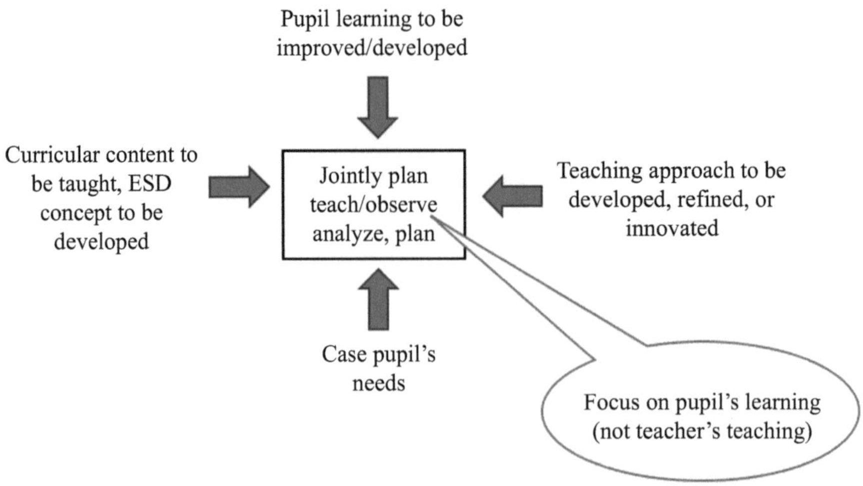

Fig. 21.2 Balanced focuses on pupil learning in ESD research lesson. (Reproduced from Dudley (2011))

3) that are jointly planned, taught/observed, and analyzed by a lesson study group. Although Dudley's lesson study process consists of three research lessons, this study was conducted with two research lessons. Using the following adapted balanced focus on pupil learning, the lesson study groups jointly plan, teach/observe, and analyze the science lessons as shown in Fig. 21.2.

21.6 Lesson Study: Integrating the ESD Concepts Into Science Lessons

The lesson study group consisted of four teacher trainers from Sagaing University of Education and 16 junior science teachers from four basic education schools. Dudley's (2011) lesson study cycle process of plan, teach, observe, and reflect/post lesson discussion was used for conducting the lesson study. Science lessons integrating with ESD were selected from grades six and eight science textbooks (Table 21.1). In the planning step, the teacher educators facilitated the participants in formulating the goals to be achieved or identifying the problems to be solved, selecting the science lesson topics, identifying the lesson objectives, preparing the teaching learning materials and creating the learning activities, constructing the post-test questions, and preparing the observation sheets. The lesson study group members conducted the trial lessons; case pupil interviews; and post-lesson discussion, reviewed the research lesson, refined it, and taught it as lesson study cycle 2 to the next class. Among them, four teachers did not participate in observing the classroom teaching because they had to teach the primary classes in parallel. Lesson study group members selected two model teachers to teach the research lessons, and

Table 21.1 Planning science lessons' integration with ESD

Grade/topic	Grade six Malleability and ductility of metals	Grade eight Food chain and food web
Learning objectives	Students understand the concepts through self-experiments and reflect on how these properties of metals can be utilized in their daily lives Think about the positive/negative effects of the use of metal products on their lives in terms of the social, economic, and environmental perspectives (sustainability)	Students understand the formation of the food chain and reasons for the formation of the food web Recognize the interdependence of living things on each other and realize how organisms try to survive Establish a social network from the food web
Activities	Study the physical properties of objects by hammering and identify them as metal or nonmetal Conduct experiments with metals to learn the concepts of malleability and ductility, such as copper wire (which is pulled into a thin and longer one) and a piece of lead (which is hammered into a thin sheet) Classify the given objects into metals and nonmetals and then name the metal objects that are used in their daily lives (e.g., gold, silver, copper, and brass)	Examine and discuss the formation of the food chain and food web as well as the original source of energy and its importance Make the food chain and food web chart Make their social network chart
Focus group discussion questions	Was the flow of the lesson coherent and did it support the student's learning of the concept? Were the activities and materials helpful in achieving the goal of the lesson? Did the group discussion help promote students' understanding? Did the teacher's questions engage and facilitate student thinking?	Were the activities and materials helpful in achieving the goal of the lesson? Did the group discussion help promote student understanding?
Pupil questions	What did you enjoy most about the lesson? What did you learn? (What can you do now that you could not do earlier? What can you do better? How is it better?) Talk about the positive/negative impacts on the social, economic, or environmental perspectives (any one)	What did you learn? (What can you do now that you could not do earlier?)
Achievement test	Textbook-based posttest questions	

the rest were observers. In addition, they identified the case pupils who might typify the different groups of learners in the class (pupils who are making good, average, or below average progress in academic achievement). The case pupils were interviewed at the end of the lesson. The lesson study groups referred to the suggested questions (Shuilleabhain, 2015) to be used for the observation and post-lesson case pupil interviews. Qualitative (case study) and quantitative methods were used.

21.6.1 Results of the Grade Six Lesson Study Group

After the lesson, the posttest (grade six) was administered immediately.

The average of the students' academic achievement, being an immediate assessment, was above 65% in cycle 1, and in cycle 2 their achievement improved to over 75% (Table 21.2). These achievement values were the outcomes of the lesson study, collaborative planning, teaching/observing, and freely making productive suggestions. In particular, pupils in the low learning groups (names were suggested by the teachers) achieved academic success even above the level of the average groups in cycle 2 (day 2). It is obvious that through lesson study, the teachers could perceive the pupil learning occurring in much shaper detail than is usually possible, and the gap between what they had assumed was happening when the pupils learned and what is actually happening (Dudley, 2011).

Interviews with the students in post lessons was conducted at the end of the lesson individually. The students responded to the question "what did you enjoy most about the lesson?" by stating that they enjoyed learning science because they had opportunities to do practical work and participate actively in the hands-on activities (learning by doing).

Specifically, one case pupil replied,

> I was very excited to pull the wire because I thought that it would get cut, but it became longer.

The above reply is a response to the question "what did you learn?" (what do you know about something that you did not know before?).

The case pupil responded (particularly, a low achiever but not a slow learner) that he could participate in the activities, share his ideas with the others in the discussion, do the group work successfully, and was satisfied with his achievement. As a consequence, the students have a sense of self-worth combined with a respect for the other individuals and cultures. This quality is also essential for students who are educated for sustainability (Henderson, 2009).

Focus Group Discussion Post Lessons The teachers interpreted that they could stimulate the interest of their students in learning experiences and participation in activities. They were satisfied with a variety of their students' answers. For example, the students responded to the question "name the objects used in your home." The answers were cables made of copper (ductility), rings, earrings, hand chains that are made of gold and silver, and some tools used in the farm. Moreover, their

Table 21.2 Evaluation of the student learning outcomes for grade six

Student achievement	Day 1 ($N = 35$)	Day 2 ($N = 48$)
Mean (total)	68.9%	76.4%
Case pupils (high achiever)	92.5%	90%
Case pupils (average)	75%	76%
Case pupils (low achiever)	62.5%	82.5%

response to the question "do you know the places where these metals can be found?" was (copper mine) Salingyi Township of Sagaing Region, Myanmar. It illustrates that the teachers could encourage the students to relate to the content they have learned in the classroom to its use in their daily life; they could make sense of the concept of sustainability. This means that the teachers could make the students aware of the natural resources in Myanmar.

The common observation of the observer teachers about the science lesson "ductility of metals" was,

> They measured the length of the wire themselves…stretched it with pliers…then, they measured the wire again…I could see their satisfaction on the conceptual understanding of the science concept of the 'ductility of metal'.

One of the less experienced junior teachers continued,

> Collaborative planning makes me more confident in and satisfied with the teaching and learning situation.

Although the middle school science teachers' content knowledge and pedagogical knowledge need to be enhanced, the improvement of their teaching skills is evident. One experienced teacher who taught the research lesson, for instance, said,

> We usually teach the content as prescribed by the textbook…and rarely prepare science activities…This is because I thought that it would be time consuming…Now I realize that the planning and collaborative planning are worth it." He continued, "I am hesitant about introducing the lesson for the first time…then my teaching becomes smooth and gains momentum.

This indicates that the teachers change their teaching techniques to better support learning and consequently discover ways to plan learning that is better matched to the pupils' needs.

Furthermore, he stated,

> I did not know about a few chemicals and simple laboratory apparatus, and I had never seen them.

21.6.2 Results for Grade Eight

Students had to examine and discuss the formation of the food chain and food web as well as the original source of energy and its importance. Then, they made the food chain and food web charts, which, in turn, encouraged the students to create their social network chart (Figs. 21.3 and 21.4). These activities enabled the students to experience the results of group accomplishment. Moreover, the students discussed the question "do you live alone and why or why not?" Accordingly, the students could establish the social network, and the concept of "all for all" made sense to them.

Fig. 21.3 Making the food chain/food web chart and feeling a sense of group accomplishment

Fig. 21.4 Making the social network chart

21.6.3 Discussion

The teachers who participated in this study realized that they need to help their students imbibe the qualities of being a sustainable person. This is because the student who is educated for sustainability has traits, such as a sense of responsibility to the environment, to other people, and to the future of both; the will, knowledge, and skills to translate this responsibility into action in both their personal and public life; and a capacity to observe the links between the individual and group actions and the external events (Henderson, 2009). They understood how the ESD teaching strategies (learner-centered) are different from their traditional methods (teacher-centered) and rote learning. Generally, the learning objectives and the overall goals identified during the planning were achieved. The teachers accepted that they could create a lively learning environment where the students could discover new things (which the students did not know before), share ideas, and improve the students' interest. Lesson study enables the teachers to achieve more in the context of a supportive teaching learning community, which is strongly committed to helping pupils to learn and to the professional learning of the members of the group (Dudley, 2011).

In group work, allowing the students to examine the materials using their senses, brainstorm ideas of what they discovered about the attributes of the materials, and asking them to note the properties of the materials led to an improvement in the students' critical thinking skills and allowed them to experience a sense of group accomplishment. Critical thinking competency is one of the key competencies for sustainability: the ability to question the norms, practices, and opinions; to reflect on one's own values, perceptions, and actions; and to take a position in the sustainability discourse. Moreover, the students gained collaborative competency. It is also the key capability for sustainability that the students must possess. Collaborative competency is the ability to learn from others; to understand and respect the needs, perspectives, and actions of others (empathy); to understand, relate to, and be sensitive to others (empathic leadership); to deal with conflicts in a group; and to facilitate collaborative and participatory problem solving (UNESCO, 2017).

21.7 Conclusion

Sustainable development encompasses the environment, economics, and society. Therefore, people need basic knowledge from the natural sciences, social sciences, and humanities to understand the principles of sustainable development, how they can be implemented, the values involved, and the ramifications of their implementation. Baseline survey results showed that the teachers' perception toward ESD is generally good (Aye et al., 2019). However, their ability to use the teaching techniques to integrate the ESD concept into the teaching subjects needs to be improved. ESD addresses learning skills, perspectives, and values that guide and motivate people to seek sustainable livelihoods and live in a sustainable manner by involving the study of local and global issues. All the teachers are collectively responsible for ESD. To address it, the teachers need to be fully prepared. This will be more easily achieved by the teachers if, in addition to having a positive attitude toward ESD, they fully understand the various aspects of the ESD concept and possess the teaching skills required in integrating the ESD concept into their teaching subjects.

In the Myanmar education system, however, examinations have been used and continue to be used in assessing student achievement. This leads teachers to teach what will be tested and students to learn only what might appear in the examination. Moreover, parents usually judge the teachers on the basis of the students' success in the examinations, and the school authorities also view the teachers' performance on the same basis. The impact of the basic education high school examination, which involves examinations across a selected set of subject combination-based examinations and is utilized for determining grades from matriculation to higher education, is a significant factor for exam-orientation teaching methods (Hayden &

Martin, 2013). The Ministry of Education, as a responsible organization, realizes the weakness of the current teacher pedagogy. It states that the emphasis on the teaching strategies needs to be shifted away from the narrow goal of merely succeeding in the examinations because most Myanmar students cannot think critically, raise questions or solve problems, and classroom instruction focuses primarily on getting the students to understand and memorize the facts in the textbooks (Myanmar Education Research Bureau, 1992).

The aim of this study is to establish the ways in which teachers can integrate the ESD concepts into science teaching. It is an unrealistic expectation that they will have sufficient teaching competency for sustainability. As mentioned above, in the current situation, it would be difficult to implement the ESD practices and reorient the educational system toward quality education. The challenges that the Myanmar education system is facing are multidimensional, and most of them deal with the key concepts in education reform, such as access, quality, and equity (Ministry of Education, 2016). However, the findings of this study are evidence that teachers can create the science lesson plans which improve the students' competencies for sustainability and that the ESD concepts can be integrated into science teaching through lesson study. Accordingly, the students will develop the key competencies for sustainability, which are essential skills for the developing citizens of the twenty-first century. No one universal approach for ESD exists. Political, social, and cultural realities and specific environmental and ecological challenges make the contextual basics of ESD essential (UNESCO, 2017). In conclusion, even though the integration of the ESD concepts into science lessons is a demanding task (for the teachers), and many challenges exist in the implementation of the integration of ESD into science teaching, we need to identify the locally and nationally relevant interpretations of ESD and implement the related forms of education.

Acknowledgments The work described in this chapter was supported by the JSPS Core-to-Core Program "Formation of International Center of Excellence to Promote Teacher Education on ESD." Moreover it would not be possible to do this work without the generosity of the junior teachers who gave so freely of their time, expertise, and knowledge through the study.

References

Aye, S., Win, Y. M., & Maw, S. S. (2019). In-service teachers' perception towards education for sustainable development in Myanmar. *Journal of Physics: Conference Series, 1280*, 032043. https://doi.org/10.1088/1742-6596/1280/3/032043

Curriculum, Syllabus and Textbook Committee, Ministry of Education. (2016). *Grade six and grade eight science textbooks*. Ministry of Education. (in Burmese).

Dudley, P. (2011). *Lesson study: A handbook*. https://lessonstudy.co.uk/wp-content/uploads/2012/03/new-handbook-revisedMay14.pdf.

Hayden, M., & Martin, R. (2013). Recovery of the education system in Myanmar. *Journal of International and Comparative Education, 2*(2), 47–57.

Henderson, R. (2009). *Education for sustainable development: Teacher training manual*. Live & Learn Environmental Education.

Mckeown, R. (2006). *Education for sustainable development toolkit*. UNESCO.
Ministry of Education. (2016). *National education strategic plan 2016–21 summary*. Ministry of Education. (in Burmese).
Myanmar Education Research Bureau. (1992). *Education sector study. Phase 1. Final report*. Myanmar Education Research Bureau. (in Burmese).
Shuilleabhain, A. N. (2015). *Lesson study as a form of in-school professional development: Case studies in two post-primary schools. Technical report*. National Council for Curriculum and Assessment.
UNESCO. (2009). *Second collection of good practice in education for sustainable development: UNESCO associated schools*. UNESCO.
UNESCO. (2010). *Education for sustainable development lens: A policy and practice review tool*. UNESCO.
UNESCO. (2012). *Education for sustainable development: Sourcebook*. UNESCO.
UNESCO. (2014). *Roadmap for implementing the global action Programme on education for sustainable development*. UNESCO.
UNESCO. (2017). *Education for sustainable development goals: Learning objectives*. UNESCO.
UNESCO. (2018). *Integrating education for sustainable development (ESD) in teacher education in South-East Asia: A guide for teacher educators*. UNESCO.
Union of Myanmar. (2008). *Constitution of the republic of the Union of Myanmar*. Ministry of Information. (in Burmese).
Wals, A. (2009). *Review of contexts and structures for education for sustainable development*. UNESCO.
Wiltshire, W. (2008). *Teachers' guide for education for sustainable development in the Caribbean*. UNESCO.

Chapter 22
Conclusion: Toward Advancing Science Education for Sustainable Development in Asia

Sun-Kyung Lee and Hiroki Fujii

22.1 Diversity and Uniqueness of Science Education for Sustainable Development in Asia

22.1.1 Seeking Social Participation and Individual Transformation

International perspectives are critical when considering science education for sustainable development. The complex topics addressed by science education for sustainable development, such as biodiversity, food security, clean water and sanitation, renewable energy, preservation of the environment, and sustainable use of natural resources, require a global approach (Kilinc & Aydin, 2011). Simultaneously, each country's approach to coping with sustainability issues is shaped by its culture, economy, and politics (UNESCO, 2018), and the nature of each country is the basis for this shaping. Therefore, science education for sustainable development inevitably involves local diversity. Considering these, this book aims to pursue science education for sustainable development in Asia by vividly depicting regional diversity in the context of nature, culture, economy, etc., while maintaining international perspectives.

The first section of the conclusion explores the diversity and commonalities in science education highlighted in its chapters. To do so, the keywords of science education connected to Education for Sustainable Development (ESD) mentioned

S.-K. Lee (✉)
Cheongju National University of Education, Cheongju, Republic of Korea
e-mail: sklee@cje.ac.kr

H. Fujii
Okayama University, Okayama, Japan

H. Fujii, S.-K. Lee (eds.), *Science Education for Sustainable Development in Asia*, Education Innovation Series, https://doi.org/10.1007/978-981-99-8711-5_22

in the introductory chapter are reintroduced. The keywords discussed first belong to science education objectives, and scientific literacy is a common keyword in this book. The idea that scientific literacy should be the primary objective of science education and that this understanding should allow individuals to effectively participate in decision-making about matters related to our world (Sadler, 2011) will have different degrees of penetration in each Asian country. Against this backdrop, the research and practices presented by the contributors to this book, developed under the umbrella of fostering science literacy among students and teachers, demonstrate rich diversity.

Chapter 8, which discusses the 2015 revised Korean national science curricula, notes that the curricula are inadequate in incorporating elements of environmental education in elementary, middle, and high schools. In particular, environmental attitudes and values, competencies, and participation and action are much less reflected in the curriculum than in environmental knowledge. Reflecting on these elements of environmental education implies promoting science education for sustainable development, which is one of the challenges in revising the national science curricula in Korea.

A discussion directly related to this challenge is presented in Chap. 2, which reports on Korean and Japanese high-school students collaborating to learn science for sustainability. Bilateral partnerships and collaborative science lessons foster students' multifaceted understanding, critical thinking, decision-making, abilities, and viewpoints to respond to the global society advocated by ESD. These lessons engage students from global and local perspectives on sustainability issues and encourage their social participation; they are a good example of how SDG 17 – Partnerships for the goals: Strengthen the means of implementation and revitalize the global partnership for sustainable development – can be embodied in science education for sustainable development.

A keyword discussed in the other chapters is individual transformation, which the following two chapters address. Chapter 6 of the Philippines proposes ESD as (i) the head, referring to competencies from global and local perspectives; (ii) the hand, incorporating a skills-oriented teaching model; and (iii) the heart, emphasizing a value-driven approach to achieve holistic attitudes toward the environment. The authors' assertion that science education has great potential for promoting individual transformation by connecting with such ESD approaches may suggest the potential for the general development of science education. Chapter 15, which introduces the ASEAN Sustainability Leadership Program, reports that through experiential and collaborative learning, researchers from various disciplines of higher education institutions who participate in the program can increase their awareness of learning about and engaging in the discourse on sustainability; it can also help explore ways to promote the study of sustainable development. For researchers who participated in the program, science education for sustainable development is an important opportunity for individual transformation.

Regarding science teacher education related to this keyword, Chap. 17 depicts the implementation of ESD in pre-service science teacher training in Indonesia. This shows that the newly established compulsory and elective university courses promote science teachers' awareness of sustainability and competencies in practicing ESD, which involve their individual transformations. Furthermore, the research discussed in this chapter has led to subsequent research identifying the competencies required of science teachers who practice ESD (Eliyawati et al., 2023). Chapter 18 from Japan highlights the domain "Continue to learn, reflect, innovate and transform" of the Asia-Pacific ESD Teacher Competency Framework (Okayama University ESD Promotion Center, 2020) that the contributors to this book participated in its creation. A science teacher training program for ESD in this framework, emphasizing "reflection," provides an excellent opportunity for them to be transformed as reflective practitioners.

Finally, Chap. 5 from Mongolia reports on a national survey, showing that the introduction of STEM education in elementary schools across the country has improved the sustainability competencies of students and teachers. The keywords that appeared were problem-solving and collaborative competencies. However, understanding that STEM education and ESD have different roots requires careful work to theorize their coupling. This is an inevitable challenge for science education for sustainable development in Asia.

22.1.2 Addressing Local and Global Issues

Natural disasters are among the most significant sustainability issues faced by Asian countries. For example, earthquakes, tsunamis, storms, and floods are frequent in East and Southeast Asia and can cause tremendous damage. The United Nations Convention on Biological Diversity states that biodiversity conservation is a pressing issue in biodiversity-rich Asia. In addition, reducing greenhouse gas emissions – which exacerbate climate change – and shifting to sustainable consumption and production are challenging issues in Asia, which is undergoing rapid industrialization and urbanization. Although these are sometimes local issues that call for specific measures, they always involve global connections. Therefore, a common keyword in the themes of research and practice in science education discussed in these chapters is how to address local and global issues.

To highlight how each chapter is addressed, Chap. 3 from Japan proposes a junior high school science lesson for post-natural disaster recovery. In this chapter, lessons exploring the geological environment of a local area affected by an earthquake will help students understand that the area's rich soil produces abundant crops, transforming their view of the local area from negative to positive. These results prove that science education for sustainable development can influence students' values. Values and ethics will need to become an integral part of education if sustainability is to be seriously addressed (Wals & Dillon, 2015).

Chapter 9, which focuses on science teacher training in Indonesia, implements a problem-solving-based volcano learning program on volcanic disasters. The program encompasses ESD from all perspectives of the curriculum; that is, it applies the characteristics of ESD to the content of subjects arranged, presented in a structured and systematic context, and is driven by the principles of ESD. This chapter provides useful suggestions for approaches that include disaster reduction in science teacher training curricula.

The theme of biodiversity is addressed in Chaps. 11 and 12. Chapter 11 describes the development of an elementary school science curriculum based on the Japanese sociocultural context of biodiversity conservation, focusing on invasive alien species. Intervention lessons based on the curriculum transformed students' one-dimensional and negative perceptions of invasive alien species into multifaceted and appropriate perceptions. This is a good example of a lesson that fosters future thinking about how children can engage in biodiversity conservation. Chapter 12 from Japan highlights ways to design educational activities and engage the public to promote awareness of the relationship between humans and wildlife, focusing on Asian elephants and Japanese oriental storks listed as endangered species on the IUCN Red List in 2018. For this purpose, university students created "Kamishibai," a paper theater of a popular literacy form in Japan, and presented it to the public near these animals exhibited in a zoo. These results will be fruitful for developing specific teaching materials on animal diversity conservation.

Other themes covered in the chapters are the utilization of solar energy, a form of renewable energy, in Chap. 2, and soil pollution and its restoration in Chap. 4. Chapter 4 reports on a high school science lesson in Beijing, China, addressing soil pollution caused by a chemical plant. This sustainability issue affects students' real lives and organizes the content of the high school biology unit "cells" from the perspective of sustainable development. The authors call this way of organizing content a holistic approach to linking the three perspectives of ESD: environment, economy, and society.

However, most chapters, including Chap. 4, do not address the themes of economic and social perspectives. Although there is criticism of the hierarchical equivalence of environmental, economic, and social perspectives because they reduce the value of the environment, a model in which these three perspectives form a single circle is a universal basis for sustainable development (Gericke et al., 2019). If so, naturally, the primary interest of science education in sustainable development is environmental perspectives; however, linking environmental perspectives with economic and social perspectives, which are inseparable from environmental perspectives, remains a challenge.

In this context, Chap. 10 uses interviews with Indonesian in-service science teachers to illustrate how the local context, as an intersection of sustainability and science curricula, is intricately intertwined with economic, social, and cultural perspectives, as well as from an environmental perspective. Therefore, painstakingly unraveling these contexts is the key to addressing the challenges discussed earlier.

22.1.3 Adopting Active Pedagogical Approaches

A common keyword in the educational practices depicted in the chapters is active pedagogical approaches, which aim to encourage behavioral changes in learners, such as learner-centered, inquiry-based, experiential, participative, collaborative, and inter- and transdisciplinary approaches (Fujii, 2022). Chapter 16 states that China's national science curricula aim to promote active pedagogical approaches and that ESD-infused science teacher education focuses on creating lessons in schools that incorporate these approaches to enable students to solve practical issues. Chapter 4 describes the development of Chinese high-school biology lessons and provides a good example of an inquiry-based pedagogical approach that addresses the issue of soil pollution and its restoration in a local community.

Regarding science teacher education, Chap. 20 in Laos presents a training program for science teachers' professional development that incorporates ESD. As a result of this training, science teachers have begun to improve their school curricula and incorporate interactive teaching strategies into their classes. Chapter 21 of the study conducted in Myanmar also presents a training program for in-service science teachers on ESD based on the lesson study model. With such training, science teachers can improve the active teaching-learning process in terms of teachers' content knowledge, teaching methodology, and their perspectives toward students' participation in science activities. Similar to the point made by Fischer et al. (2022), science teacher education for sustainable development can facilitate innovation in science teacher education in general.

Chapters addressing informal science education include participation and collaboration as keywords among active pedagogical approaches. The Japanese practice in Chap. 12 is a collaboration between a university and a zoo – notably, not only the citizens but also the university students themselves acquire sustainability competencies through science communication with the public using paper theaters of a popular literacy form as a tool. The study in Chap. 14 is a university-school collaboration to improve the quality of science education in the United Nations University-registered ESD Regional Center of Expertise in Penang, Malaysia. This project, a competition for young inventors to promote sustainable development, embodies quality science education that sustains a science community capable of tackling global issues (Eijck & Roth, 2007).

Participation and collaboration should include the entire school and the local community. Chapter 7, which highlights the activities of the Southeast Asian Ministers of Education Organization (SEAMEO) with its secretariat in Thailand, shows that whole-school ESD practices transcending school science lessons can encourage student and teacher participation and collaboration and enrich learning on topics such as disaster risk reduction, saving energy, local wisdom for environmental conservation, and plastic problems for transforming communities. As the OFSTED (2009) points out, using the whole school to teach sustainability can sustain student behavior in adolescence and adulthood. Chapter 13, which presents ESD in a Japanese junior high school, illustrates the process of forming a multifaceted, collaborative initiative that includes students, teachers, and local community members while enhancing a multilayered community of practice. These findings can be applied to science education for sustainable development in collaboration with local communities. In addition, Chap. 19 from Kazakhstan pitches steps to achieve ESD goals in the country, including private investment in education, digitalization of education, and creating an educational hub. These steps are prerequisites for educating and inspiring science teachers, other teachers, and local community members to infuse ESD into their educational practices.

Furthermore, the keywords for inter- and transdisciplinary approaches are noteworthy. In the ASEAN Sustainability Leadership Program introduced in Chap. 15, researchers from various disciplines in higher education institutions participate in learning about local community issues in Siem Reap, Cambodia, through their experiences in the field. Participants are stimulated by collaborative activities with those in different specialties, leading to the emergence of ideas and the search for better solutions. This study provides suggestions for developing interdisciplinary and transdisciplinary education for sustainable development as an emerging field, with science education as a starting point.

To summarize this subsection, the discussions in the chapters depict how the keywords of science education connected to ESD are understood and embodied as educational practice in the Asian context. Common to the arguments included in them are the pursuit of social participation and individual transformation through the development of scientific literacy, addressing local and global socio-scientific issues (SSI) and applying active pedagogical approaches, which are closely interrelated. At the same time, the included critical discussion addresses the weak link between environmental perspectives and economic and social perspectives in sustainability, the easy combination of STEM education and ESD, and the lack of verification of the effectiveness of active pedagogical approaches. In addition, it addresses the need for a holistic design in science education curriculum, pedagogy, and assessment based on the idea of sustainable development. These discussions argue for a fundamental reorientation of science education. Science education for sustainable development offers a direction for how to innovate science education in general in the context of current and future social changes.

22.2 Toward Advancing Science Education for Sustainable Development in Asia

As mentioned earlier, the cases of ESD in science education presented in Chaps. 2–21 show various and unique science education for sustainable development in Asia. This information is expected to provide diverse possibilities regarding the role of science education in a sustainable society and in the future. However, the cases illustrated in this book have limitations to be overcome and issues to be discussed. Considering them, the following recommendations are discussed and suggested toward advancing science education for sustainable development in Asia:

The purpose and significance of science education for sustainable development need to be further clarified. The cases of science education presented in the previous chapters have yet to comprehensively analyze the exact meaning or purpose of ESD in science education or science education for sustainable development as recognized in each country, which should be investigated and clarified. The cases included in this book were collected via a research project titled “Teacher Education for ESD in Asia-Pacific Region,” a collaboration for science educators in Asia. This project explores strategies for integrating ESD into science education and links them to improve teachers' expertise. To this end, lesson study was used, in which teachers in schools or communities worked together to prepare research questions, study textbooks, design lessons, and conduct demonstration classes before exchanging information via communication. These processes help teachers enhance their teaching capabilities and expertise. The project played a crucial role in identifying significant cases in Laos, Myanmar, and the Philippines. However, ESD in science education, recognized in participating countries, has not been sufficiently investigated and explored.

In many cases, science education aims at cultivating scientific literacy. This also applies to science education in the Republic of Korea (Ministry of Education, 2015). Scientific literacy refers to the ability to read and write scientifically; it includes not only the understanding and explanation of scientific facts but also the critical literacy and thinking regarding them, which can be called secondary literacy. Critical thinking is one of cross-cutting key competencies for achieving the Sustainable Development Goals (SDGs) with competencies such as systems thinking, anticipatory, problem-solving, normative, strategic, collaborative, self-awareness, and integrative problem-solving. Buntting and Jones (2015) emphasized that future thinking should function as one of the main pillars of future science education. Therefore, integrating sustainability into science education is deemed particularly important in contributing to a sustainable society and future and improving the quality of science education. Thus, when similar joint projects or research are implemented in the future, the potential and meaning of science education for sustainable development and teachers' recognition should be reviewed and examined. It is also necessary to discuss the significance and critical elements of lesson study to improve teachers' expertise and bring about changes in schools.

Science education should contribute to understanding, exploring, and tackling sustainability issues including climate change. Climate change dubbed "the climate crisis" or "climate emergency" has become a major global issue (Ripple et al., 2020), which is threatening the sustainability of modern mankind on Earth. On November 5, 2019, 5 people including Ripple declared a climate emergency in a joint statement published in Bioscience with 11,258 scientists from 153 countries (Ripple et al., 2020). They wrote that "the world's population continues to grow by 80 million annually, and the destruction of the Amazon rainforest is increasing again. Analyzing data on population growth, meat consumption, deforestation, energy consumption, and economic losses due to extreme weather events, It is clear that humanity is facing a global emergency." They also emphasized that "the crisis is already in front of us and is progressing faster and more seriously than many scientists expected, threatening the fate of mankind and ecosystems" and "if we do not take immediate action to preserve the earth, the climate crisis will bring great suffering to mankind" and "there is no time to waste now."

The Intergovernmental Panel on Climate Change (IPCC) that periodically releases reports based on scientific facts. The IPCC report published in 2021 showed that climate change has accelerated faster than expected by scientists (IPCC, 2021). The sixth assessment report of IPCC (IPCC AR6), passed unanimously by participants and released in 2023, emphasizes the urgent need for the international community to tackle climate change (IPCC, 2023). According to the IPCC AR6 (2023), widespread and rapid changes in the atmosphere, ocean, cryosphere, and biosphere has occurred, and it is unequivocal that human influence has warmed the atmosphere, ocean, and land. The scale of recent changes across the climate system as a whole and the present state of many aspects of the climate system are unprecedented over many centuries to many thousands of years.

It is also worth noting here the way science deals with sustainability issues such as climate change based on evidences and scientists' consensus. First, science describes and reports current conditions, such as climate change, that are closely linked to the physical sustainability of the earth. As mentioned earlier, the IPCC, which has published six reports on climate change, is an intergovernmental panel composed of scientists and government representatives. Although it does not conduct separate research, it reviews peer-reviewed articles and other references related to climate change that has been published through existing research, and diagnoses the situation of the earth in relation to climate change based on the evidence contained in them. It predicts the impact of climate change and the damage that will occur due to climate change. In the case of the fifth assessment report, more than 800 scientists from more than 80 countries were selected as a team of authors, 1000 contributing authors, more than 1000 expert reviewers participated, and more than 30,000 scientific papers were evaluated. About 3000 scientists around the world address the current progress of climate change based on evidence based on existing researches and publish the agreed results in the IPCC reports. As a result, the sustainability of the Earth is seriously threatened due to climate change, and greenhouse gas emissions such as carbon dioxide must be drastically reduced in order to prevent an increase in global surface temperature (1.5 °C), which will cause irreversible changes. It was suggested that real greenhouse gas reductions should be implemented promptly. Based on this, each country is developing efforts for greenhouse gas reduction and carbon neutrality by 2050.

As regards to climate change, what is happening worldwide in the context of climate change should be understood scientifically to take practical measures to mitigate or adapt to climate change. It is essential to understand greenhouse gases (CO_2, methane, etc.), fossil fuels, renewable energy, and carbon sinks, as well as the causes of climate change and its impact on ecosystems. Understanding the connection between climate change, biodiversity, and pandemics is a starting point for resolving this issue. Science and technology have played a pivotal role in combating and tackling climate change. For instance, when discussing energy conversion for wind power generation, the direction and strength of the wind in regions or nations must first be understood scientifically. Solar power, geothermal power, zero-energy buildings, and climate-change mitigation plans also require scientific understanding. Against this backdrop, science education must address environmental sustainability-related issues more fundamentally.

Mitigation and adaptation are important in climate change, and science and technology can contribute to both. In the process of reducing greenhouse gas emissions to mitigate climate change, science and technology can be key, such as the development and application of renewable energy, the development and operation of electric vehicles, the development and application of new processes, and buildings considering energy efficiency. Science and technology are also key to capturing greenhouse gases that have already been released. Even in the process of climate change adaptation, science and technology are needed to identify vulnerable areas and vulnerable groups and prepare countermeasures for them. However, science and technology alone cannot solve everything, and sometimes there are always problems that the development and application of new climate technology can bring. Therefore, in the field of science and science education, opportunities to constantly review and discuss issues of sustainability and their relevance to science and technology should be provided. In addition, complex discussions on social, economic, and political aspects other than science and technology are necessary so that a just transition can be made in the process toward sustainability.

Practical issues related to social or global sustainability should be explored in a multidisciplinary, interdisciplinary, and/or transdisciplinary way in environmental, economic, and social contexts. Most issues facing modern society – including climate change, can be referred to as SSI or Socially Acute Questions (SAQ) – are science-related social problems that should be tackled based on science. Therefore, they require a comprehensive understanding of and response to the connection between science and society. A clear understanding of the relationships among the environment, economy, society, and the environment is also deemed very significant. In other words, scientific knowledge and ethical and political aspects should be explored together, which can play an important role in resolving these issues. As described in Chap. 1, the key topics in science education for sustainable development include climate change, renewable energy, biodiversity, natural disaster prevention and mitigation, sustainable production/consumption, and health promotion. The topics closely relate to the SDG 13 and the other SDGs proposed by the UN. Diverse issues, including climate change, need to be tackled based on an understanding of scientific knowledge. However, they cannot be easily defined or coped with, because they are wicked problems which have no simple solutions and connected to various realms, such as the environment, economy, society, politics, and culture (Incropera, 2015; Block et al., 2019).

The cases introduced in chapters fail to reveal a clear connection between science education and economics, which can be one of limitations of this book. As mentioned in Chaps. 10 and 15, community issues are explored meaningfully in connection with science education, linking them to the environmental, social, and cultural spheres. However, their connection to the economic realm has not been significantly highlighted. Economic features are important elements affecting citizens' daily lives. Therefore, when discussing science education for sustainable development, its connection with economic and environmental, social, and cultural realms should be considered, explored, and addressed, followed by making necessary decisions.

Science education containing practical measures that can contribute to resolving problems in real situations should be provided. Moreover, it is necessary to offer science education that makes it possible to implement innovative measures at

individual and social levels. Most approaches introduced herein mainly relate to improving awareness, learning, exploration, and others but fail to lead to practical solutions and real actions. To prevent science education from being limited to laboratories, real-world issues should be dealt with and discussed, leading learners to experience the process of producing solutions and bringing about change. This reflects Vision III for scientific literacy, science education for change and innovation, as proposed by Sjöström and Eilks (2018). Roberts (2007, 2011) has classified scientific literacy into Vision I, which focuses on scientific content and processes for later application, and Vision II, which aims to understand the usefulness of scientific knowledge in life and society. Vision III, which was dubbed "critical scientific literacy" by Hodson (2009), implies a politicized science education aiming at dialogical emancipation and socio-ecojustice and emphasizes transdisciplinarity, philosophical values, and praxis-oriented global citizenship (Sjöström & Eilks, 2018). Recent discussions on education have focused on change and innovation. The OECD Learning Framework for 2030 (OECD, 2018) emphasizes that future education aims to achieve collective well-being characterized by innovative capabilities and that student agency should be ensured in school environments. The roadmap for ESD for 2030 (UNESCO, 2020) highlights that individual and social innovation and new technologies are required for a sustainable future and that ESD must contribute to conducting the initiative. In this context, science education must contribute to bringing change and innovation to the real world, which features unsustainability.

Indigenous knowledge needs to apply to science education. In most cases, indigenous knowledge is based on biocultural diversity. Lee et al. (2015) identified environmental wisdom in traditional culture in China, Japan, and Korea in terms of costumes, food, and housing, which was conducted while promoting cooperation through the Tripartite Environmental Education Network (TEEN) among the three countries. Based on this, teaching materials were developed for elementary school students. Such environmental wisdom in traditional culture reflects regional and environmental characteristics and cultural diversity. Biocultural diversity refers to cases in which community members identify and use specific indigenous biological resources daily. Knowledge of traditions or biocultural diversity can become crucial to achieving a sustainable future. Zidny et al. (2020) discussed strategies for applying traditional knowledge in science education. Kim and Dionne (2014) categorized knowledge into indigenous knowledge, indigenous science, and traditional ecological knowledge. Kim et al. (2017) proposed a traditional ecological knowledge Polygon framework consisting of place-based teaching approaches, environmental education for sustainability, promoting multiculturalism in science education, relevant curricula for indigenous students, and the cosmology of traditional ecological knowledge. Based on these findings, we explored ways to apply these elements to science education. In other words, place-based teaching approaches connected to traditional ecological knowledge can be applied to the social, cultural, economic, political, and natural realms in regional environments (Smith, 2002), while in the context of science education, traditional ecological knowledge facilitating discussions on sustainability and communities, thereby enabling environmental education for sustainability (McConney et al., 2011; Kimmerer, 2012). In connection with science education, it is necessary to recognize that various learning activities based on cultural perspectives have contributed to the development of modern science (Eijck & Roth, 2007; Mueller & Tippins, 2010). Traditional ecological knowledge offers culturally relevant curricula to indigenous students, facilitating science learning and strengthening their cultural identities (Kim et al., 2017). In addition, traditional ecological knowledge can establish a basic framework for the cosmology of indigenous knowledge systems, relying on the interconnectivity of all objects and their relations with the Earth (McGregor, 2006; Kim et al., 2017). These principles should be considered the main elements of science education for sustainable development.

Science education should provide opportunities for democratic participation for a sustainable society and for cultivating scientific knowledge and research capabilities. In connection with democratic participation, many studies emphasize that science education aims to nurture citizens to boost democratic participation through scientific literacy. However, Levinson (2010) points out the difficulties in cultivating such citizens and classifies feasible democratic participation in school environments into the following four processes: (1) a democratic deficit enabling limited participation, (2) deliberative democracy, (3) science education as a practice, and (4) science education for conflicts and objections. It proposes that inquiring about, discussing, and making decisions on authentic issues in real-world situations be introduced to school environments via science education as a practice and science education for conflicts and objections. Professionals do not deliver sustainability; a sustainable world is created by members of specific groups, regions, or nations discussing and reaching agreements on various issues via conflict resolution. Therefore, the democratic participation of members and the organization and operation of governance are deemed essential elements. To this end, communities where members learn together must be established. Therefore, social learning should be enabled in this environment. Scott and Gough (2003) claim that one-way teaching, communication, and mediation are vital in education and learning in a sustainable society.

Science education for sustainable development needs teachers equipped with literacy and expertise in sustainability. Therefore, programs for cultivating literacy and expertise in sustainability should be included in the training of pre-service and in-service teachers. In particular, the Asia-Pacific ESD Teacher Competency framework presented in Chap. 1 contains the following three domains: Competency I: facilitate learning; Competency II: connect, collaborate, and engage; and Competency III: continue to learn and create (Okayama University ESD Promotion Center, 2020). Therefore, opportunities should be provided to enhance these competencies. This is because teachers should be able to shape their own future.

Teacher cultivation programs need to be connected to and integrated into the pre-service teacher training process. Most importantly, pre-service teachers should be exposed to a process of cultivating literacy and expertise in sustainability. One university of education in Korea operates programs in which topics and activities related to environmental education and sustainability are integrated into courses for subjects such as Korean, Social Studies, and Science. Consequently, teachers can determine the connections between general subjects and environmental education/ESD and secure teaching and learning repertoires for offering ESD via diverse courses. In-service teacher training programs should also provide opportunities to secure repertoires that can be used in various teaching courses through cases related to subjects and school activities. Teachers should be given opportunities to understand ESD and sustainable development better.

Professional learning communities (PLC) that are set up and run in teacher training processes can lead diverse teaching activities for "Science" and other subjects to promote sustainability and can also help bring change to schools. Teachers must be supported in establishing and operating PLC inside and outside schools.

Science education for sustainable development can be offered in connection with physical facilities in schools and ultimately requires a whole-school approach. The fundamental pillar of educational activities in schools is teaching and learning based on curricula. Schools function as a microcosm and model of a sustainable society and contain diverse elements besides curricula, teaching, and learning activities. In 2005, researchers included in the Environment and School Initiatives (ENSI), a global network for environmental education and ESD, collected cases of eco-schools, environmental schools, schools for ESD, and green schools in 13 nations, including Korea, Austria, and Australia, to explore the selection criteria, operational processes, and results, and compiled them into a book (Mogensen & Mayer, 2005). They identified quality criteria that schools should consider when facilitating ESD (Breiting et al., 2005). The quality criteria include three domains: (i) teaching and learning, (ii) school management and operation, and (iii) relations between schools and the outside world, and proposed standards that need to be considered by domain. When all the activities in these domains promote sustainability, schools can be reshaped into those for ESD or become closer to sustainability. Offering science education can allow students to consider sustainability and facilitate future critical and systems thinking. It can also create visible changes inside and outside the school by applying diverse and innovative teaching and learning methods. These educational systems enable students to learn about sustainable development through content and processes. These need to be connected to other types of education and the school's curricula as a whole, school operation/decision-making, school facility management, and communities, thereby changing the school's culture as a whole. In particular, when attempting to improve environmental sustainability via specific actions for climate change mitigation and carbon neutrality in the real world, including in schools and communities, physical facilities in schools can be connected to science curricula in terms of their educational content and methodology. If this happens, science education will become closely linked to the real world, including students' lives, thus enhancing its relevance. In this context, experiencing the process of creating diverse types of change and improving sustainability will help students resolve practical issues and bring about future changes.

The significance of research regarding science education for sustainable development should be stressed. Topics and pedagogies for science education for sustainable development need to be reviewed and examined. How the aforementioned realms are managed in terms of science education for sustainable development and whether they achieve their goals can be studied by determining the meaning of diverse endeavors, monitoring specific actions, and evaluating their performance. Based on reality analysis, issues related to science education for sustainable development can be explored, thereby presenting new goals. In connection with UNESCO-ASPnet schools in Japan, flagship schools for ESD promotion and evaluation reports are released periodically. These results can be reflected in educational activities, program operations, and student performance evaluations in newly designated schools, presenting a new direction for the future operation of schools. However, for science education in sustainable development, assessing socio-emotional spheres leading to practical actions is more significant than cognitive achievement. Therefore, adequate research on this topic should be conducted. Moreover, research on the entire system and changes in the school as a whole will help study the process of change and innovation, as well as the elements that facilitate them. These studies will enable science education to reflect more effectively on sustainable development in the future.

Finally, the importance of science communication should be emphasized in science education for sustainable development. In Ripple et al. (2020), scientists warned the urgency of climate change and propose six recommendations to cope with it, which include energy transition and high carbon tax, reduction of short-lived pollutants such as methane, ecosystem protection and restoration increasing carbon sinks, and more vegetable food less animal products, pursuit of a carbon-free economy, less development of the biosphere, stabilization of the global population while considering social and economic justice. Based on these recommendations, scientists urged policy makers to collaborate with scientists on climate action. But in the real world, climate action by policy makers and the general public has not been swift. While there is nearly universal consensus among scientists, the same cannot be said about the general population. When the Journal of Environmental Education was first published in 1969, environmental education and communication were considered two sides of the same coin. In the first issue, editor Schoenfeld (1969) wrote an article titled "What's New about Environmental Education?" According to him, environmental education is defined as "communication to nurture citizens who know the fact that there are environmental problems and how to solve them, and who have the will to implement the solutions." Similarly, Cox (2010) referred to environmental communication as "crisis discipline" in the sense that it is an academic field that will contribute to the diagnosis and resolution of environmental crises, and also called it communication about and for the environment (Kim, 2022). As such, science education alone is insufficient to convey the impending crisis: that ways in which scientific ideas, findings and impact can be effectively communicated to the general population are important. While this is not seen as science among some people, it is an important area of science education that is missing. This broadly mean that science education needs a fairly fundamental re-orientation to reconnect with the population in order for its work to be recognized to be relevant, and essentially for society and beyond.

References

Block, T., Poeck, K., & Östman, L. (2019). Tackling wicked problems in teaching and learning. *Sustainability issues as knowledge, ethical and political challenges*. https://doi.org/10.4324/9781351124348-3

Breiting, S., Mayer, M., & Mogensen, F. (2005). *"Quality criteria for ESD schools" Guidelines to enhance the quality of Education for Sustainable Development*. Austrian Federal Ministry of Education, Science and Culture.

Buntting, C., & Jones, A. (2015). Futures thinking in the future of science education. In D. Corrigan, C. Buntting, J. Dillon, A. Jones, & R. Gunstone (Eds.), *The future in learning science: What's in it for the learner?* (pp. 229–244). Springer.

Cox, R. (2010). *Environmental communication and the public sphere* (3rd ed.). Sage.

Eijck, M., & Roth, W.-M. (2007). Keeping the local local: Recalibrating the status of science and traditional ecological knowledge (TEK) in education. *Science Education, 91*(6), 926–947.

Eijck, M., & Roth, W.-M. (2008). Improving science education for sustainable development. *PLoS Biology, 5*(12), 2763–2769.

Eliyawati, Widodo, A., Kaniawati, I., & Fujii, H. (2023). The development and validation of an instrument for assessing science teacher competency to teach ESD. *Sustainability, 15*(4), 3276.

Fischer, D., King, J., Rieckmann, M., Barth, M., Büssing, A., Hemmer, I., & Lindau-Bank, D. (2022). Teacher education for sustainable development: A review of an emerging research field. *Journal of Teacher Education, 73*(5), 509–524.

Fujii, H. (2022). Trends and perspectives of climate change education in the Asia-Pacific. In W. O. Lee, P. Brown, A. L. Goodwin, & A. Green (Eds.), *International handbook on education development in Asia-Pacific*. Springer Nature.

Gericke, N., Pauw, J. B., Berglund, T., & Olsson, D. (2019). The sustainability consciousness questionnaire: The theoretical development and empirical validation of an evaluation instrument for stakeholders working with sustainable development. *Sustainable Development, 27*(1), 35–49.

Hodson, D. (2009). *Teaching and learning about science: Language, theories, methods, history, traditions and values*. Sense.

Incropera, F. (2015). *Climate change: A wicked problem. Complexity and uncertainty at the intersection of science, economics, politics, and human behavior*. Cambridge University Press.

Intergovernmental Panel on Climate Change (IPCC). (2021). *Climate change 2021: The physical basis*. Cambridge University Press.

Intergovernmental Panel on Climate Change (IPCC). (2023). *Climate change 2023: Synthesis report. Summary for policymakers*. IPCC.

Kilinc, A., & Aydin, A. (2011). Turkish student science teachers' conceptions of sustainable development: A phenomenography. *International Journal of Science Education, 35*(5), 731–752.

Kim, N. (2022). Environmental communication. In Ministry of Environment (Ed.), *Common textbook for level 3 environmental educators (Volume 2. Practical course)* (pp. 293–303). Ministry of Environment.

Kim, E.-J. A., & Dionne, L. (2014). Traditional ecological knowledge in science education and its integration in grades 7 and 8 Canadian science curriculum documents. *Canadian Journal of Science, Mathematics and Technology Education, 14*(3), 311–329.

Kim, E.-J. A., Asghar, A., & Jordan, S. (2017). A critical review of traditional ecological knowledge (TEK) in science education. *Canadian Journal of Science, Mathematics and Technology Education, 17*(4), 258–270.

Kimmerer, R. W. (2012). Searching for synergy: Integrating traditional and scientific ecological knowledge in environmental science education. *Journal of Environmental Studies and Sciences, 2*, 317–323.

Lee, S., Lee, J., Jung, W., & Cho, E. (2015). *Our common future: Environmental wisdom in traditional culture. Tripartite environmental education network among China, Japan and Korea*. Ministry of Environment in Republic of Korea.

Levinson, R. (2010). Science education and democratic participation: An uneasy congruence? *Studies in Science Education, 46*(1), 69–119.
McConney, A., Oliver, M., Woods-McConney, A., & Schibeci, R. (2011). Bridging the gap? A comparative, retrospective analysis of science literacy and interests in science for indigenous and non-indigenous Australian students. *International Journal of Science Education, 33*(14), 2017–2035.
McGregor, D. (2006). Traditional ecological knowledge. *Ideas: The Arts and Science Review, 3*. Retrieved from https://doi.org/www.silvafor.org
Ministry of Education. (2015). *2015 revised science curriculum*. Ministry of Education, Republic of Korea.
Mogensen, F., & Mayer, M. (2005). *Eco-schools: Trends and divergencies. A comparative study on ECO-school development processes in 13 countries*. Austrian Federal Ministry of Education, Science and Culture.
Mueller, M., & Tippins, D. (2010). Van Eijck and Roth's utilitarian science education: Why the recalibration of science and traditional ecological knowledge invokes multiple perspectives to protect science education from being exclusive. *Cultural Studies in Science Education, 5*(4), 993–1007.
OECD. (2018). *The future of education and skills: Education 2030. Position paper*. OECD.
Office for Standards in Education (OFSTED). (2009). *Education for sustainable development: Improving schools – Improving lives*. Office for Standards. in Education.
Okayama University ESD Promotion Center. (2020). *Guide for the effective dissemination of the Asia-Pacific ESD teacher competency framework*. Okayama University.
Ripple, W. J., Wolf, C., Newsome, T. M., Barnard, P., Moomaw, W. R., & 11,258 Scientist Signatories from 153 Countries. (2020). World scientists' warning of a climate emergency. *Bioscience, 70*(1), 8–12.
Roberts, D. A. (2007). Scientific literacy/science literacy. In S. K. Abell & N. G. Lederman (Eds.), *Handbook of research on science education* (pp. 729–780). Lawrence Erlbaum.
Roberts, D. A. (2011). Competing visions of scientific literacy: The influence of a science curriculum policy image. In C. Linder, L. Ostman, D. A. Roberts, P.-O. Wickman, G. Erickson, & A. MacKinnon (Eds.), *Exploring the landscape of scientific literacy* (pp. 11–27). Routledge.
Sadler, T. (2011). Situating socio-scientific issues in classrooms as a means of achieving goals of science education. In T. Sadler (Ed.), *Socio-scientific issues in classroom: Teaching, learning, and research* (pp. 1–10). Springer.
Schoenfeld, C. (1969). What's new about environmental education? *Environmental Education, 1*(1), 1–4.
Scott, W., & Gough, S. (2003). *Sustainable development and learning: Framing the issues*. Routledge Falmer.
Sjöström, J., & Eilks, I. (2018). Reconsidering different visions of scientific literacy and science education based on the concept of *Bildung*. In Y. J. Dori, Z. R. Mevarech, & D. R. Baker (Eds.), *Cognition, metacognition, and culture in STEM Education: Learning, teaching and assessment* (pp. 65–88). Springer.
Smith, G. A. (2002). Place-based education: Learning to be where we are. *Phi Delta Kappan, 83*(8), 584–594.
UNESCO. (2018). *Issue and trends in education for sustainable development*. UNESCO.
UNESCO. (2020). *Education for Sustainable Development: A roadmap. ESD for 2030*. UNESCO.
Wals, A., & Dillon, J. (2015). Foreword. In S. K. Stratton, R. Hagevik, A. Feldmann, & M. Bloom (Eds.), *Educating science teachers for sustainability* (pp. v–viii). Springer.
Zidny, R., Sjöström, J., & Eilks, I. (2020). A multi-perspective reflection on how indigenous knowledge and related ideas can improve science education for sustainability. *Science & Education, 29*, 145–185.

Index

H. Fujii, S.-K. Lee (eds.), *Science Education for Sustainable Development in Asia*, Education Innovation Series, https://doi.org/10.1007/978-981-99-8711-5

MIX
Papier aus verantwortungsvollen Quellen
Paper from responsible sources
FSC® C105338

FSC
www.fsc.org

If you have any concerns about our products,
you can contact us on
ProductSafety@springernature.com

In case Publisher is established outside the EU,
the EU authorized representative is:
Springer Nature Customer Service Center GmbH
Europaplatz 3, 69115 Heidelberg, Germany

Printed by Libri Plureos GmbH
in Hamburg, Germany